Microwave Absorbing Materials

Microwave Absorbing Materials

Yupin Duan
Hongtao Guan

Pan Stanford Publishing

Published by

Pan Stanford Publishing Pte. Ltd.
Penthouse Level, Suntec Tower 3
8 Temasek Boulevard
Singapore 038988

Email: editorial@panstanford.com
Web: www.panstanford.com

British Library Cataloguing-in-Publication Data
A catalogue record for this book is available from the British Library.

Microwave Absorbing Materials

ISBN 978-981-4745-10-9 (Hardcover)
ISBN 978-981-4745-11-6 (eBook)

Printed in the USA

Contents

Preface

Due to their extensive applications in military stealth technology, most of the research on microwave absorbing materials has been kept secret and classified over the years. In the recent past, with increasing requirements for microwave absorbing performances of these materials and their prosperity in civil applications, new kinds of microwave absorbing materials have emerged, and either their absorbing mechanisms or their applications have attracted considerable attention and made pronounced progress.

This book presents a concise scope of modern microwave absorbing materials, also known as electromagnetic absorbing materials, and their absorption characterizations. The objective is to provide a sound understanding of the fundamentals and concepts of microwave absorbing theories, which also form the basis of the principles of microwave absorbing materials and their absorbing mechanisms.

The content in this book is presented in eight chapters. Chapter 1 is devoted to the fundamental aspects of interactions between electromagnetic waves and microwave absorbing materials. On the basis of principle theory, the crucial factors which may influence the absorbing performances of microwave absorbing materials, such as density, particle size, shape, chemical compositions, and stability, are also included. Chapters 2 to 5 discuss traditional microwave absorbing materials based on manganese oxides, iron matrix alloys, conductive polyanilines, and barium titanates. The preparation techniques and their electromagnetic characterizations are also dealt with. Chapters 6 to 8 give a description of hybrid microwave absorbers, cement matrix absorbing materials, and structural pyramidal materials. Chapter 6 also gives an overview of two main absorbers, absorbing coatings and absorbing structures. Several representative absorbing coatings and structures based on epoxide resin, polyurethane (PU) varnish, silicon rubber, and acrylonitrile-butadiene-styrene (ABS) are introduced briefly. Chapter 7 elaborates on the electrical and electromagnetic properties of cement-based composite materials filled with carbon materials, metal fillers, and porous fillers. On the basis of the microwave absorbing properties of cement composites filled with expanded polystyrene (EPS), the energy conservation law in electromagnetic fields has been proposed. In Chapter 8, we present the design philosophy of the pyramid absorbers widely used in most anechoic chambers. And also, we propose a new kind of resonant absorber based on carbon-coated EPS and discuss its absorbing mechanism in detail.

To give a more intuitive understanding of the materials in each chapter, we give a full list of references related to the main contents in that chapter. The readers can refer to these lists to get more information.

I would like, first, to thank gratefully my colleagues and students for their assistance and contribution to this book. These include Prof. Liu, Prof. Guan, Huifang Pang, Wei Liu, Yahong Zhang, Jin Liu, Qun Xi, LuLu Song, Gaihua He, Liyang Chen, Lidong Liu, Baoyi Li, Shuping Lv, Guangli Wu, Shuchao Gu, Jia Zhang, Hui Jin, He Ma, Zhuo Liu, Ming Wen, Long Wang, Junlei Chen, Jizhu Du, and Xiaodong Chen, who provided excellent expertise and support for the language and pictures, especially Prof. Guan, who gave many good suggestions on the design and polishing of the content. I am also deeply indebted to my family for their patience, encouragement, and support and for contributing so much to my confidence in dealing with the writing of this book.

Yuping Duan

May 2016

Chapter 1

Fundamentals of Electromagnetic Wave Absorbing Theory

Maxwell's equations indicate that the time-varying electromagnetic (EM) field is a rotational solenoidal field in the source-free space ($\rho = 0, \vec{J} = 0$). In other words, electric force lines and magnetic field lines are closed without any endpoints. The electric field and magnetic field cross-link and excite each other to generate EM waves, which are the main form for EM energy to propagate forward.

According to the shape of the phase surface (also known as the wave front plane), waves are classified into plane waves, spherical waves, and cylindrical waves. Among them, a plane wave is the simplest one in structure. In a small region far away from the radiating source, the spherical wave may be approximated as a plane wave. All the field quantities are in a plane normal to the direction of their propagation. The other forms of EM waves are created by mixing plane waves. If the amplitude and direction of the plane-wave electric field are identical on the whole equiphase surface, it can be defined as a uniform plane wave. Thus, it is very important to comprehend and master the transmission characteristics and basic properties of plane waves[1-5].

Microwave Absorbing Materials
Yuping Duan and Hongtao Guan

ISBN 978-981-4745-10-9 (Hardcover), 978-981-4745-11-6 (eBook)
www.panstanford.com

1.1 Plane Electromagnetic Wave in Lossy Medium Space

In practice, there is more of the space consisting of lost media, although most of the time we think the space is lossy as an approximation. Next we will discuss the planar transmission characteristics of the source-free boundless plane waves in a lossy medium.

We assume the characteristic parameters of lossy medium are σ, ε, and μ in source-free boundless space, and the time-varying EM field satisfies Maxwell's equations as

$$\nabla \times \vec{H} = \sigma \vec{E} + j\varepsilon\omega \vec{E} \tag{1.1a}$$

$$\nabla \times \vec{E} = -j\omega\mu \vec{H} \tag{1.1b}$$

$$\nabla \cdot \vec{H} = 0 \tag{1.1c}$$

$$\nabla \cdot \vec{E} = 0 \tag{1.1d}$$

Rewrite Eq. 1.1 as

$$\nabla \times \vec{H} = j\omega\left(\varepsilon - j\frac{\sigma}{\omega}\right)\vec{E}$$

and

$$\varepsilon_c = \varepsilon - j\frac{\sigma}{\omega} = \varepsilon(1 - j\tan\delta) \tag{1.2}$$

ε_c is complex permittivity, and it is related to the frequency. $\tan\delta = \sigma/\varepsilon\omega$ is the loss angle tangent; then

$$\nabla \times \vec{H} = j\omega\varepsilon_c \vec{E} \tag{1.3}$$

This equation is in the same form as Maxwell's equation in an ideal medium. This means that the solution of plane EM waves can be obtained if ε in the relative equations of an ideal medium was replaced by ε_c.

Wave equations are

$$\nabla^2 \vec{E} + \omega^2 \mu\varepsilon_c \vec{E} = 0 \tag{1.4}$$

$$\nabla^2 \vec{H} + \omega^2 \mu\varepsilon_c \vec{H} = 0 \tag{1.5}$$

If $k_c^2 = \omega^2\mu\varepsilon_c$, then

$$k_c = \omega\sqrt{\mu\left(\varepsilon - j\frac{\sigma}{\omega}\right)} = \omega\sqrt{\mu\varepsilon}\sqrt{1 - j\tan\delta} = \beta - j\alpha$$

$$\alpha = \omega\sqrt{\frac{\mu\varepsilon}{2}\left(\sqrt{1+\tan^2\delta}-1\right)} \tag{1.6}$$

$$\beta = \omega\sqrt{\frac{\mu\varepsilon}{2}\left(\sqrt{1+\tan^2\delta}+1\right)} \tag{1.7}$$

If $\vec{E} = E_x\vec{a}_x$, we can obtain the values of the electric field and the magnetic field as

$$E_x = E_{xm}e^{-jk_z z} = E_{xm}e^{-\alpha z}e^{-j\beta z} \tag{1.8}$$

$$H_y = \frac{j}{\omega\mu}\frac{\partial E_x}{\partial z} = \sqrt{\frac{\varepsilon}{\mu}}(1-j\tan\delta)E_{xm}e^{-\alpha z}e^{-j\beta z} \tag{1.9}$$

On the basis of the equations above, in a lossy medium:

- As the transmitting range increases, the amplitudes of this field decreases. The attenuation constant is α.
- As the phase-moving constant β is related to frequency ω, we know the phase velocity also relates to it. This phenomenon is called dispersion.
- The phase of electric field intensity is different from that of magnetic field intensity, and the ratio of these two is known as complex wave impedance.

$$\eta_c = \sqrt{\frac{\mu}{\varepsilon_c}} = \sqrt{\frac{\mu}{\varepsilon(1-j\tan\delta)}} = \frac{\eta}{\sqrt{\varepsilon(1-j\tan\delta)}} \tag{1.10}$$

In this equation, $\eta = \sqrt{\mu/\varepsilon}$. In practical applications, the following two special situations always can be seen[6]: low-loss medium and high-loss medium.

1.1.1 Low-Loss Medium

Under this condition ($\tan\delta << 1$, that is, $\sigma << \omega\varepsilon$), using the binomial theorem

$$\begin{aligned} k_c &= \omega\sqrt{\mu\varepsilon}(1-j\tan\delta)^{1/2} \\ &= \omega\sqrt{\mu\varepsilon}[1-j\tan\delta/2+\cdots] \end{aligned}$$

And ignoring the higher-order term behind the quadratic term, we get

$$k_c \approx \omega\sqrt{\mu\varepsilon} - j\frac{\sigma}{2}\sqrt{\frac{\mu}{\varepsilon}} = \beta - j\alpha \tag{1.11}$$

Then

$$\beta \approx \omega\sqrt{\mu\varepsilon} \tag{1.12}$$

$$\alpha \approx \frac{\sigma}{2}\sqrt{\frac{\mu}{\varepsilon}} \tag{1.13}$$

$$\eta_c \approx \sqrt{\frac{\mu}{\varepsilon}} = \eta \tag{1.14}$$

Thus it can be seen that β and η_c have almost the same results as in an ideal medium; only they have an attenuation on amplitude, which is not serious. Then we can approximately take the low-loss medium as the ideal medium in practice.

1.1.2 High-Loss Medium

In this case $\tan\delta >> 1$, that is, $\sigma >> \omega\varepsilon$

$$k_c = \omega\sqrt{\mu\varepsilon}(1 - j\tan\delta)^{1/2} \approx \sqrt{\omega\mu\sigma}e^{-j\pi/4} \tag{1.15}$$

So we can get

$$\alpha = \beta = \sqrt{\frac{\omega\mu\sigma}{2}} \tag{1.16}$$

$$\eta_c = \sqrt{\frac{\mu}{\left(\varepsilon - j\frac{\sigma}{\omega}\right)}} \approx \sqrt{\frac{\omega\mu}{\sigma}}e^{j\pi/4} \tag{1.17}$$

From the above it is obvious that the higher the frequency of EM waves, the faster the transmission attenuation of EM waves in the medium, and the shorter the distance of transmission. Therefore, high-frequency EM waves only exist in the thin layer which is near the conductor surface. This phenomenon is said to be the skin effect whose level is characterized by skin depth (or penetrating depth).

Skin depth is defined as the distance traveled by waves in a conducting medium at which its amplitude falls to $1/e$ of its value on the surface of that conducting medium. It is represented as h, and obviously there is

$$h = \frac{1}{\alpha} = \sqrt{\frac{2}{\omega\mu\sigma}} = \frac{1}{\sqrt{\pi f \mu\sigma}} \tag{1.18}$$

As a good conductor has a high conductivity, there is a very small skin depth for high-frequency EM waves. Therefore, the metal has an excellent function of shielding radio waves.

1.2 Reflection and Refraction of Uniform Plane Waves

In practice, EM waves inevitably encounter the interfaces of various media where they are reflected and transmitted. In this section we start out from simple boundary problems to introduce the characteristics of reflection and refraction[7–10].

1.2.1 Vertically Incident, Uniform Plane Electromagnetic Waves on the Interface

As shown in Fig. 1.1, uniform plane waves are transmitted from medium 1 to medium 2 with the interface of infinite plane $z = 0$. The parameters of medium 1 are ε_1, μ_1, and σ_1 and those of medium 2 are ε_2, μ_2, and σ_2.

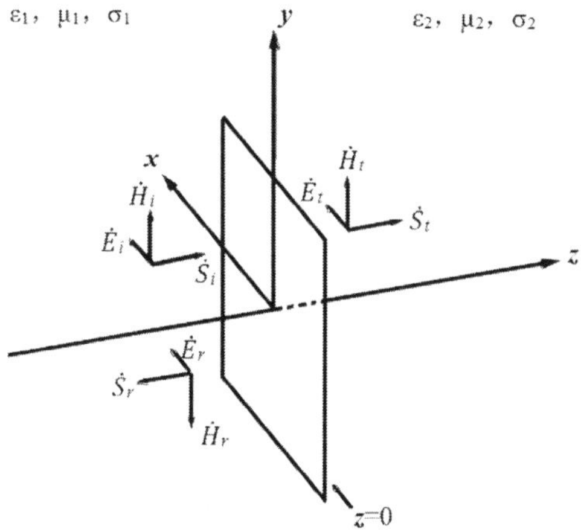

Figure 1.1 Vertical incidence of the boundary.

According to the characteristics of plane waves in lossy medium space mentioned in Section 1.1.1, we assume the electric field of incident waves in common media is

$$\vec{E}_{\mathrm{i}}(z)=\vec{a}_x E_{\mathrm{im}} e^{-jk_{\mathrm{c1}}z} \tag{1.19}$$

Then their magnetic fields are

$$\vec{H}_{\mathrm{i}}(z)=\vec{a}_y \frac{E_{\mathrm{im}}}{\eta_{\mathrm{c1}}} e^{-jk_{\mathrm{c1}}z} \tag{1.20}$$

And the reflected electric and magnetic fields are

$$\vec{E}_{\mathrm{r}}(z)=\vec{a}_x E_{\mathrm{rm}} e^{jk_{\mathrm{c1}}z} \tag{1.21}$$

$$\vec{H}_{\mathrm{r}}(z)=-\vec{a}_y \frac{E_{\mathrm{rm}}}{\eta_{\mathrm{c1}}} e^{jk_{\mathrm{c1}}z} \tag{1.22}$$

Electric and magnetic fields of transmitted waves in medium 2 are

$$\vec{E}_{\mathrm{t}}(z)=\vec{a}_x E_{\mathrm{tm}} e^{-jk_{\mathrm{c2}}z} \tag{1.23}$$

$$\vec{H}_{\mathrm{t}}(z)=\vec{a}_y \frac{E_{\mathrm{tm}}}{\eta_{\mathrm{c2}}} e^{-jk_{\mathrm{c2}}z} \tag{1.24}$$

In Eqs. 1.19–1.24

$$k_{\mathrm{c1}}=\omega\sqrt{\mu_1\varepsilon_1}(1-j\tan\delta_1)^{1/2}$$

$$\eta_{\mathrm{c1}}=\sqrt{\frac{\mu_1}{\varepsilon_1}}(1-j\tan\delta_1)^{-1/2}$$

$$k_{\mathrm{c2}}=\omega\sqrt{\mu_2\varepsilon_2}(1-j\tan\delta_2)^{1/2}$$

$$\eta_{\mathrm{c2}}=\sqrt{\frac{\mu_2}{\varepsilon_2}}(1-j\tan\delta_2)^{-1/2}$$

The composite electric and magnetic fields in medium 1 are

$$\vec{E}_1(z)=\vec{E}_{\mathrm{i}}(z)+\vec{E}_{\mathrm{r}}(z)=\vec{a}_x(E_{\mathrm{im}}e^{-jk_{\mathrm{c1}}z}+E_{\mathrm{rm}}e^{jk_{\mathrm{c1}}z}) \tag{1.25}$$

$$\vec{H}_1(z)=\vec{H}_{\mathrm{i}}(z)+\vec{H}_{\mathrm{r}}(z)=\vec{a}_y\left(\frac{E_{\mathrm{im}}}{\eta_{\mathrm{c1}}}e^{-jk_{\mathrm{c1}}z}-\frac{E_{\mathrm{rm}}}{\eta_{\mathrm{c1}}}e^{jk_{\mathrm{c1}}z}\right) \tag{1.26}$$

According to Eqs. 1.23–1.26 and the boundary conditions of the tangential electric field and the tangential magnetic field on $z = 0$

$$\vec{E}_1(z=0)=\vec{a}_x(E_{\mathrm{im}}+E_{\mathrm{rm}})=\vec{E}_{\mathrm{t}}(z=0)=\vec{a}_x E_{\mathrm{tm}}$$

$$\vec{H}_1(z=0)=\vec{a}_y\left(\frac{E_{\text{im}}}{\eta_{\text{c1}}}-\frac{E_{\text{rm}}}{\eta_{\text{c1}}}\right)=\vec{H}_t(z=0)=\vec{a}_y\frac{E_{\text{tm}}}{\eta_{\text{c2}}}$$

Combining these two equations above, we can get

$$E_{\text{rm}}=\frac{\eta_{\text{c2}}-\eta_{\text{c1}}}{\eta_{\text{c2}}+\eta_{\text{c1}}}E_{\text{im}} \tag{1.27}$$

$$E_{\text{tm}}=\frac{2\eta_{c2}}{\eta_{c2}+\eta_{c1}}E_{\text{im}} \tag{1.28}$$

For the convenience of analysis, we define the rate of reflection wave fields' amplitudes and incident wave fields' amplitudes as the reflection coefficient, in terms of Γ. Then there is

$$\Gamma=\frac{E_{\text{rm}}}{E_{\text{im}}}=\frac{\eta_{\text{c2}}-\eta_{\text{c1}}}{\eta_{\text{c2}}+\eta_{\text{c1}}} \tag{1.29}$$

Defining the rate of transmitted and incident wave fields' amplitudes as the reflection coefficient, in terms of τ, we get

$$\tau=\frac{E_{\text{tm}}}{E_{\text{im}}}=\frac{2\eta_{\text{c2}}}{\eta_{\text{c2}}+\eta_{\text{c1}}} \tag{1.30}$$

The reflection coefficient and the transmission coefficient are two of the most important parameters to analyze the transmission characteristics of EM waves in bounded space, and there is

$$1+\Gamma=\tau \tag{1.31}$$

Let's substitute Γ and τ in Eqs. 1.23 and 1.25. Then the electric fields in medium 1 and medium 2 are

$$\vec{E}_1(z)=\vec{a}_xE_{\text{im}}(e^{-jk_{c1}z}+\Gamma e^{jk_{c1}z}) \tag{1.32}$$

$$\vec{E}_t(z)=\vec{a}_x\tau E_{\text{im}}e^{-jk_{c2}z} \tag{1.33}$$

For the common media, as η_{c1} and η_{c2} are complex, so the reflection coefficient and the transmission coefficient also should be complex items. It indicates that reflected and transmitted waves through the interface vary not only in the amplitudes but also in the phases and their variation is related to the medium parameters.

Next let's analyze two special conditions:

- Interface between an ideal medium and an ideal conductor
- Interface of an ideal medium

1.2.1.1 Interface between an ideal medium and an ideal conductor

We assume medium 1 is an ideal medium and medium 2 is an ideal conductor; that is, $\sigma_1 = 0$ and $\sigma_2 = \infty$. Then

$$k_{c1} = \omega\sqrt{\mu_1\varepsilon_1} = k_1$$

$$\eta_{c1} = \sqrt{\frac{\mu_1}{\varepsilon_1}} = \eta_1$$

$$k_{c2} = \infty$$

$$\eta_{c2} = 0$$

Let's substitute them in Eqs. 1.29, 1.30, 1.32, and 1.33; we obtain

$$\Gamma = -1 \tag{1.34}$$

$$\tau = 0 \tag{1.35}$$

$$\vec{E}_1(z) = -\vec{a}_x 2jE_{\text{im}} \sin k_1 z \tag{1.36}$$

$$\vec{H}_1(z) = \vec{a}_y \frac{2E_{\text{im}}}{\eta_1} \cos k_1 z \tag{1.37}$$

$$\vec{E}_{\text{t}}(z) = 0 \tag{1.38}$$

$$\vec{H}_{\text{t}}(z) = 0 \tag{1.39}$$

These equations state that there is no time-varying electric field in a perfect conductor. Besides, according to the boundary conditions of $\vec{n} \cdot \vec{E} = \rho_{\text{s}}$ and $\vec{n} \times \vec{H} = \vec{J}_{\text{s}}$, it can be known that there is no inductive charge on the surface of a perfect conductor. The surface inductive charge is

$$\vec{J}_{\text{s}} = -\vec{a}_z \times \vec{H}_1(z=0) = \vec{a}_x \frac{2E_{\text{im}}}{\eta_1} \tag{1.40}$$

To analyze the transmission characteristics in medium 1, we express the total electric field and the total magnetic field in the form of an instant. These are

$$\vec{E}_1(z,t) = \text{Re}[\vec{E}_1(z)e^{j\omega t}] = \vec{a}_x 2E_{\text{im}} \sin k_1 z \sin \omega t \tag{1.41}$$

$$\vec{H}_1(z,t) = \text{Re}[\vec{H}_1(z)e^{j\omega t}] = \vec{a}_y \frac{2E_{\text{im}}}{\eta_1} \cos k_1 z \cos \omega t \tag{1.42}$$

The space-time curves of Eqs. 1.41 and 1.42 are shown in Fig. 1.2. They indicate that electric and magnetic fields pulsate with time, but there are a fixed max point (wave antinode points) and zero point (wave nodal points) in the z direction. This kind of waves are called pure standing waves. The locations of electric field wave antinode points (magnetic wave nodal points) are $z=-\frac{(2n+1)\lambda}{4}$, $n = 0, 1, 2, \ldots$ The locations of electric wave nodal points (magnetic antinode point) are $z=-\frac{n\lambda}{2}$, $n = 0, 1, 2, \ldots$

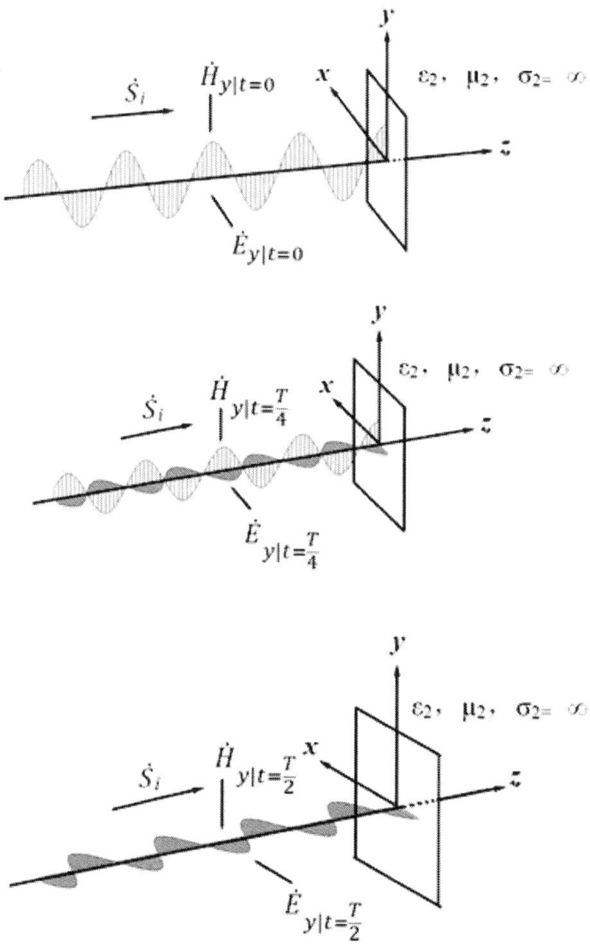

Figure 1.2 Space and time distributions of electric fields and magnetic fields in the case of perpendicular incidence.

The average synthesized energy flux density of magnetic waves in medium 1 is

$$\vec{S}_{av} = \frac{1}{2}\mathrm{Re}[\vec{E}_1(z) \times \vec{H}_1^*(z)] = 0 \tag{1.43}$$

This equation indicates that pure standing waves can't be the method to transmit magnetic energy. Therefore, electric energy and magnetic energy mutually convert between the two nodal points with a 1/4 cycle.

1.2.1.2 Interface of an ideal medium

We assume $\sigma_1 = \sigma_2 = 0$, that is, media 1 and 2 are both ideal media. Then

$$k_{c1} = \omega\sqrt{\mu_1\varepsilon_1} = k_1$$

$$\eta_{c1} = \sqrt{\frac{\mu_1}{\varepsilon_1}} = \eta_1$$

$$k_{c2} = \omega\sqrt{\mu_2\varepsilon_2} = k_2$$

Substituting them in Eqs. 1.29, 1.30, 1.32, and 1.33,

$$\Gamma = \frac{\eta_2 - \eta_1}{\eta_2 + \eta_1} \tag{1.44}$$

$$\tau = \frac{2\eta_2}{\eta_2 + \eta_1} \tag{1.45}$$

$$\vec{E}_1(z) = \vec{a}_x E_{\mathrm{im}}[(1+\Gamma)e^{-jk_1 z} + j\Gamma 2\sin k_1 z] \tag{1.46}$$

The first item in Eq. 1.46 expresses the magnetic waves with their amplitude of E_{im} $(1 + \Gamma)$ transmitting along the y direction; these kind of waves are called travelling waves. The second item expresses that of standing waves. In consequence, the waves $\vec{E}_1(z)$ containing both travelling wave components and standing wave components are named standing traveling waves (mixed waves). Standing traveling waves have fixed antinode points and nodal points. But compared to pure standing waves, their field intensity on antinode points is not zero. The transmitted waves are unidirectional, and their characteristics are similar to plane waves'. So it won't be discussed here.

For the standing traveling wave, the standing wave coefficient S (or standing wave ratio) is introduced to describe its characteristics, and it is defined as the rate of the maximum value and the minimum value in the standing wave electric field. It can be expressed as

$$S = \frac{|E|_{\max}}{|E|_{\min}} \tag{1.47}$$

To analyze the characteristics of a standing wave in medium 1, it can be rewritten as

$$\vec{E}_1(z) = \vec{a}_x E_{\text{im}} e^{-jk_1 z}(1 + \Gamma e^{j2k_1 z}) \tag{1.48}$$

For the ideal medium, η_1 and η_2 are both positive real numbers; so are Γ and τ. But as the relationship of η_1 and η_2 changes, Γ can be a positive real number or a negative one.

- Condition 1: $\Gamma > 0$, that is, $\eta_2 > \eta_1$.

On the basis of what is mentioned above, the maximum value of the electric field is

$$|E_1|_{\max} = E_{\text{im}}(1 + \Gamma) \tag{1.49}$$

The locations of the maximum value in the electric field are $2k_1 z = -2n\pi$:

$$z = -\frac{n\lambda_1}{2} \quad n = 0, 1, 2, \ldots \tag{1.50}$$

The minimum value in the electric field is

$$|E_1|_{\min} = E_{\text{im}}(1 - \Gamma) \tag{1.51}$$

The locations of the minimum value in the electric field are $2k_1 z = -(2n + 1)\pi$:

$$z = -\frac{(2n+1)\lambda_1}{4} \quad n = 0, 1, 2, \ldots \tag{1.52}$$

- Condition 2: $\Gamma < 0$, that is, $\eta_2 < \eta_1$.

Obviously, amplitudes of the maximum and minimum point are the same as those in condition 1, but they are at the opposite locations. Substituting Eqs. 1.49 and 1.51 in Eq. 1.47, we get

$$S = \frac{1 + |\Gamma|}{1 - |\Gamma|} \tag{1.53}$$

and

$$|\Gamma| = \frac{S+1}{S-1} \tag{1.54}$$

The standing wave ratio S is a real number and convenient to be measured; therefore we often get the reflection coefficient by computing S, which can be given from measurement. The EM energy flow densities in these two media are

$$\vec{S}_{1\text{av}} = \frac{1}{2}\text{Re}[\vec{E}_1(z)\times\vec{H}_1^*(z)] = \frac{E_{\text{im}}^2}{2\eta_1}(1-\Gamma^2)\vec{a}_z \tag{1.55}$$

$$\vec{S}_{\text{tav}} = \frac{1}{2}\text{Re}[\vec{E}_\text{t}(z)\times\vec{H}_\text{t}^*(z)] = \frac{\tau E_{\text{im}}^2}{2\eta_2}\vec{a}_z \tag{1.56}$$

It can be proved that the sum of the transmitted waves' and reflected waves' energy flow equals the incident waves' energy flow, which satisfies the law of conservation of energy.

1.2.2 Normal Incidence on the Interface of Multilayered Media

Multilayered media are very common in practical applications, such as composite materials, surface shielding materials, and absorbing coated materials. Taking the case of the three-layered medium let's discuss the problem of normal incidence of plane waves. The analysis model is shown in Fig. 1.3.

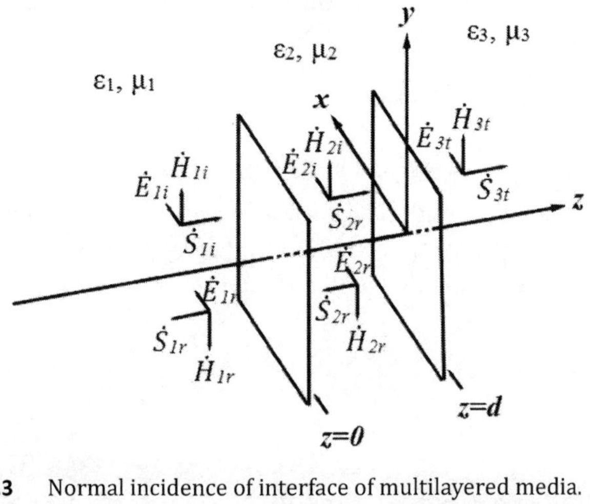

Figure 1.3 Normal incidence of interface of multilayered media.

According to the characteristics of plane waves, we can present the formulas of the superposed electric field and magnetic field. In medium 1

$$\vec{E}_1(z)=\vec{a}_x E_{1\mathrm{im}}(e^{-jk_1 z}+\Gamma_1 e^{jk_1 z}) \tag{1.57a}$$

$$\vec{H}_1(z)=\vec{a}_y \frac{E_{1\mathrm{im}}}{\eta_1}(e^{-jk_1 z}-\Gamma_1 e^{jk_1 z}) \tag{1.57b}$$

In medium 2

$$\vec{E}_2(z)=\vec{a}_x E_{2\mathrm{im}}(e^{-jk_2(z-d)}+\Gamma_2 e^{jk_2(z-d)}) \tag{1.58a}$$

$$\vec{H}_2(z)=\vec{a}_y \frac{E_{2\mathrm{im}}}{\eta_2}(e^{-jk_2(z-d)}-\Gamma_2 e^{jk_2(z-d)}) \tag{1.58b}$$

In medium 3

$$\vec{E}_3(z)=\vec{a}_x E_{3\mathrm{im}} e^{-jk_3(z-d)} \tag{1.59a}$$

$$\vec{H}_3(z)=\vec{a}_y \frac{E_{3\mathrm{im}}}{\eta_3} e^{-jk_3(z-d)} \tag{1.59b}$$

For interface 2, there is

$$\Gamma_2=\frac{\eta_3-\eta_2}{\eta_3+\eta_2} \tag{1.60}$$

When $z = 0$ on interface 1, the tangential components of the electric field and the magnetic field are continuous. They are

$$\vec{E}_1(0)=\vec{E}_2(0)$$

$$\vec{H}_1(0)=\vec{H}_2(0)$$

Dividing both sides of the two equations above and substituting the result in Eqs. 1.57 and 1.58, we can get

$$\eta_1 \frac{1+\Gamma_1}{1-\Gamma_1}=\eta_2 \frac{e^{jk_2 d}+\Gamma_2 e^{-jk_2 d}}{e^{jk_2 d}-\Gamma_2 e^{-jk_2 d}} \tag{1.61}$$

If there is

$$Z_\mathrm{p}=\eta_2 \frac{e^{jk_2 d}+\Gamma_2 e^{-jk_2 d}}{e^{jk_2 d}-\Gamma_2 e^{-jk_2 d}}$$

Then substitute Eq. 1.60 into the equations above. Combining those with Euler's formula, we can get

$$Z_\mathrm{p}=\eta_2 \frac{\eta_3+j\eta_2 \tan k_2 d}{\eta_2+j\eta_3 \tan k_2 d} \tag{1.62}$$

Z_p is the ratio of the electric and magnetic fields on $z = 0$ in medium 2. It has the same dimension as impedance; therefore we define it as the equivalent wave impedance on plane 1 in medium 2. Substituting Z_p in Eq. 1.61, one can get

$$\Gamma_1 = \frac{Z_p - \eta_1}{Z_p + \eta_1} \tag{1.63}$$

Equations 1.62 and 1.63 are very useful in the analysis of reflection and transmission characteristics. When there are n layers in the medium, starting from the layer $n - 1$ we can compute the reflection coefficient of these layers by recycling Eqs. 1.62 and 1.63. Then the electric and magnetic fields of all the layers can be computed.

Next we discuss two special cases of multilayers:

- Quarter-wave matching layer
- Half-wavelength dielectric window

1.2.2.1 Quarter-wave matching layer

As shown in Fig. 1.4, assuming the thickness of medium 2 is $\frac{\lambda_2}{4}$, $k_2 d = \frac{2\pi}{\lambda_2}\frac{\lambda_2}{4} = \frac{\pi}{2}$ and $\tan k_2 d = \infty$. Substituting them in Eqs. 1.62 and 1.63, one can get that the equivalent wave impedance and reflection coefficients are

$$Z_p = \frac{\eta_2^2}{\eta_3} \tag{1.64}$$

$$\Gamma_1 = \frac{\eta_2^2 - \eta_1\eta_3}{\eta_2^2 + \eta_1\eta_3} \tag{1.65}$$

That is to say there is no reflection on interface 1. It means that medium 2 transforms η_3 to $Z_p = \eta_1$ of interface 1, which counteracts the reflection and plays the role of impedance conversion. Thus medium 2 is termed the "quarter-wave matching layer." This property can be used in the design of absorbing coatings to improve the effects of absorbing materials.

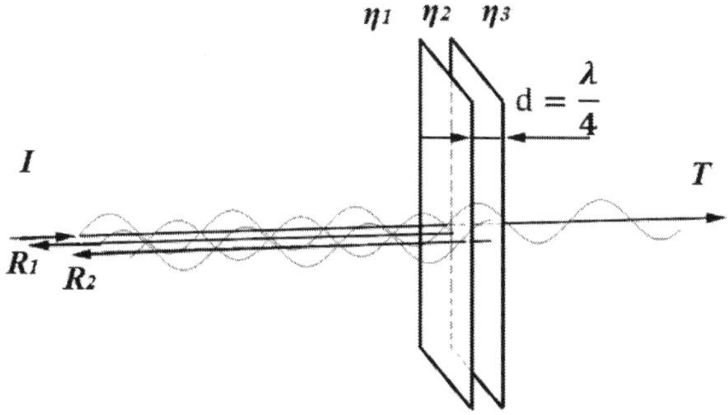

Figure 1.4 Quarter-wave matching layer.

1.2.2.2 Half-wavelength dielectric window

If the thickness of medium 2 is $\frac{\lambda_2}{2}$, $k_2 d = \frac{2\pi}{\lambda_2}\frac{\lambda_2}{2} = \pi$ and $\tan k_2 d = 0$. Then $Z_p = \eta_3$. If we assume $\eta_3 = \eta_1$, medium 1 is the same as medium 3; then there is

$$\Gamma_1 = \frac{Z_p - \eta_1}{Z_p + \eta_1} = 0 \tag{1.66}$$

There is no reflection on interface 1 and it is easy to conclude $E_{3tm} = E_{1im}$. It indicates that incident waves transmit from medium 1 to medium 3 just like there is no medium 2. Thus we call it a "half-wavelength dielectric window." This property is extensively used in the design of radomes.

1.2.3 Oblique Incidence of Uniform Plane Electromagnetic Waves on the Interface

When an EM wave strikes a plane boundary with any arbitrary angle, we refer to it as oblique incidence. In this case, as the plane of electric and magnetic fields is not horizontal with the plane of the interface, the analysis is much more difficult, although the overall thought of analytical method is the same as the oblique incident before.

For the sake of convenient analysis, first we introduce the concept of an incidence plane and a reflection plane. Then we classify oblique incidence plane waves into parallel polarized waves and vertically polarized waves. As shown in Fig. 1.5, the plane which includes the normal to the boundary $\vec{n}$ and the incident ray is called the plane of incidence, θ_i is called the angle of incidence, the plane that includes the normal to the boundary $\vec{n}$ and the reflected ray is called the plane of incidence, θ_r is called the angle of reflection, and θ_t is the angle of transmission. When the field of the incident wave is normal to the incident plane, we called it the perpendicularly polarized wave; when the field of the incident wave is parallel to the incident plane, we called it the parallel polarized wave. The polarized waves of any orientation can be decomposed into parallel polarized waves and perpendicularly polarized waves.

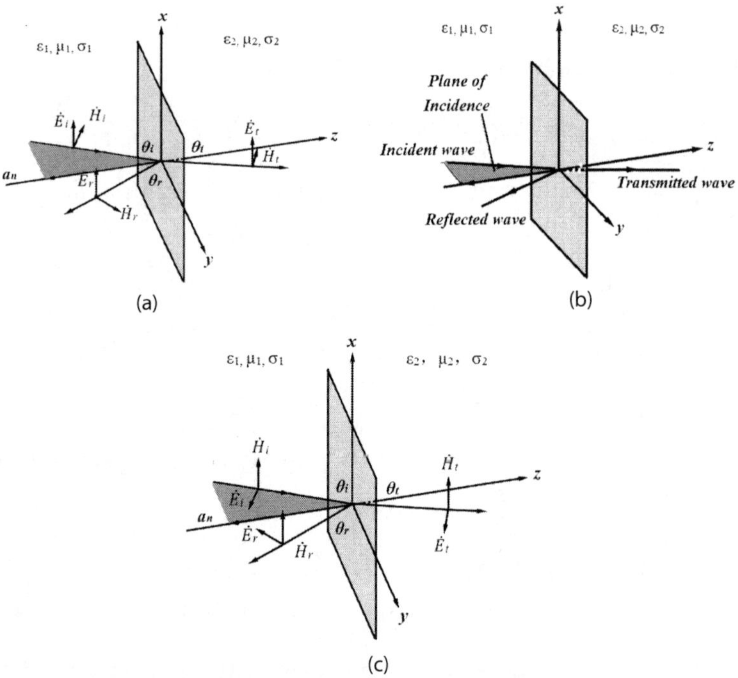

Figure 1.5 (a) Basic definition and the sketch map for (b) perpendicularly and (c) parallel polarized waves.

1.2.3.1 Oblique incidence on the surface of an ideal medium plane

For the sake of convenient analysis and in general, we assume $z = 0$ as the interface of the ideal medium and there is a coincidence between the incident plane and the coordinate plane. They are shown in Fig. 1.6. According to the boundary conditions, the components of the tangential electric and magnetic fields are continuous. This relationship must be satisfied on the whole interface. In other words, the transmit velocity components of incident, reflected, and transmitted waves along the interface must be equivalent. That is,

$$\frac{v_1}{\sin\theta_\mathrm{i}} = \frac{v_1}{\sin\theta_\mathrm{r}} = \frac{v_2}{\sin\theta_\mathrm{t}}$$

So

$$\theta_\mathrm{r} = \theta_\mathrm{i} \tag{1.67}$$

$$\frac{\sin\theta_\mathrm{i}}{\sin\theta_\mathrm{t}} = \frac{v_1}{v_2} = \frac{\sqrt{\mu_2\varepsilon_2}}{\sqrt{\mu_1\varepsilon 1}} \tag{1.68}$$

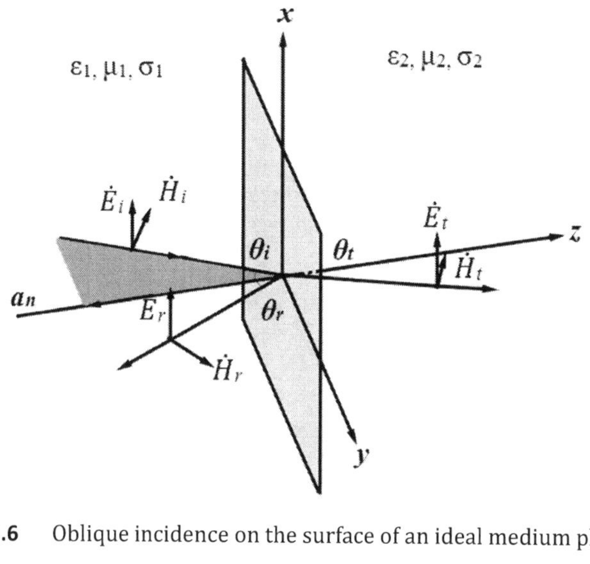

Figure 1.6 Oblique incidence on the surface of an ideal medium plane.

Equations 1.67 and 1.68 are Snell's laws of refraction and refraction, respectively. For general media, $\mu_1 = \mu_2 = \mu_0$. $n = \sqrt{\varepsilon_\mathrm{r}}$ is called the refractive index. Then Eq. 1.68 can be written as

$$n_\mathrm{t}\sin\theta_\mathrm{t} = n_\mathrm{i}\sin\theta_\mathrm{i} \tag{1.69}$$

That is the mathematical expression for the optic refraction law. For parallel polarized waves, boundary conditions of the tangential field are

$$E_{\mathrm{i}}\cos\theta_{\mathrm{i}} - E_{\mathrm{r}}\cos\theta_{\mathrm{r}} = E_{\mathrm{t}}\cos\theta_{\mathrm{t}} \tag{1.70a}$$

$$\frac{E_{\mathrm{i}}}{\eta_1} + \frac{E_{\mathrm{r}}}{\eta_1} = \frac{E_{\mathrm{t}}}{\eta_{\mathrm{t}}} \tag{1.70b}$$

Combining these two equations above, we can get

$$E_{\mathrm{r}} = \frac{\eta_1\cos\theta_{\mathrm{i}} - \eta_2\cos\theta_{\mathrm{t}}}{\eta_1\cos\theta_{\mathrm{i}} + \eta_2\cos\theta_{\mathrm{t}}} E_{\mathrm{i}} = \Gamma_{||}E_{\mathrm{i}} \tag{1.71}$$

$$E_{\mathrm{t}} = \frac{2\eta_2\cos\theta_{\mathrm{i}}}{\eta_1\cos\theta_{\mathrm{i}} + \eta_2\cos\theta_{\mathrm{t}}} E_{\mathrm{i}} = \tau_{||}E_{\mathrm{i}} \tag{1.72}$$

$\Gamma_{||}$, $\tau_{||}$ in the above equations are the reflection coefficient and transmission coefficient of parallel polarized waves, respectively. And

$$\Gamma_{||} = \frac{\eta_1\cos\theta_{\mathrm{i}} - \eta_2\cos\theta_{\mathrm{t}}}{\eta_1\cos\theta_{\mathrm{i}} + \eta_2\cos\theta_{\mathrm{t}}} \tag{1.73}$$

$$\tau_{||} = \frac{2\eta_2\cos\theta_{\mathrm{i}}}{\eta_1\cos\theta_{\mathrm{i}} + \eta_2\cos\theta_{\mathrm{t}}} \tag{1.74}$$

$$\tau_{||} = (1 + \Gamma_{||})\,\eta_2/\eta_1 \tag{1.75}$$

η_1 and η_2 are the wave impedance of medium 1 and medium 2, respectively. Similarly, for vertically polarized waves

$$\Gamma_{\perp} = \frac{\eta_2\cos\theta_{\mathrm{i}} - \eta_1\cos\theta_{\mathrm{t}}}{\eta_2\cos\theta_{\mathrm{i}} + \eta_1\cos\theta_{\mathrm{t}}} \tag{1.76}$$

$$\tau_{\perp} = \frac{2\eta_2\cos\theta_{\mathrm{i}}}{\eta_2\cos\theta_{\mathrm{i}} + \eta_1\cos\theta_{\mathrm{t}}} \tag{1.77}$$

$$\tau_{\perp} = 1 + \Gamma_{\perp} \tag{1.78}$$

- Condition 1: Total reflection and critical angle

  Total reflection is the special case for $|\Gamma| = 1$. When $\theta_{\mathrm{t}} = \pi/2$, from Eqs. 1.73 and 1.76, it can be known that $\Gamma_{||} = \Gamma_{\perp} = 1$. The incident angle of total reflection is named the critical angle; it can be expressed as θ_{C}. Thus it can be concluded from Eq. 1.65 that

  $$\theta_{\mathrm{c}} = \arcsin\sqrt{\varepsilon_2/\varepsilon_1} \tag{1.79}$$

  Obviously, only when $\varepsilon_1 > \varepsilon_2$, that is, EM waves move from a high-optical-density medium to a lower one, total reflection can be observed.

- Condition 2: Total transmission and Brewster angle
  The total transmitted wave is the special case for $|\Gamma| = 0$. For a parallel polarized wave, it can be concluded from Eq. 1.73 that $|\Gamma| = 0$, that is,
  $\eta_1 \cos\theta_i = \eta_2 \cos\theta_t$
  Substituting it into Eq. 1.69, we get

$$\theta_i = \theta_B = \arcsin\sqrt{\frac{\varepsilon_2}{\varepsilon_1 + \varepsilon_2}} \tag{1.80}$$

In the formula above, θ_B is known as the Brewster angle. We can also conclude from it that when parallel polarized waves are incident on the surface of an ideal medium with an angle of θ_B, the EM wave energy will pass totally into medium 2. Then the reflected energy is zero.

For a vertical polarization wave, the analysis above is equally applicable. If $\Gamma_\perp = 0$, there must be $\varepsilon_1 = \varepsilon_2$. Then we know these two media are the same, which means that the total transmission can't happen to vertical polarization waves.

1.2.3.2 Oblique incidence of an ideal conductor plane

Assume medium 1 is an ideal medium, μ_1, ε_1, $\sigma_1 = 0$ and medium 2 is an ideal conductor with $\sigma_2 = \infty$. The uniform plane wave is incident on the surface of the conductor with the incident angle of θ_i.

As shown in Fig. 1.7, for vertical polarization waves as $\sigma_2 = \infty$, $\eta_2 = 0$. From Eqs. 1.76 and 1.77 we can conclude that $\Gamma_\perp = -1$, $\tau_\perp = 0$, that is, there is total reflection. From the analysis, the composite electric and magnetic fields in the medium can be expressed as

$$E_1 = \vec{E}_i + \vec{E}_r = -\vec{a}_y j2E_{im} \sin(k_1 z\cos\theta_i)e^{-jk_1 x\sin\theta_i} \tag{1.81a}$$

$$\vec{H}_1 = \vec{H}_i + \vec{H}_r = -\vec{a}_x \frac{2E_{im}}{\eta_1}\cos\theta_i \cos(k_1 z\cos\theta_i)e^{-jk_1 x\sin\theta_i}$$

$$-\vec{a}_z \frac{j2E_{im}}{\eta_1}\sin\theta_i \sin(k_1 z\cos\theta_i)e^{-jk_1 x\sin\theta_i} \tag{1.81b}$$

It can be known from these equations that when a vertically polarized wave is incident on an ideal conductor surface, the EM wave in the medium area has the following features:

- In the z direction on the surface of the vertical conductor, composite waves distribute as pure standing waves with no EM energy transmission along this direction.
- The composite wave in the x direction is a traveling wave. From $e^{-jk_1 x \sin\theta_\mathrm{i}}$, we can know the phase shift constant in the x direction is $k_1 \sin \theta_1$. Then the phase velocity in this direction is

  $$v_{\mathrm{p}x} = \frac{\omega}{k_1 \sin\theta_\mathrm{i}} = \frac{v_\mathrm{p}}{\sin\theta_\mathrm{i}} > v_\mathrm{p}$$

  In this equation, there is

  $$v_\mathrm{p} = \frac{1}{\sqrt{\mu_1 \varepsilon_1}}$$

- As the uniform amplitude plane z and equal phase plane x are both constants, the composite wave is a nonuniform plane wave.
- There are components of the magnetic field but no component of the electric field in the transmission direction (x direction). Such kind of wave is called a transverse electric (TE) wave.

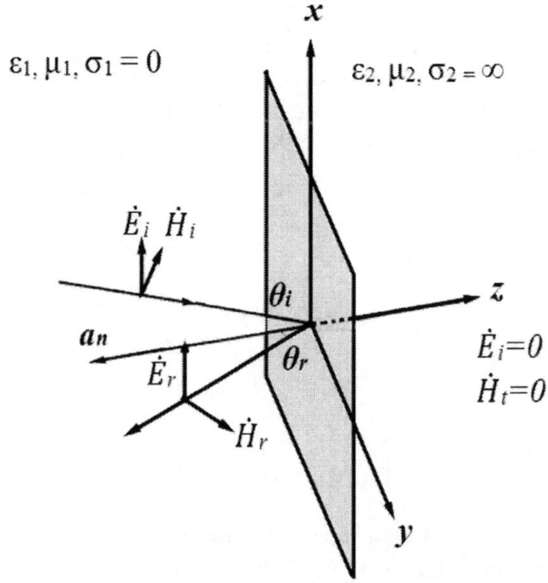

Figure 1.7 Oblique incidence of the ideal conductor plane.

For a parallel polarized wave, $\sigma_2 = \infty$ and $\eta_2 = 0$. From Eqs. 1.75 and 1.76 we can obtain that $\Gamma_{//} = 1$ and $\tau_{//} = 0$. Therefore

$$\vec{E}_1 = -\vec{a}_x j2E_{\text{im}} \cos\theta_{\text{i}} \sin(k_1 z\cos\theta_{\text{i}})e^{-jk_1 x\sin\theta_{\text{i}}} - \vec{a}_z 2E_{\text{im}} \sin\theta_{\text{i}} \cos(k_1 z\cos\theta_{\text{i}})e^{-jk_1 x\sin\theta_{\text{i}}} \quad (1.82\text{a})$$

$$\vec{H}_1 = \vec{a}_y \frac{2E_{\text{im}}}{\eta_1} \cos(k_1 z\cos\theta_{\text{i}})e^{-jk_1 x\sin\theta_{\text{i}}} \quad (1.82\text{b})$$

It can be found from Eq. 1.82 that when a parallel polarized wave is under oblique incidence to the surface of an ideal conductor, the features of EM waves in the medium area are the same as those of vertically polarized waves. There are components of the electric field but no component of the magnetic field in the transmission direction (x direction) at the moment. Such kind of wave is called a TE wave.

1.3 Theoretical Fundamentals of Absorbing Materials

According to the absorbing mechanism, absorbents can be divided into three classes: resistive-type absorbents, dielectric-type absorbents, and magnetic-type absorbents. Resistive-type absorbents absorb EM waves mainly by the interaction with the EM wave. The absorption rate of resistive-type absorbents depends on the conductance and permittivity of material. Resistive-type absorbents are primarily represented by carbon black (CB), metal powders, and silicon carbide. Dielectric-type absorbents absorb EM waves mainly through the dielectric polarization relaxed loss, such as barium titanate and ferroelectric ceramics. Magnetic-type absorbents play the role mainly from resonance and hysteresis loss, such as ferrite and carbonyl-iron (CI). In addition, absorbing coatings contain radioisotopes of high-energy particles can also attenuate EM waves by ionizing the air nearby. But, there are many difficulties to overcome in practical use of radioisotopes, such as complex process control and high cost.

The most important requirement of conventional absorbing materials is the high absorption property. But new absorbing materials are required for better properties, such as thinner thickness, less weight, wider absorption band, and higher strength.

For futuristic absorbing materials, more intensified requirements are raised. The restrictions of absorbents limit the development of EM absorbing materials. Therefore, it's very important to develop new absorbing materials.

With rapid development of modern science and technology, the advent of various increasing mature methods of new materials and new preparation methods is a beneficial guarantee for new absorbing materials. The current study of new absorbents includes nanometer-sized material, chiral materials, and conducting polymers.

1.3.1 Property Characterization of Absorbents

The study of absorbents is an important part of developing radar absorption materials (RAMs), and it is also the material base for studying the absorbing material and improving properties. A good understanding and an apt description of absorbents are of great importance and necessity for the research, production, and selection of RAMs.

1.3.1.1 Electromagnetic parameters and absorbing properties

EM parameters include permittivity (ε) and permeability (μ). They are both important parameters to represent EM properties. A RAM can absorb EM waves as much as possible by adjusting the EM parameters. From the point of a medium, the bigger the ε (ε', ε''), the bigger the μ (μ', μ''), and the better the properties of the RAM. But the characteristic impedance matching must be considered in the design. Therefore, there is an optimum value of ε and μ when we choose a proper material. It is important to balance the characteristic impedance matching and EM wave absorbing.

According to transmission line theory, for a single-layer absorbing material backed by a perfect conductor, the input impedance (Z_{in}) at the air–material interface is given by

$$Z_{in} = \sqrt{\frac{\mu}{\varepsilon}} \tanh\left(j\frac{2\pi}{\lambda} d\sqrt{\mu\varepsilon} \right) \tag{1.83}$$

where μ is the complex permeability ($\mu' - j\mu''$) and ε is the complex permittivity ($\varepsilon' - j\varepsilon''$).

When the incident wave is perpendicularly transmitted to the absorber, the reflection loss R is expressed as follows:

$$R = \frac{Z - 1}{Z + 1} \tag{1.84}$$

The propagation constant γ can be expressed as

$$\gamma = \alpha + j\beta = j\frac{2\pi f}{c}\sqrt{\varepsilon \cdot \mu} \tag{1.85}$$

$$\alpha = \frac{\pi f}{c}(\mu'\varepsilon')^{1/2}$$

$$\left\{2\left[\tan\delta_\varepsilon \tan\delta_{\mathrm{m}} - 1 + (1 + \tan^2\delta_\varepsilon + \tan^2\delta_m + \tan^2\delta_\varepsilon\delta_m)^{\frac{1}{2}}\right]\right\}^{\frac{1}{2}} \tag{1.86}$$

where α, β, c, and f denote the attenuation coefficient, the phase coefficient, the velocity of light, and the frequency of wave, respectively. $\tan\delta_\varepsilon = \varepsilon''/\varepsilon'$, and $\tan\delta_m = \mu''/\mu'$, are the dielectric and magnetic loss factors, respectively.

From Eqs. 1.83–1.86, you can draw this conclusion: To satisfy impedance matching and high absorption properties, a reasonable design of (ε, μ) must be chosen[13–15].

1.3.1.2 Confirmation of electromagnetic parameters

Representations of EM parameter determination are calculation methods and direct measurement methods. The calculation methods of EM parameters are the direct calculation method and the indirect calculation method. The direct calculation method estimates EM parameters on the basis of the magnetic polarization intensity and electric field intensity of the absorbent in the EM field[16]. The main indirect calculation methods are the transmission/reflection method and the multistate, multithickness method[17–22]. Measuring waveguides of samples and the S parameters of coaxial lines are widely employed in the transmission/reflection method; then complex permittivity ε and complex permeability μ are calculated according to the related formulas. Because measuring the amplitudes and phases of S_{11} and S_{21} simultaneously is difficult, changing the terminal states and thicknesses of samples is employed to measure the corresponding complex reflection coefficient and to obtain the complex permittivity and complex permeability through calculation.

There are two main direct testing methods. The first method is that the absorbent and binder are made into coatings or block samples to measure ε and μ. The measured parameters are actually the constants of the complex medium, that is, the relative dielectric constant and magnetic permittivity. Therefore, the quality, mixing ratio, and size dimension should remain the same in this method. The second method is the measurement of absorbent EM parameters ε and μ. The ε and μ measurement of a powder absorbent is the intuitional method to represent its EM properties. Therefore, a sample frame waveguide should be prepared. The two ends of the waveguide sealed with a high-transmission material slice form a rectangular vessel, and the above cover of the waveguide is opened to form the waveguide sample frame. The absorbent powder is filled into the frame according to apparent density and tap density, and the complex permittivity ε and complex permeability μ of the sample frame filled with absorbent are measured using the waveguide measuring system.

1.3.1.3 Electromagnetic parameters of absorbing materials with different absorbent content

Absorbing materials with different absorbents should be measured, and relational expressions should be summarized to obtain the accurate relationship between EM parameters of an absorbing material and the volume percentage of absorbents. μ', μ'', ε', and ε'' are the EM parameters of the absorbing material; μ_1', μ_1'', ε_1', and ε_1'' are the EM parameters of the absorbent; ε'_g and ε''_g are the dielectric constants of the matrix; V_1 and V_g are the volume percentages of the absorbent and the matrix, respectively; and f is the operating frequency. There are no dependencies among the above parameters[23]:

$$\begin{aligned}
\mu' &= f_1(\mu_1', V_1, f) \\
\mu'' &= f_2(\mu_1'', V_1, f) \\
\varepsilon' &= f_3(\varepsilon_1', \varepsilon_g', V_1, V_g, f) \\
\varepsilon'' &= f_4(\varepsilon_1'', \varepsilon_g'', V_1, V_g, f)
\end{aligned} \tag{1.87}$$

where μ_1', μ_1'', ε_1', and ε_1'' are the EM parameters when the quantity of the absorbent is 100% in the absorbing material. Absorbents are usually powder materials and can't be pressed into pure samples; therefore the values of μ_1', μ_1'', ε_1', and ε_1'' cannot be obtained through testing directly. It is feasible for us to calculate the values of V_1 and V_g and test the values of μ', μ'', ε', and ε'' as well as ε'_g and ε''_g for each absorbing material sample. Taking no account of the explicit function expression of the arguments in the above formulas, the numerical values of the parameters at different frequencies can be measured, and therefore, μ_1', μ_1'', ε_1', and ε_1'' can be calculated through four formulas.

However, as the functional relationships in the above formulas are unknown, multiple samples are needed to confirm the values of μ_1', μ_1'', ε_1', and ε_1''. If the given functions are correct, the obtained μ_1', μ_1'', ε_1', and ε_1'' from each sample should reach the same values. Actually, for the existence of measuring error and sample inhomogeneity, these values are just closed and a few values may have a larger difference. In the actual processing, the values with relatively larger deviations are rejected and then the average value is calculated. When these values are brought into the above functions, the relation curve between μ', μ'', ε', and ε'' and V_1 can be determined.

In the actual test, the volume fraction can be determined on the basis of the ratio of the density and weight of each component. If the densities of the absorbent and matrix are ρ_1 and ρ_g, respectively, the density of the absorbing material sample is ρ_s; the compound dielectric of the matrix is $\varepsilon_g' - j\varepsilon_g''$; and the weight ratios of the absorbent and matrix are G_1 and G_g, respectively, and $G_1 + G_g = 1$. The volume percentages of the absorbent, matrix, and air are V_1, V_g, and V_k, respectively, and they can be calculated according to the following formulas:

$$\begin{aligned} V_1 &= \rho_s G_1/\rho_1 \\ V_g &= \rho_s G_g/\rho_g \\ V_k &= 1 - V_1 - V_g \end{aligned} \tag{1.88}$$

1.3.2 Density of Absorbents

The density of an absorbent includes apparent density, tap density, and true density. Apparent density is the density obtained when the powder fills up the specified standard container freely. If the container is shaken to make it tight during the powder-filling process, then the obtained density is called tap density. The true density is measured by a pycnometer. The EM parameters measured with different densities are generally varied; hence, the referring EM parameters must be a numerical value under specific test conditions. In addition, the density of the absorbent will have a great influence on the entire EM wave absorption effect. In a composite material, the density of the absorbent is the percentage composition of the absorbent. On the basis of the EM parameter and impedance matching, the density of an absorbent has an optimum value on microwave absorption performance.

1.3.3 Particle Size of Absorbents

The particle size of an absorbent has a large influence on the EM wave absorbing performance and the selection of absorbing frequency. Nowadays there are two choice trends of the absorbent grain size: First, the particle size of absorbing agents trending toward miniaturization and nanocrystallization is the research focus at present. When a particle is refined to a nanoparticle, owing to the small size and large specific area, the atoms on the nanoparticle's surface and the dangling chemical bonds are numerous and the activity of the nanomaterial is enhanced. Interfacial polarization and multiple scattering are the main reasons that nanomaterials show the microwave absorption property[24]. Second, the absorbing units are discontinuous. The percolation point of the absorbent in the matrix appeared relatively early after refining, and the absorbent formed a conductive network, which is effective to reflect the EM wave strongly and make it difficult for the EM wave to enter the material. It is insufficient to fully absorb EM wave when the absorbent content is controlled under the percolation point. Therefore, the discontinuous millimeter-level absorbing unit should be formed in the absorber, and the content of the absorbent in each absorbing unit should be increased as much as possible. The absorber matches the impedance of free space favorably, the EM wave enters the material maximally in this way, and then the absorbing band is broadened and the absorption efficiency is improved enormously.

1.3.4 Shapes of Absorbents

An absorbent is the crux to obtain a high-performance absorbing material. Besides the particle content, particle size, and aggregate state, the absorbent particle shape will influence the absorbing properties undoubtedly. The main shapes of absorbents are sphere, rhombus, arborization, flake, acicular, etc. Researchers think that when the particles contain a certain number of disks and an acicular structure, the absorbing property is much better than an absorbing material with other shapes[25]. The various shapes of absorbent structures influence the EM parameters and scattering effect of the absorbent immediately, thus influencing its absorbing property.

1.3.5 Technological Properties

An absorbent is generally not used alone and should be combined with other matrix materials to form a certain structure form first. Therefore, a good technological property is required in order to mix with or dope with other materials. The method of combining several absorbents together is always employed to fulfill the purpose of broadening the absorbing band and enhancing absorbing performance. The combination methods include simple mixing, coating, cladding, and modifying.

1.3.6 Chemical Stability and Environmental Performance

An absorbent should be mixed with a solvent or other material during its preparation process, and it is inevitable to use high temperature during the preparation or application process. In addition, when the absorbent is used in a weapon system it must undergo harsh conditions, including the atmosphere, seawater, oil pollution, acid/base, etc; hence, resistance to corrosion is required. The absorbent must possess favorable chemical and environmental stability to guarantee design performance of the material in all kinds of application conditions.

In conclusion, the performance, frequency band, and environmental conditions must be taken into account when choosing an absorbent. In addition, on the basis of guaranteed performance, the production cost should be reduced as much as possible and volume production should be fulfilled to make it widely used.

References

1. Chen GR (1998). *Engineering Electromagnetic Field and Electromagnetic Waves*, Northwest University Press, Xi'an.
2. Yang XQ, Zhao JS, Wang Y (2003). *Electromagnetic Field and Electromagnetic Waves*, Defense Industry Press, Peking.
3. Yang RG (2003). *Electromagnetic Field and Electromagnetic Waves*, Higher Education Press, Peking.
4. Zhang KQ, Li DJ (2001). *Electromagnetic Theory for Microwaves and Optoelectronics*, Publishing House of Electronics Industry, Peking.
5. Jin AK (2003). *Electromagnetic wave theory*, Beijing Publishing House of Electronics Industry, Peking.
6. Zhao KH, Chen XM (2003). *Electromagnetism*, Higher Education Press, Peking.
7. Jia QM, Zheng YL, Chen JY (2001). *Electromagnetism (2nd edition)*, Higher Education Press, Peking.
8. Niu XQ, Zhu MZ, Lu ZY, Lu HM (2001). *Foundations of Electromagnetic Fields Theory*, Publishing House of Electronics Industry, Peking.
9. Yin JJ (2004). *Microwave Technology and Antenna*, Publishing House of Electronics Industry, Peking.
10. Guo HP, Liu XG (2002). *Electromagnetic Field and Electromagnetic Waves*, Xi'an Electronic Science and Technology University Press, Xi'an.
11. Peng PF (2007). *Microwave Technology and Experiment*, Tsinghua University Press, Peking.
12. Liao CE (2004). *Fundamentals of Microwave Technology*, Electronic Science and Technology University Press, Xi'an.
13. Xing L (ed) (2004). *Stealth Materials*, Chemical Industry Press, Beijing.
14. Wu M (1997). Present status and developingtrend of radar absorbing materials, *J Magn Mater Devices*, **28**(2), 26–30.
15. Shi J (1993). Microwave absorbents and evaluation of their properties, *Aerosp Mater Technol*, (5), 1–4.
16. Wu Z, Zhao Z, He H (1999). Theoretical calculations of microwave complex permeability and complex permittivity of polycrystalline iron fiber absorbers, *J Adv Mater*, **30**(1), 91–94.
17. Ni E (1991). An qlnncertainty analysis for the measurement of intrinsic properties of materials by the combined transmission-reflection method, *J Zhejiang Uni*, **25**(5), 523–531.

18. Nicolson AM, Ross GF (1970). Measurement of the intrinsic properties of materials by time-domain techniques, *IEEE Trans Instrum Meas,* **19**(4), 377–382.
19. Weir WB (1974). Automatic measurement of complex dielectric constant and permeability at microwave frequencies, *Proc IEEE,* **62**(1), 33–36.
20. Ligthart LP (1983). A fast computational technique for accurate permittivity determination using transmission line methods. *IEEE Trans Microw Theory Techn,* **31**(3), 249–254.
21. Henaux JC (1990). Dimensional correction of high dielectric and magnetic constants determined by sparameters measurements, *Electron Lett,* **26**(15), 1151–1153.
22. Cao J (1994). A review on measurement of electromagnetic parameters of materials, *J Astronaut Metrol Meas,* **13**(3), 30–34.
23. Wang X, Sheng W, Qiu Z, Qian J, Lu H, Hu G (1992). The dependence of ε^* and μ^* of microwave absorbing material on the percentage volume of its constituent, *J Nanjing Uni: Nat Sci Ed,* **28**(4), 551–555.
24. Guo F, Xu Z (2004). Recent developments of new nanometer microwave-absorbing agents, *Adv Ceram,* **25**(3), 23–26.
25. Ge F, Zhu J, Chen L (1996). Effects of inclusion-shape on absorbing ability of microwave absorbing material, *Aerosp Mater Technol,* (5), 42–49.

Chapter 2

Manganese Dioxide Absorbents

EM wave absorbents are desirable because they possess both excellent dielectric loss and magnetic loss properties. Manganese dioxides have received great interest owing to the flexibility with which their physical or chemical properties can be readily controlled through size, shape, and dimension in the synthesis stage, which makes them attractive candidates in fields ranging from catalysts[1] and electrode materials[2] to water treatment[3].

MnO_2 exists in different structural forms in nature, such as the α-, β-, γ-, and δ-types. The crystallographic forms are generally believed to be responsible for their properties. Bach et al.[4] have especially pointed out that the electrochemical properties of MnO_2 strongly depend on parameters such as powder morphology, crystalline structure, and bulk density. Meanwhile, many newly derived performances and functions of manganese dioxide have been explored by doping with flexible ions, such as Al, Ni, Co, Cr, Fe, and Li, resulting in continually broadening applications[5–11]. Jana et al.[10] reported the synthesis of flower-like Ag-doped MnO_2; they implied that a red shift in the optical response and a concomitant reduction of the band-gap energy were observed after the incorporation of Ag into MnO_2. Yu et al.[11] reported the effects of metal ions Fe^{3+} and Al^{3+} on the structures, morphologies, and electrochemical properties of

Microwave Absorbing Materials
Yuping Duan and Hongtao Guan

ISBN 978-981-4745-10-9 (Hardcover), 978-981-4745-11-6 (eBook)
www.panstanford.com

MnO_2, and their analysis also showed that the addition of Fe^{3+} and Al^{3+} affected the hydrothermal growth of MnO_2 by changing the chemical potential of the solution.

Further research has shown that both the morphologies and structures can be readily controlled in terms of ion doping, as the large tunnel cavity and octahedral space of manganese dioxide offer a possibility to accommodate extrinsic ions. Manganese oxide with the todorokite structure reported by Shen et al.[12] was found to accommodate some inorganic cations such as Mg^{2+}, Co^{2+}, Ni^{2+}, Cu^{2+}, and Zn^{2+} in different contents, and the synthesized samples had three kinds of shapes—plates, needles, and fibers—depending on the nature of the cations. Moreover, from a fundamental point of view, knowledge of the influence of doping on the dielectric and its derived properties has shown its potential in further understanding the basic mechanisms of dielectric absorption[13].

2.1 Different Crystalline Structures and Composition of MnO_2

Crystallized MnO_2 materials have several crystalline structures, including α-, β-, γ-, and δ-MnO_2. Among them, α-, β-, and δ-MnO_2 have a tunnel structure (2 × 2 octahedral units for α-MnO_2, the relatively large tunnel structured phase; 1 × 1 octahedral units for β-MnO_2, the more compact and dense phase), and δ-MnO_2 has a relatively open layered structure (as shown in Fig. 2.1 by birnessite).

The crystallographic forms are generally believed to be responsible for their properties. The dielectric and magnetic characteristics and microwave absorption properties of manganese dioxide with different crystallographic structure are different.

2.1.1 Polymorphism

It has been recognized that different preparation conditions can result in different MnO_x structures. For example, gradually increasing the precursor acidity can lead to a progression from layered δ-MnO_2, through the relatively large tunnel-structured phase α-MnO_2, to a more compact and dense β-MnO_2. In a NaOH or KOH solution, the product is mainly a δ-MnO_2 phase. These structural changes in MnO_2

give rise to significant changes in electronic and ionic conductivity, affecting the materials' electromagnetic (EM) behavior[15].

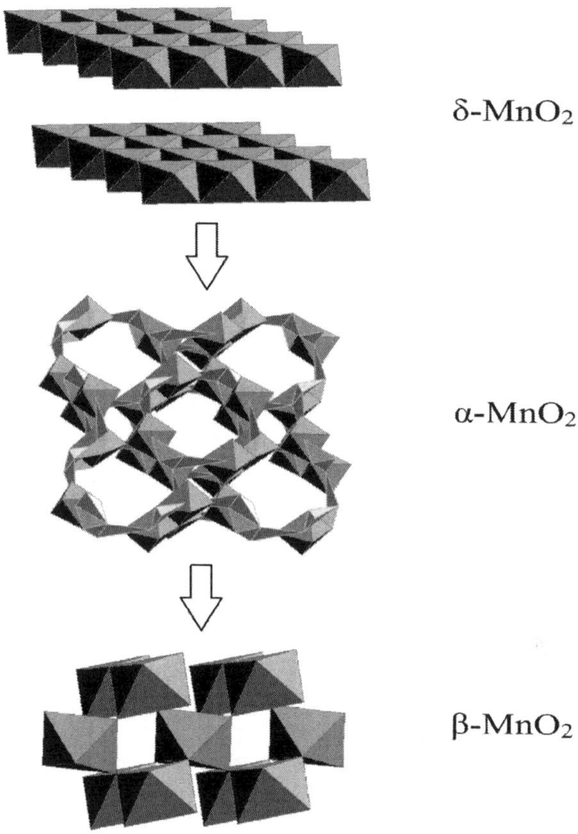

Figure 2.1 Manganese dioxide structural transitions induced during material synthesis[14].

MnO_2 was synthesized successfully by $MnSO_4{\cdot}H_2O$ and $(NH_4)_2S_2O_8$ powders, which were maintained at 120°C for 1, 2, 4, 12, 24, and 48 h. The X-ray power diffraction (XRD) pattern reveals the phase and purity of the products. The typical XRD patterns of the products are shown in Fig. 2.2. It can be seen that in the case of MnO_2 prepared at 120°C for 1 h, the system showed evolution of the nucleation of the nanocrystalline nature of MnO_2 particles (Fig. 2.2a), with peaks in the γ-MnO_2 (JCPDS card no. 14-0644) structure. There was an increase in crystallinity with respect to the increase of the hydrothermal dwell time from 1 to 48 h, as evidenced by the appearance of sharper peaks. Figures 2.2a–d shows gradual transformation of γ-MnO_2 to α-MnO_2 (JCPDS 44-0141) and β-MnO_2 (JCPDS 24-0735). However, when the hydrothermal dwell time was prolonged to 24 and 48 h, all the peaks of the sample could be indexed to a pure tetragonal phase of β-MnO_2 (JCPDS 24-0735). This is shown in Figs. 2.2e and 2.2f. In addition, the sharp reflection peaks indicate that the products are well crystallized. Figure 2.2 also shows that the purity of the MnO_2 crystalloid in the product increased significantly with the increasing reaction time, which illustrates that the purity and crystallinity of the as-prepared MnO_2 depend on the reaction time.

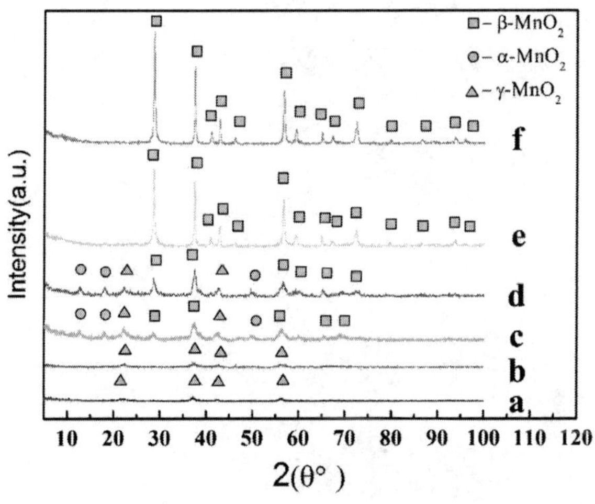

Figure 2.2 XRD patterns of MnO_2 synthesized hydrothermally at 120°C for the following times: a, 1 h; b, 2 h; c, 4 h; d, 12 h; e, 24 h; and f, 48 h.

Figure 2.3 shows scanning electron microscopy (SEM) images of MnO_2 obtained at 120°C with different hydrothermal dwell times. It clearly shows that many smooth-faced microspheres about 2–3 μm in size were observed at an early reaction stage (1 h), as shown in Fig. 2.3a. It is noteworthy that when the reaction times were prolonged to 2, 4, or 12 h, growth of flocculus on the surface of MnO_2 microspheres was observed. The growth increased with the increasing reaction time, as shown in Fig. 2.3b–d. Finally, when the reaction time was further prolonged to 24 and 48 h, more than 3 μm dispersive nanorods were formed, as shown in Figs. 2.3e and 2.3f.

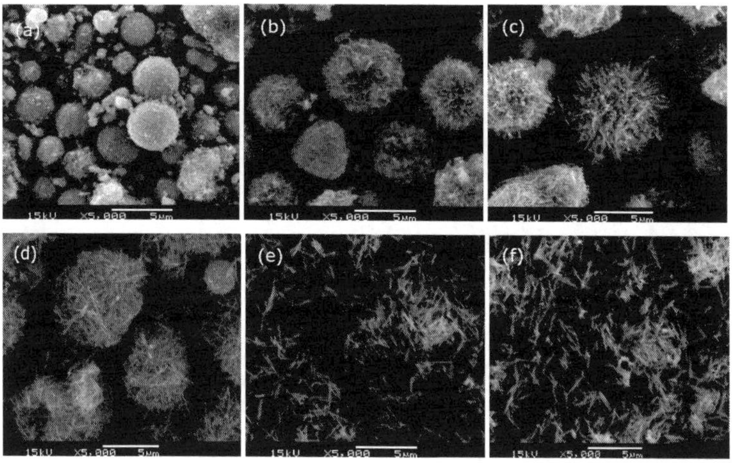

Figure 2.3 SEM images of MnO_2 synthesized hydrothermally at 120°C with different dwell times: (a) 1 h, (b) 2 h, (c) 4 h, (d) 12 h, (e) 24 h, and (f) 48 h.

The morphologies and microstructures of the as-prepared MnO_2 samples were further investigated with transmission electron microscopy (TEM), high-resolution TEM (HRTEM), and selected area electron diffraction (SAED). Figure 2.4 shows typical TEM images of the samples with hydrothermal dwell times of 2, 12, and 48 h. In Fig. 2.4a, most of the products with a reaction time of 2 h are composed of microspheres with diameters of 2–3 mm. The surfaces of these microspheres are composed of large quantities of acicular nanocrystallites that interweave together. However, as seen in the SAED pattern of the as-prepared MnO_2 microspheres in the inset of Fig. 2.4a, the purity and crystallinity of the products are low, which is in agreement with the XRD results in Fig. 2.2b. Clearly, we can also see that the individual MnO_2 microsphere in Fig. 2.4b is in the sea urchin shape with a diameter of several micrometers. The microsphere is formed by nanorods radiating from its center whose purity and crystallinity were greatly improved. This can be seen on the SAED pattern on the inset in Fig. 2.4b and the XRD in Fig. 2.2d. Figure 2.4c shows that the samples of β-MnO_2 consisted of straight and smooth nanorods with diameters in the range of

50–70 nm and lengths of up to several micrometers. The SAED pattern of the β-MnO_2 nanorods (inset in Fig. 2.4c) taken from a single rod indicates that these crystalline β-MnO_2 nanorods are of high quality.

On the basis of the above statement, the morphology, crystal structure, and purity of the as-prepared MnO_2 samples are dependent on the reaction time. The formation mechanism of the MnO_2 particles may be deduced as follows: First, MnO_2 colloids were produced in the solution by the redox reaction between $MnSO_4$ and $(NH_4)_2S_2O_8$. The resulting MnO_2 colloids formed spherical aggregates, owing to the irrelatively high surface energies. As the reaction proceeded, the reactants $MnSO_4$ and $(NH_4)_2S_2O_8$ were gradually consumed. Thereafter, a thermodynamically stable system was arrived at. In the next step, the MnO_2 nanorods gradually evolved from these spherical aggregates, owing to their one-dimensional growth habit. This resulted in the formation of a large quantity of sea urchin–like MnO_2 microspheres and the straight and smooth MnO_2 nanorods[16].

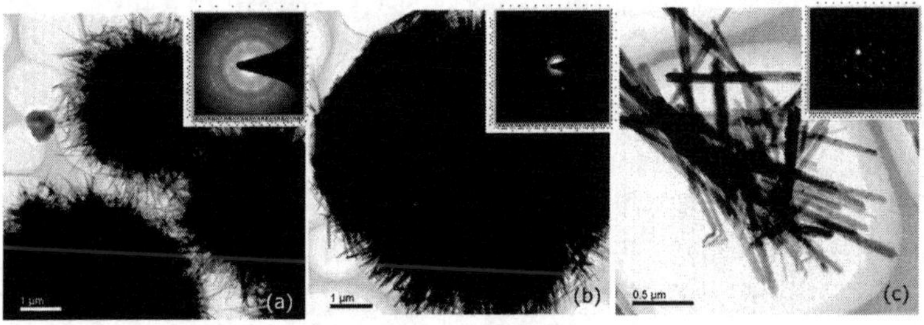

Figure 2.4 Typical TEM images of the as-prepared MnO_2 with the following hydrothermal dwell times: (a) 2 h, (b) 12 h, and (c) 48 h. Insets are selected area electron diffraction (SAED) patterns.

2.1.2 Microwave Properties

Permittivity, ε, of a dielectric material is a quantified response of molecules polarizing themselves along the electric field caused by attraction or repulsion between a set of charges. Normally, $\varepsilon = \varepsilon_0\varepsilon_r$, where ε_0 ($1/36\pi \times 10^{-9}$ F/m) is the absolute permittivity of a free space (vacuum), and ε_r is a dimensionless number that specifies the relative permittivity of the material. Permittivity can then be written as $\varepsilon_r = \varepsilon_r' - j\varepsilon_r''$,where ε_r' and ε_r'' are the real and imaginary parts of ε_r, respectively, and ε_r'', associated with electric field loss, are results of the electric-dipole polarization at microwave frequencies.

Permeability, μ, of a magnetic material is characterized by $\mu = \mu_0\mu_r$, where μ_0 ($4\pi \times 10^{-7}$ H/m) is the absolute permeability of the free space and μ_r is a dimensionless number that specifies the relative permeability of the material. Thus, permeability can be expressed as $\mu_r = \mu_r' - j\mu_r''$. Here, μ_r' and μ_r'' are the real and the imaginary parts of μ_r, respectively, and μ_r'', associated with magnetic field loss is a result of the magnetic-dipole magnetization. A material can absorb radiation, as long as its ε_r'' and μ_r'' values are not negligible[4].

The electric loss tangent (tan δ_ε) and the magnetic loss tangent (tan δ_μ) of a material can be written as tan $\delta_\varepsilon = \varepsilon_r''/\varepsilon_r'$ and tan δ_μ μ_r''/μ_r', respectively. This means that the ratio of the energy loss/unit radian in the dielectric to the energy storage in the dielectric can be determined[17].

Figure 2.5 illustrates the real and imaginary parts of the complex relative dielectric permittivity and permeability versus the frequency of the prepared MnO_2 powder–paraffin wax composites. Paraffin wax is an insulator and a nonmagnetic material, thus, impermeable to EM waves. In the MnO_2/paraffin wax composite, paraffin wax only functions as a matrix. New substances and chemical bonds, which can provide electrical dipoles for the composite, will be hardly formed from the interactions between MnO_2 and paraffin wax. Thus, the electric loss of the composite is mainly the contribution of MnO_2.

It can be seen from Fig. 2.5a,b that the values of ε_r' decrease with an increase in the frequency. In contrast, an increasing trend of the ε_r' values in relation to the frequency, except for the 2 h sample, can be seen. The possible reason for the change in permittivity is the faster change of the external alternating EM field. This is due to the dispersive character of the MnO_2 particles. That is why MnO_2 in the paraffin wax can be polarized repeatedly in a high-frequency electric field, causing a change of the electric energy into other forms, mainly heat energy. The heat is then dissipated. Moreover, MnO_2 has relatively low conductivity, of the magnitude order of 10^3 Ω·cm. Consequently, its permittivity is higher than its permeability. Likewise, its temperature can increase from 298 to 1378 K in

100 s at the speed of 10.80 K/s when subjected to EM wave irradiation. In view of these, it can be deduced that MnO_2 is a promising dielectric loss absorbent[18]. In addition, it is noteworthy that the real part values of the complex relative dielectric permittivity of different MnO_2 products, as shown in Fig. 2.5a, are comparatively smaller than the values of metal powder, as reported by Wu et al. and Wen et al[19,20]. The low real part value of the complex relative dielectric permittivity is a great advantage in striking a balance between permeability and permittivity. This can decrease the reflection coefficient of the absorber. Other metal magnetic materials used for microwave absorbing application may not have this ability[21].

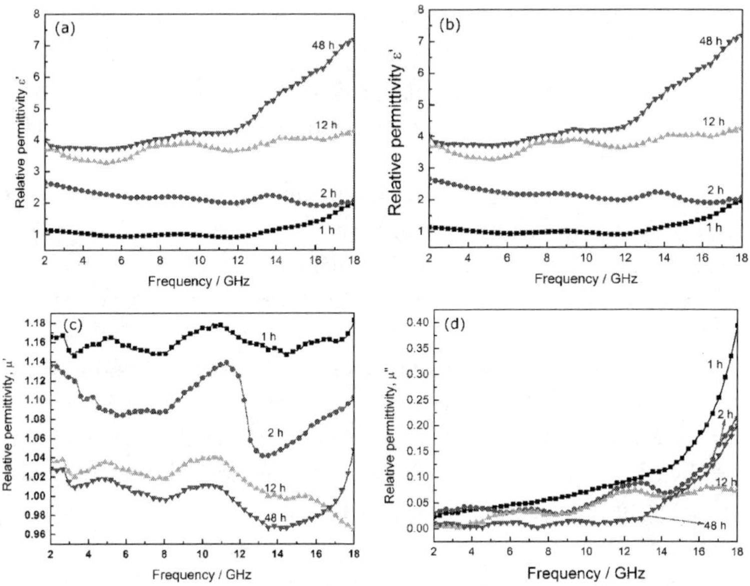

Figure 2.5 (a, b) Complex relative permittivity and (c, d) permeability of MnO_2 synthesized hydrothermally at 120°C for different dwell times (1 h, 2 h, 12 h, and 48 h) as a function of frequency.

In Figs. 2.5c,d, the real parts μ_r' of the MnO_2/paraffin wax composite have several distinct resonance peaks with values of more than 1. In addition, the values of μ_r'' generally showed increases in the frequency range 2–18 GHz. These results indicate that MnO_2 with different crystalloid structures has paramagnetism and magnetic loss.

The magnetic loss of the paramagnetism material mainly depends on paramagnetic resonance. When subjected to radiation from an external EM wave at resonant frequency, an electron transition from low-spin energy level to high-spin energy level occurs. This induces magnetic loss of MnO_2 materials. Therefore, MnO_2 can provide magnetic dipoles that allow a high-frequency magnetic field to cause repeated magnetization and produce magnetic loss[22].

Thus, it can be concluded that MnO_2 is both a dielectric lossy material and a magnetic loss of microwave radiation material. However, the dielectric loss plays a dominant role in the total loss part. This can also be proven effectively by the loss tangent, as shown in Fig. 2.6. It can be seen that with the increase in frequency, the dielectric loss tangent of the different MnO_2 products ranges from 0.2 to 0.5, while the magnetic loss tangent ranges from 0.002 to 0.15. Obviously, the value of magnetic permeability ($\tan\,\delta_\mu$) is less than that of the dielectric loss tangent ($\tan\,\delta_\varepsilon$). Therefore, it can be concluded that the magnetic loss of MnO_2 with different crystal structures is small or even negligible when the frequency is low compared to that with the dielectric loss.

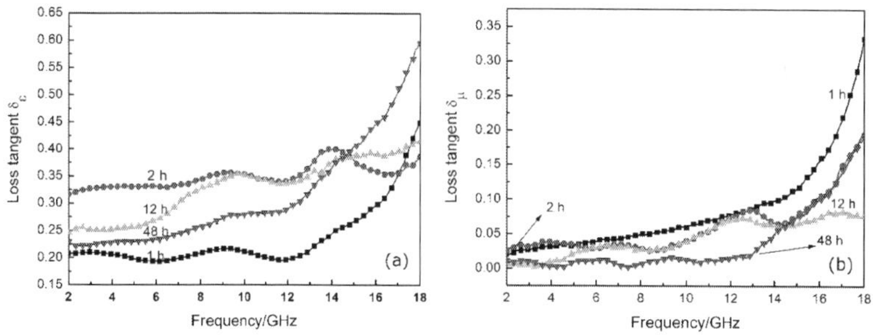

Figure 2.6 (a) The electric loss tangent and (b) magnetic loss tangent of MnO_2 synthesized hydrothermally at 120°C for different dwell times (1 h, 2 h, 12 h, and 48 h) as a function of frequency.

It is interesting to note that the dielectric permittivity and magnetic permeability of the as-prepared MnO_2 changed systematically at different reaction times (Fig. 2.5). The values of both the real and imaginary parts of the complex permittivity for different MnO_2 products increased systematically with increasing reaction times, as can be seen in Fig. 2.5a,b. The values of the 12 h and 48 h samples are significantly higher than those of the 1 h and 2 h samples. In contrast, the values of both the real and imaginary parts of the complex magnetic permeability for different MnO_2 products decreased systematically with the increase in reaction times, as can be seen in Fig. 2.5c,d.

The probable reasons are the variety of the crystalloid state and purity of the as-obtained MnO_2, resulting in the change of the crystal lattice, the crystal defect, the hole, and so on. These all play key roles on the difference of the EM parameter and wave loss. Second, the morphology of the as-prepared MnO_2 transformed from microspheres to nanorods with the increase in reaction times. The transformation is very important for the formation of the microwave EM characteristics. Compared to the spherical particulate, the flaky or strip-shaped structure can form more dipole moments, so it has more scattering cross-sections when subjected to EM radiation. This enhances dielectric permittivity. The decrease in complex magnetic permeability with the increasing reaction times can be attributed to the frequency dispersion effect of MnO_2 products[23].

2.1.3 Reflection Loss

To further prove the dependence of microwave absorption properties on permittivity and permeability, the reflection losses (RLs) (dB) of different MnO_2 samples were calculated according to transmission line theory. The RL of EM radiation under normal incidence of the EM field on the surface of a single-layer material backed with a perfect conductor can be defined as[24]

$$\mathrm{RL} = 20\log\left|\frac{Z-1}{Z+1}\right| \tag{2.1}$$

where Z is the normalized impedance between input impedance Z_{in} of the single-layered absorber and the impedance of free space Z_0, expressed as

$$Z = \frac{Z_{in}}{Z_o} = \sqrt{\frac{\mu_r}{\varepsilon_r}}\tanh\left(j\frac{2\pi d}{\lambda_0}\cdot\sqrt{\mu_r\varepsilon_r}\right) \tag{2.2}$$

where μ_r and ε_r represent the relative complex dielectric permittivity and magnetic permeability, respectively, of the composite medium; d is the thickness of an absorber; and λ_0 is the wavelength of the incident wave in free space. Thus, the RL of an absorber is a function of six characteristic parameters: ε_r', ε_r'', μ_r', μ_r'', f, and d.

The microwave absorption properties of MnO_2–paraffin wax have been calculated from a computer simulation using Eq. 2.1. The measured values of ε_r', ε_r'', μ_r', and μ_r'' previously obtained (Fig. 2.5) were calculated with the assumption that $d = 2.0$ mm. The RLs as a function of the frequency for samples with different crystal structures of MnO_2 were also calculated and are shown in Fig. 2.7.

As can be seen, the maximum RL of the samples with 1 h MnO_2 filling reached –24.21 dB at 17.36 GHz, and the effective absorption band below –5 dB is from 13 to 18 GHz. The sample with 2 h MnO_2 inclusion shows a maximum RL of –20.95 dB at 16.56 GHz, and the absorption range below –5 dB is from 11.3 to 18 GHz. On the other hand, when the reaction time of as-prepared MnO_2 reached 12 h, the maximum RL (–35.46 dB at 11.44 GHz) and absorption band (8.8–16.8 GHz) below –5 dB of the samples were greatly improved. In comparison, the sample with the 48 h MnO_2 filler weakened and the RL was –25 dB at 9.84 GHz, while the absorption band ranged from 8 to 12.7 GHz.

The possible reason for this can be explained as follows: At the 2 h or 12 h reaction time, the products were MnO_2 composites or mixtures with different morphologies (as shown in Fig. 2.2). The microwave absorptions of these MnO_2 composites were high since these were the sums of attenuations by different phases. Meanwhile, when the reaction time was 1 or 48 h, the MnO_2 products were pure; thus, the microwave attenuation weakened accordingly.

In addition, it is of interest that the increasing reaction times shifted the absorbing peak value to the lower-frequency band, as shown in Fig. 2.7. The shift can be attributed mainly to the real parts of the corresponding complex relative permittivity (see Fig. 2.5). The more real parts the complex permittivity has, the higher the frequency is. Hence, the absorbing peaks shifted to the lower-frequency band when the real parts of the complex permittivity and the reaction times of the MnO_2 products increased. Furthermore, dielectric polarization loss versus frequency of the MnO_2 with different crystalline structures also played an important role in the shifts of the absorption peaks.

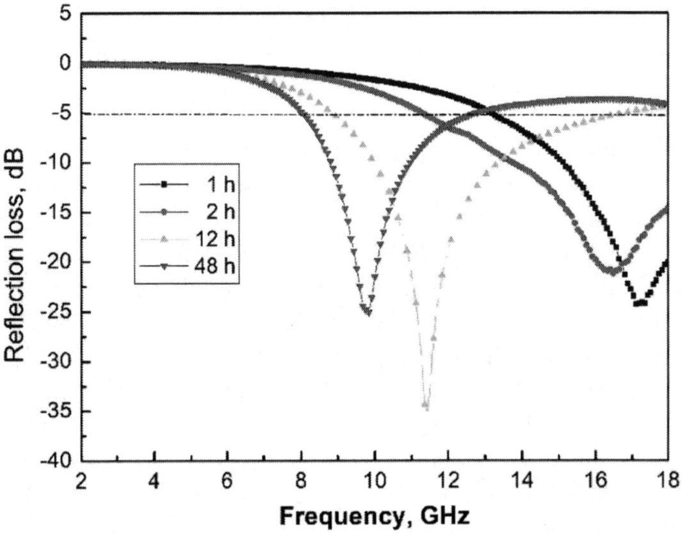

Figure 2.7 Calculated reflection loss of the 60 wt% MnO_2–paraffin wax composites with the same thickness (2 mm) versus frequency. The filling MnO_2 are products with different hydrothermal dwell times: 1 h, 2 h, 12 h, and 48 h.

2.2 Effects of a High Magnetic Field on MnO_2

External magnetic fields can be viewed as important components of synthetic processes for generation of unusually shaped materials. Manganese oxide with the sea urchin–like ball-chain shape can be synthesized in a high magnetic field, and the microwave EM characteristics of the prepared MnO_2 were satisfactory[25].

2.2.1 Morphology of MnO_2 Synthesized in a Different Magnetic Field

$MnSO_4$ and $(NH_4)_2S_2O_8$ was used to synthesize MnO_2 in a high magnetic field, at selected magnetic field strengths of 1, 3, 5, or 8 T. Figures 2.8 and 2.9 show the SEM and TEM images with inserted diffraction patterns and the HRTEM images prepared in the present study. The five samples corresponding to Fig. 2.8 and Fig. 2.9a–e were obtained at 70°C for 4 h with the magnetic field strengths of

0, 1, 3, 5, and 8 T, respectively. As can be seen in Figs. 2.8 and 2.9a, sea urchin–like microspheres can be observed in the precipitated sample. At the magnetic field strength of 1 T, the microspheres began to form sea urchin–like ball-chain structures (Figs. 2.8 and 2.9b), but this was not very pronounced. With magnetic field strengths of 3 to 8 T, the sea urchin–like ball-chain pattern became more evident (Figs. 2.8 and 2.9c–e). From Fig. 2.8, it is clear that the effect of a high magnetic field on the microspheres size was only slight, but it is obvious from Fig. 2.9 that the particle size of the sample 3 T is smaller and more uniform than others. The inserted diffraction pattern images in Fig. 2.9 show that the crystallization degree of MnO_2 in a high magnetic field was improved effectively compared to that without the high magnetic field. And the HRTEM images provide more detailed structural information about the stick on the surface of the microsphere, showing the apparent lattice fringes of the crystal. But the influence of the high magnetic field on lattice fringes of the crystal is not very evident.

One possible mechanism for the formation of the sea urchin-like ball-chain shape is as follows. The original sea urchin–like microspheres were arrayed randomly in the absence of a high magnetic field. Nevertheless, the microspheres arrayed themselves along a given direction under the external magnetic field (Fig. 2.8c–e). This is because, for any kind of crystals, differences in the arrangement and density of the atomic structure will result in differences in the magnetic susceptibilities of the same crystal at different crystallographic orientations. Small-size microspheres arrayed along the direction of their maximum magnetic susceptibility (which is also along the magnetic field) were equivalent to small magnetic bodies. Consequently, they assembled to form the sea urchin–like ball-chain shape.

2.2.2 Phase Structure and Compositional Analysis

According to Fig. 2.10, α-MnO_2 and γ-MnO_2 are the major MnO_2 products, and all of the diffraction peaks can be indexed to tetragonal and orthorhombic structures (JCPDS 72-1982, 82-2169, 72-1983, and 73-1539). The broadened diffraction peaks indicate that the crystalline size of the samples is small and that the crystallinity is not yet ideal. Furthermore, it is obvious that the crystalline nature and percentage of the α- and γ-phase MnO_2 in each sample are different from the X-ray results, which show that the high magnetic field has a significant effect on the crystalline nature and the phase structure.

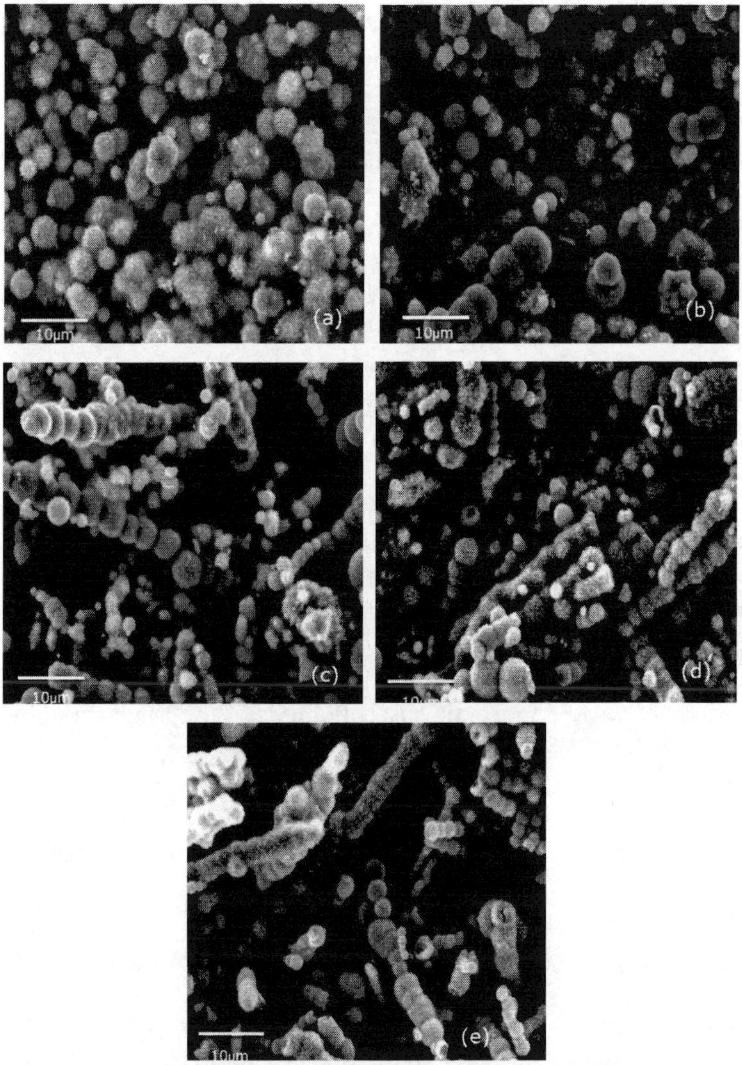

Figure 2.8 SEM images of MnO_2 samples synthesized at 70°C for 4 h (a) without a high magnetic field, (b) with a magnetic field strength of 1 T, (c) with a magnetic field strength of 3 T, (d) with a magnetic field strength of 5 T, and (e) with a magnetic field strength of 8 T.

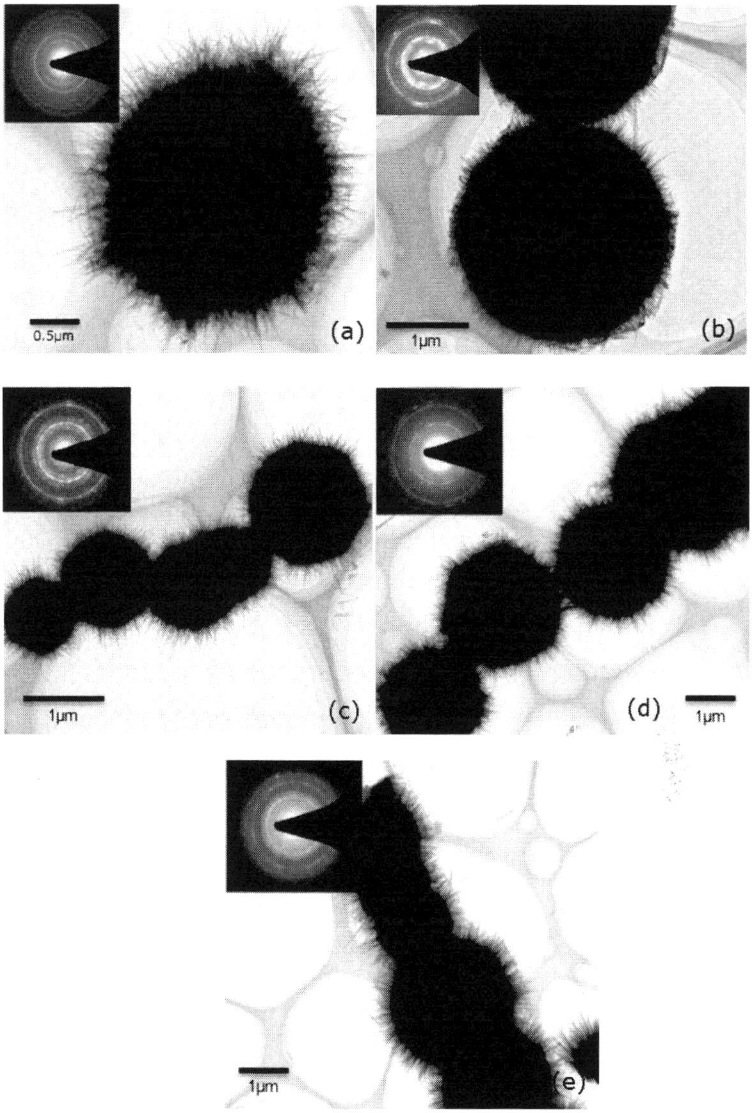

Figure 2.9 TEM images of MnO_2 samples synthesized at 70°C for 4 h (a) without a high magnetic field, (b) with a magnetic field strength of 1 T, (c) with a magnetic field strength of 3 T, (d) with a magnetic field strength of 5 T, and (e) with a magnetic field strength of 8 T.

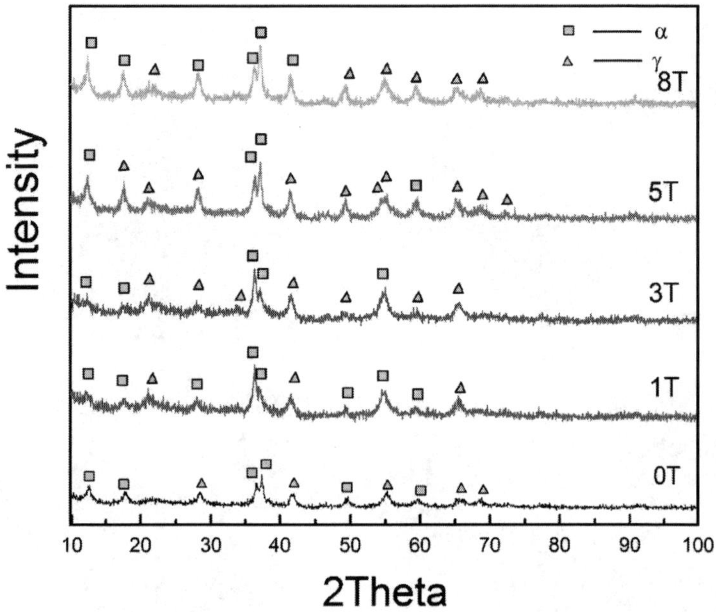

Figure 2.10 XRD patterns of as-synthesized samples with 0 T, 1 T, 3 T, 5 T, and 8 T.

2.2.3 Electromagnetic Properties

Complex permittivity, permeability, and their loss tangent represent the dielectric and magnetic properties of an absorbing material. The real parts (ε' and μ') of complex permittivity and permeability symbolize the storage capability of electric and magnetic energy. The imaginary parts (ε'' and μ'') represent the loss of electric and magnetic energy[26]. The loss tangent (tan δ) represents the loss properties of incidence EM wave in the microwave absorber. In terms of microwave absorption, the imaginary parts and the loss tangent are expected to be larger.

Figure 2.11 shows the relatively complex permittivity and permeability of the five samples in the frequency range of 2–18 GHz. As shown in Fig. 2.11a, the ε' values for the five samples declined from 9.29, 9.39, 7.77, 4.64, and 5.06 to 5.26, 5.29, 5.03, 3.38, and 3.72, respectively, with increasing frequency in the 2–18 GHz range. All of the ε'' values for the five samples also exhibited a decrease from 3.43, 3.72, 2.54, 0.93, and 1.04 to 2.20, 2.27, 1.64, 0.75, and 0.83, respectively, as the frequency was increased from 2 to 18

GHz. Increases in magnetic field strength from 0 to 8 T resulted in decreases in ε' values of the samples, except for 8 T and ε'' values, which decreased. Figure 2.11b shows that the μ' and μ'' values of samples, except for 3 and 5 T, were almost constant, with less variation throughout the whole frequency range. All $\mu' \approx 1.04$ and $\mu'' \approx 0.017$. When the external magnetic field strength was 3 T, the values of μ' and μ'' appeared as a large peak at 5.76 GHz, which is expected when magnetic loss improves.

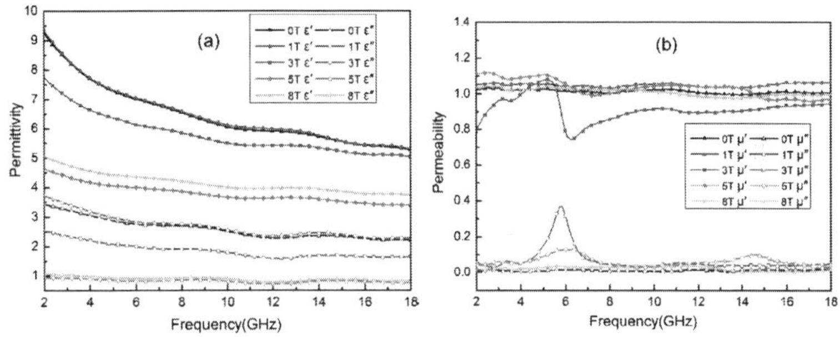

Figure 2.11 Relative complex permittivity and permeability of the five samples.

Figure 2.12 shows the frequency dependence of the electric loss tangent and magnetic loss tangent of the five samples. In Fig. 2.12a, the electric loss tangent values of the five samples are fluctuant with increasing frequency in the 2–18 GHz range. When the magnetic field strength increases, the electric loss tangent values for the five samples decrease, except for the sample under 1 T. The electric loss tangent values of the five samples were 0.37–0.42, 0.39–0.43, 0.29–0.33, 0.20–0.24, and 0.20–0.23. As shown in Fig. 2.12b, the magnetic loss tangent values of 0 and 1 T were almost constant, with less variation throughout the whole frequency range. However, the magnetic loss tangent values of 3, 5, and 8 T had clearly increased compared to those of 0 and 1 T. At 3 T, the magnetic loss tangent value reached an expected improvement in the whole frequency range and a maximum peak at 5.76 GHz, which corresponded to the data depicted in Fig. 2.11b. However, with further sequential increases in magnetic field strength, the magnetic loss tangent values gradually decreased. The magnetic loss tangent values of the five samples were 0.001–0.025, 0.002–0.035, 0.029–0.420, 0.020–0.126, and 0.001–0.047. Therefore, the EM loss property of the 3 T sample was more favorable among the five samples. The possible reasons are as follows:

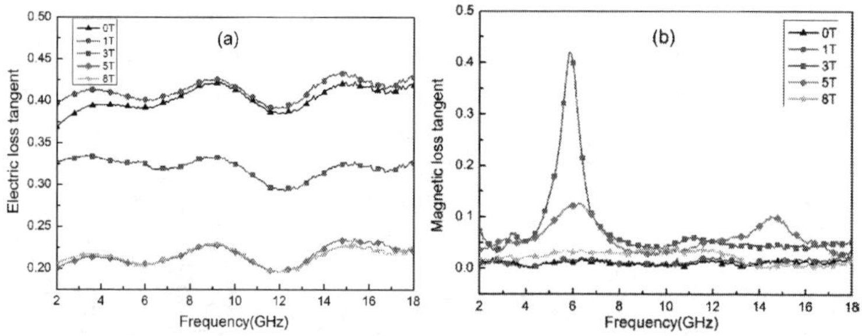

Figure 2.12 Electric loss tangent and magnetic loss tangent of the five samples.

The attenuation of EM wave in the x department $I_{(x)}$ is given by

$$I_{(x)} = I_0 \exp(-n\sigma_{ex} x) \tag{2.3}$$

where $I_{(0)}$ is the incident energy of EM waves, n is the number of absorbent particles per unit volume, and σ_{ex} the extinction cross section of the particles. According to Eq. 2.3, for a larger n, a larger extinction cross section s_{ex} is obtained, so there is greater EM wave attenuation of the particles. Due to the smaller particle size of the 3 T sample, compared to the other samples, the extinction cross section σ_{ex} is larger, which aids in the improvement of the magnetic loss. Furthermore, according to the phase structure and composition in Fig. 2.10, the crystalline nature and percentage of the α- and γ-phase MnO_2 in each sample will play an important role in their EM properties.

On the basis of this analysis, an external high magnetic field can be seen to have substantial effects on the EM properties of a prepared MnO_2 powder. The dielectric constant and the loss tangent clearly decreased under the magnetic field. The imaginary part of the magnetic permeability and magnetic loss tangent substantially increased, while the real part of the magnetic permeability, except for 3 T, was almost constant, with less variation, compared to MnO_2 prepared in the absence of an external magnetic field.

2.2.4 Electromagnetic Wave Absorption Properties

To further confirm the dependence of microwave absorption properties on permittivity and permeability, the RLs (dB) were calculated according to transmission line theory.

Figure 2.13 shows the theoretically calculated values of RL versus frequency, according to Eqs. 2.1 and 2.2. Clearly, when the magnetic field strength increases, except for 1 T, the absorption peaks become smoother and shift to a higher frequency. Over the whole range from 2 GHz to 18 GHz, the RL values of 0, 1, and 8 T only had one absorption peak. However, the RL values for 3 and 5 T have two absorption peaks, which are attributed mainly to the resonance peak frequency of the corresponding complex relative permeability (see Fig. 2.11) and wave interference on the fore- and after-surface of the sample. In the frequency range of 2–18 GHz, the 0 and 1 T samples show a maximum RL of about –20.3 dB at 17.12 and 16.56 GHz, and the effective absorption bands below –10 dB are from 14.24 to 18 GHz and 13.84 to 18 GHz, respectively. The more the real parts of the complex permittivity, the higher the frequency, and the more the wave reflection on the samples. So the absorbing peaks shift to a lower-frequency band with increasing real parts of the complex permittivity. Furthermore, dielectric polarization loss versus frequency of MnO_2 with a different crystalline structure also plays an important role in the shifts of the absorption peaks.

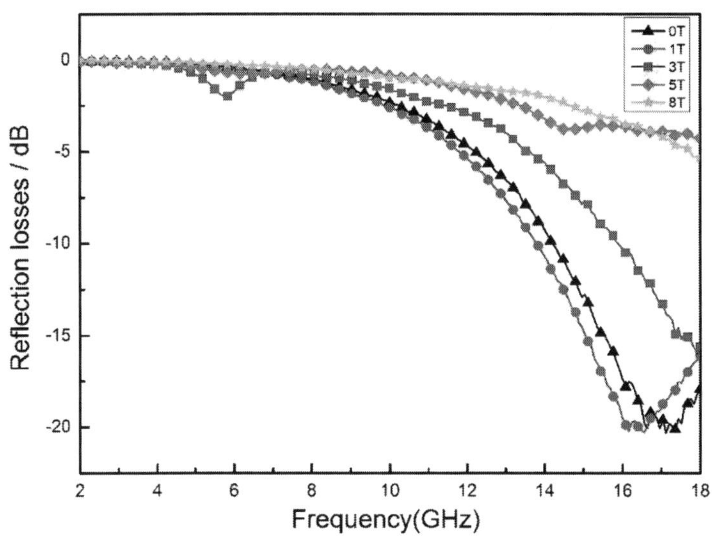

Figure 2.13 Calculated reflection loss of the five samples versus frequency.

2.3 Doped MnO_2

During the past few decades, controllable synthesis of components with specific microstructures has received considerable attention of researchers for their unique properties and potential applications in functional materials. Many newly derived performances and functions of manganese dioxide have been explored by doping of flexible ions, such as Al, Ni, Co, Cr, Fe, and Li, resulting in continually broadening applications. Jana et al. reported the synthesis of flower-like Ag doped MnO_2[10]; they implied that a red shift in the optical response and a concomitant reduction of the band-gap energy were observed after incorporation of Ag into MnO_2. Yu et al.[11] reported the effects of metal ions Fe^{3+} and Al^{3+} on the structures, morphologies, and electrochemical properties of MnO_2, and their analysis also showed that the addition of Fe^{3+} and Al^{3+} affected the hydrothermal growth of MnO_2 by changing the chemical potential of the solution. Further studies show that both the morphologies and phases can be controlled in terms of ion doping, as the large tunnel cavity and octahedral space of manganese dioxide offer a possibility to accommodate extrinsic ions. Manganese oxide with the todorokite structure reported by Shen et al. was found to accommodate some inorganic cations such as Mg^{2+}, Co^{2+}, Ni^{2+}, Cu^{2+}, and Zn^{2+} in different content, and the synthesized samples had three kinds of shapes, plates, needles, and fibers, depending on the nature of the cations[12]. Despite the interest in the above applications of ion-doped manganese oxides, few studies have been focused on the wave absorbing properties of doped manganese dioxides.

2.3.1 Fe-Doping Manganese Oxides

2.3.1.1 Synthesis

All reagents were of analytical purity and used without further purification. In our procedure, the initial aqueous solutions were prepared by dissolving mixtures of $MnSO_4 \cdot H_2O$ and $FeSO_4 \cdot 7H_2O$ with different ratios in 100 mL distilled water. Note that the concentrations of SO_4^{2-} remain constant to be 0.84 M in a series of experiments. Subsequently, 0.084 mol of $(NH_4)_2S_2O_8$ was added into the solutions while stirring. The color of the solutions turned to orange immediately as the Fe^{2+} was oxidized into Fe^{3+}. Afterward, the solutions were stirred for 5 min, heated to boiling point for 1.5 h, and then cooled naturally to room temperature. The resulting precipitates were filtered and washed several times with distilled water until the pH rose to near 7, and then dried for 12 h at 80°C prior to being characterized. The molar ratios of Fe^{3+} to the total metal ions (Mn^{2+} and Fe^{3+}) in the original solutions were increased gradually from 0% to 90%. The samples were designated as S1, S2, S3, S4, S5, and S6, corresponding to different proportions of Fe^{3+} (0, 10, 20, 30, 60, and 90, respectively).

2.3.1.2 Phase structure and composition analysis

Crystallographic structures of the samples are schematized in Fig. 2.14. The diffraction peaks can be easily assigned to the tetragonal phase of α-MnO_2 (JCPDS 44–0141) in the pure MnO_2 sample (S1) as well as the relatively less Fe-containing ones (S2 and S3). Subtle changes are present that the Bragg reflection widths enhance and the peaks intensities decrease with increasing iron ion concentration, indicating the degradation of crystallinity of MnO_2 with the introduction of iron ions. Nevertheless, the presence of a small amount of iron does not alter the phase structure of MnO_2. However, when the iron amount is increased to 30 mol%, the crystallinity of the sample converts into the hexagonal phase of delta-iron manganese oxide hydroxide with formula of $(Fe_{0.67}Mn_{0.33})OOH$ (JCPDS 14-557). With increasing iron ion concentration from 30 to 90 mol%, the peaks of the samples become broader and weaker.

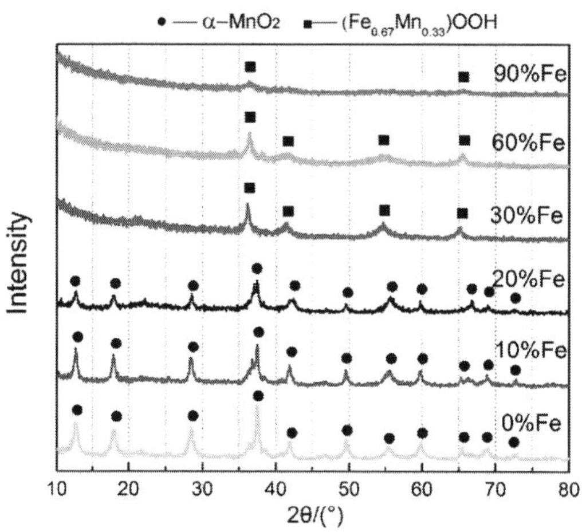

Figure 2.14 X-ray diffraction patterns of samples with different iron amounts: S1 (without iron), S2 (10 mol%), S3 (20 mol%), S4 (30 mol%), S5 (60 mol%), and S6 (90 mol%).

Accordingly, for Fe-containing samples S4–S6, it is obvious that iron ions exist in the new phase of $(Fe_{0.67}Mn_{0.33})OOH$. But the existence of iron introduced in the samples S2 and S3 cannot be determined simply by XRD patterns and hereby can be further investigated by surface composition analysis image for the three elements Fe, Mn, and O, as shown in Fig. 2.15. It can be seen that the Fe element is uniformly distributed, though relatively rare, in the samples S2 and S3, suggesting incorporation of Fe in the MnO_2 lattice. Thereby, for the samples S2 and S3 containing less iron ions, two possible deductions could be taken into consideration, which can be depicted in Fig. 2.16. On the one hand, the basic structural unit of manganese dioxide is MnO_6-octahedra, with a manganese atom in the center and six oxygen atoms in the vertices shown in Fig. 2.16b. Due to similar atomic radius, Fe ions might enter the MnO_2 crystal structure and substitute a small part of Mn ions, as shown in Fig. 2.16c, leading to the formation of a substitutional solid solution. On the other hand, particularly for α-MnO_2, which has a typical 2 × 2 tunnel structure[27], iron ions are likely located at the tunnel cavity and surrounded by eight oxygen ions, as shown in Fig. 2.14a. But the located iron ions could not fill too much in respect that positive ion repulsion caused by adjacent iron ions could upset the chemical potential balance of crystal structure. Hence, two possible substitutions of Mn^{3+} for Mn^{4+} and OH^- for O^{2-} could be considered to adjust charge balance and result in further lattice distortions. From whichever deduction described above, for the samples S2 and S3, which contain a limited amount of iron ions, the substitutions are not enough to bring out an extra phase but could, to some extent, cause geometrical frustration of the lattice constant corresponding to the calculated cell volume, as shown in Table 2.1. These distortions are responsible for poor crystallization and coincided with the decreasing intensity of XRD. However, when it comes to S4, S5, and S6 in which more iron ions are introduced, the original α-MnO_2 structures are destroyed by overloading lattice distortions and transformed into a new phase in formula of $(Fe_{0.67}Mn_{0.33})OOH$, striking a new chemical potential balance of Fe, Mn, O, and OH. More experiments are needed to verify this proposal.

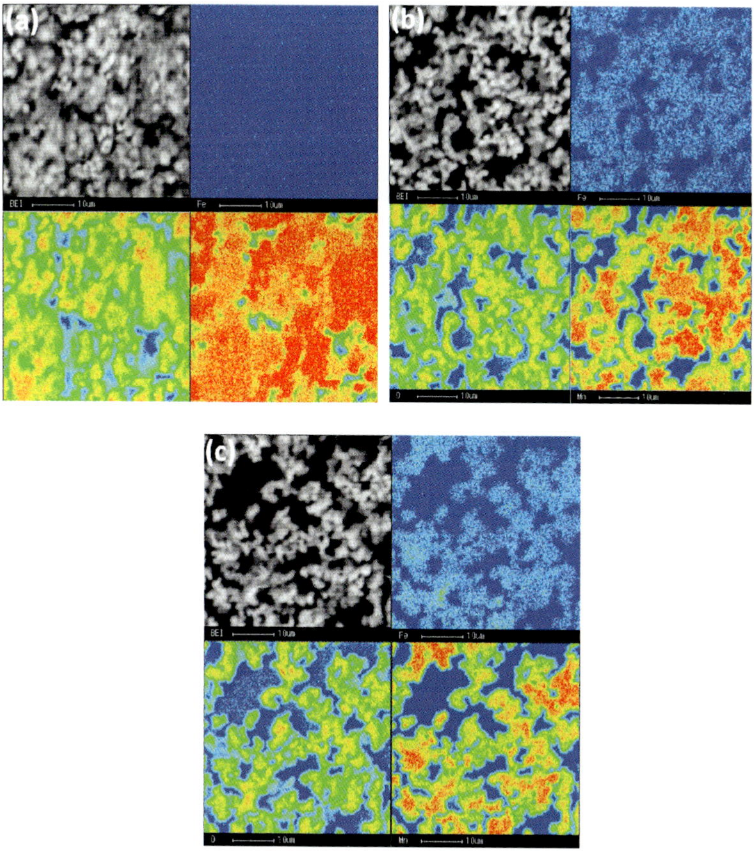

Figure 2.15 Surface composition analysis images of samples synthesized with different amounts of iron: (a) S1: without iron, (b) S2: 10 mol%, and (c) S3: 20 mol%.

Table 2.1 Calculated cell volume and c/a values of MnO_2 in S1, S2, and S3

Sample	Iron content	Cell volume	c/a
S1	0	279.93	0.28822
S2	10 mol%	280.70	0.28842
S3	20 mol%	281.10	0.29003

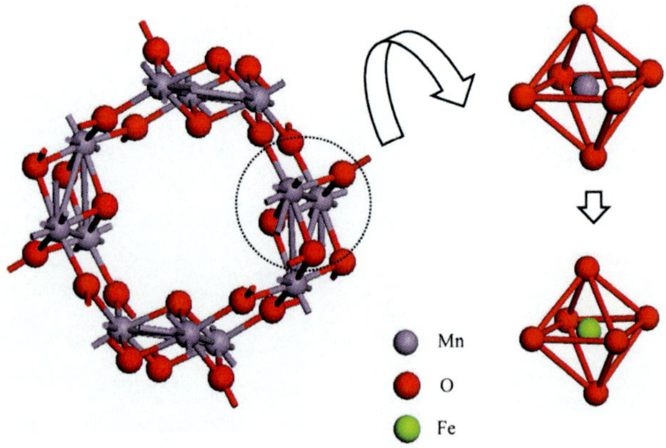

Figure 2.16 Schematic diagram of the crystal structure for Fe-containing α-MnO_2: (a) 2 × 2 tunnels structure, (b) the basic MnO_6-octahedra, and (c) a possible substitution of Fe for Mn.

2.3.1.3 Morphology analysis

The SEM study shown in Fig. 2.17 indicates that the microstructure of the as-obtained products varies remarkably with different iron ion concentration. As shown in Fig. 2.17a, pure MnO_2 exhibits the morphology of sea urchin–shaped hollow microspheres with uniform diameters of about 2 mm, which are constructed of many nanoneedles radiating from spherical shells. When a certain amount of iron ions were introduced, in S2 and S3, the radiating nanoneedles severely degraded and hollow structures with litchi-like surfaces are generated. The cavities of these slightly irregular, hollow microspheres with smaller diameters can be clearly seen from the SEM observation in Fig. 2.17b,c. For S4 and S5, with more iron ions introduced in molar percentage of 30 mol% and 60 mol%, respectively, a novel morphology of clew-like microspheres with decreasing diameters are observed in Fig. 2.17d,e. In the 90 mol% Fe-doped sample, a dramatic morphology change can be observed from microspheres into irregularly shaped micro-/nanoparticles with diameters of about 0.5 μm in Fig. 2.17f.

In the cases of S4, S5, and S6, the chemical composition and crystal structure change as a result of excess Fe doping should be predominantly responsible for the morphology change. It is worthy to mention that hollow MnO_2 structures, as exhibited by S1, S2, and S3, deserve great attention owing to their potential distinctive properties[28–30]. In previous work, α-MnO_2 hollow urchins have been prepared by Li et al. through a low-temperature mild reaction in an acidic solution and the addition of a Cu foil was found to be a crucial factor[28]. A rapid hydrothermal method used by He et al. is able to synthesize hollow α-MnO_2 spheres, which could be attributed to the appending $FeCl_3$[29]. On the basis of our experimental results, hollow-structured α-MnO_2 can be formed at the boiling temperature, no matter whether Fe ions exist or not. In particular, to a certain extent, the incorporation of iron ions can promote the formation of a hollow α-MnO_2 structure. The doped samples S2 and S3, as observed in Fig. 2.17b,c, exhibit slightly more obvious hollow structural microspheres than the undoped sample S1 in Fig. 2.17a.

Figure 2.17 SEM images of the samples synthesized with different iron amounts: (a) without iron, (b) 10 mol%, (c) 20 mol%, (d) 30 mol%, (e) 60 mol%, and (f) 90 mol%.

Almost all the reports on the hollow MnO_2 structures have reached an agreement that the underlying mechanism for the formation of hollow urchins is the well-known "Ostwald ripening process," as illustrated in Fig. 2.18[28]. In the initial stage, the reaction between $(NH_4)_2S_2O_8$ and $MnSO_4$ proceeds rapidly at the boiling temperature to produce small MnO_2 particles. These small particles aggregate to form larger solid spheres, as a virtue of the free-energy decrease of the new nuclei surface. Then as a result of the one-dimensional growth habit of MnO_2 crystals[31], the particles in the outer surfaces are epitaxially grown into nanoneedles from the initial spheres along the [110] direction and solid urchins will be formed. Meanwhile, the small MnO_2 particles located in the inner cores, which can be visualized as smaller spheres having higher curvature compared to those on the outside, have higher surface energies and can be readily dissolved. The outside nanoneedles can serve as nucleation seeds for the subsequent crystallization process of the cores. According to the Ostwald ripening process, the outside nanoneedles grow at the expense of the inside particles because of the energy difference between them. Finally, the hollow urchins are formed and enlarged with depletion of the cores, and both the diameters and lengths of the nanoneedles outside become longer and larger.

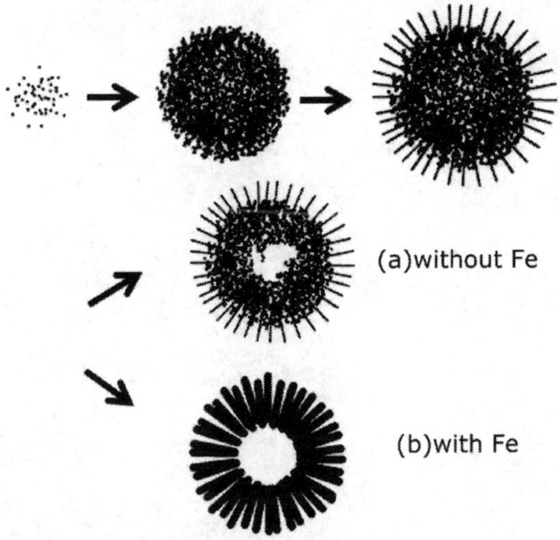

Figure 2.18 Schematic illustration of the formation of α-MnO_2 hollow urchins via the Ostwald ripening process.

During the forming process, the impact of Fe^{3+} ions on MnO_2 microspheres can be speculated in several ways. On the one hand, the addition of Fe^{3+} increases the ionic strength and the chemical potential of the reaction system and thermodynamically aggravates the Ostwald ripening process, which might make the cavities of MnO_2 microspheres more available. On the other hand, stabilization by Fe^{3+} ions chemisorptions on the small MnO_2 particles could limit the external growth stage. Besides, according to Portehault's research on growth mechanisms of MnO_2 nanowires[32], the solubility of Mn species can be increased in a highly acidic medium and can thus promote the growth stage. The chemical equations of the synthesis of Fe-doped MnO_2 in our experiments are described below:

$$2Fe^{2+} + S2O_8{}^{2-} \rightarrow 2Fe^{3+} + 2SO_4{}^{2-} \quad (2.4)$$

$$Mn^{2+} + S_2O_8{}^{2-} + 2H_2O \rightarrow MnO_2 + 2SO_4{}^{2-} + 4H^+ \quad (2.5)$$

Herein, Eq. 2.4 occurs prior to the Eq. 2.5. Meanwhile, the total amount of metal ions Mn^{2+} and Fe^{2+} remains unchanged and the oxidant $(NH_4)_2S_2O_8$ is relatively insufficient. Appropriately increasing Fe^{2+} content decreases the amount of Mn^{2+} and promotes the formation of H^+. The reduction of the acidity of reaction solution could limit the growth as well. Both of the above factors might result in coarsening of assembling nanoneedles and promote the formation of hollow structure, as observed in Fig. 2.17b,c. However, the detailed reason for the formation of a hollow α-MnO_2 structure has been not completely understood so far, and further investigations and discussions are still desperately necessary.

2.3.1.4 Electromagnetic properties

The frequency dependence of the relative complex permittivity in the 2–7 GHz range of various samples is presented in Fig. 2.19. It can be found that with increase of frequency from 2 to 7 GHz, the real (ε') and imaginary parts (ε'') of relative complex permittivity of pure MnO_2 decline from 6.55 to 5.12 and from 1.84 to 1.61, respectively. When the reactions are carried out with iron ions introduced, as in the cases of S2 and S3, the value of complex permittivity reduces with Fe content, and the curves of both ε' and ε'' demonstrate simple and slight decreases in the frequency range of 2–7 GHz. This trend suggests that the dielectric properties exhibited by MnO_2 can be dominantly attributed to relaxation polarization. Moreover, both ε' and ε'' values of S2 are relatively higher in contrast to that of S3 throughout the whole investigated frequency range. When the molar percentage of iron ions is increased to 30 mol%, the ε' value of the composite falls to about 2.6, and the ε'' value falls to about 0.03. Further enhancing the amount of iron ions seems to show no obvious effect on the permittivity, as shown in the samples S5 and S6.

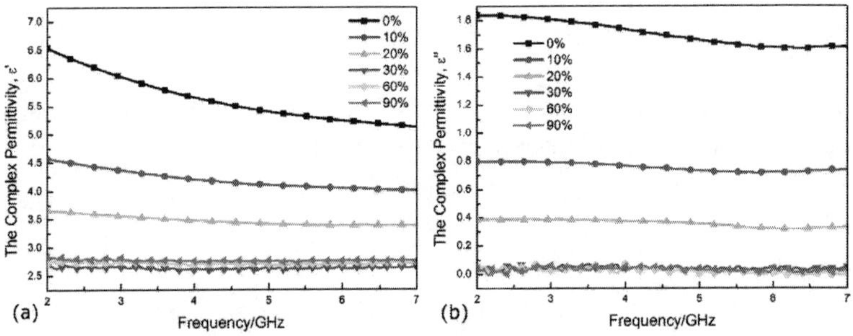

Figure 2.19 Complex relative permittivity of samples synthesized with different iron amounts: 0, 10, 20, 30, 60, and 90 mol%.

Figure 2.20 shows relative complex permeability variation of a series of samples in the 2–7 GHz range. For the cases with Fe doping below 20 mol%, the values of the real parts (μ') and the imaginary parts (μ'') increase approximately with the increase in the Fe amount. It is worthy pointing out that the sample S3 exhibits extraordinary permeability and remarkable variation in comparison with the other samples. However, when more iron ions are introduced into the reaction, the complex permeability of the sample S4 drops evidently and shows slight variation that the μ' remains about 1.02 and the μ'' is close to 0, respectively. In addition, no obvious change of the complex permeability is observed in the samples S4, S5, and S6.

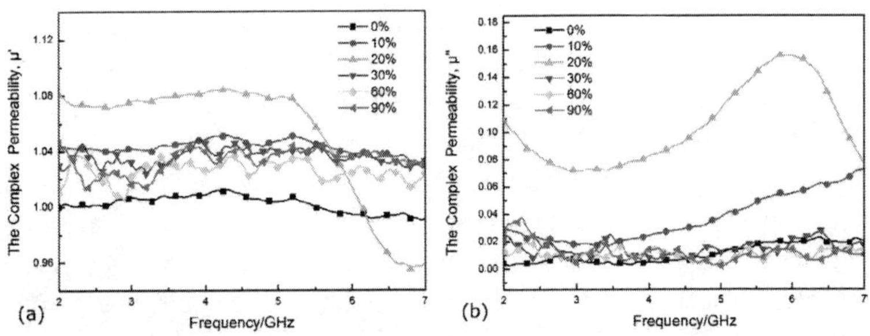

Figure 2.20 Complex relative permeability of samples synthesized with different iron amounts: 0, 10, 20, 30, 60, and 90 mol%.

Figures 2.21a and 2.21b show the frequency dependence of the electric loss tangent and magnetic loss tangent of the samples. Obviously, the pure MnO_2 sample exhibits the largest dielectric loss, while the 20 mol% Fe-doped MnO_2 one exhibits the largest magnetic loss. It can be concluded that when the amount of iron ions ranges from 30 to 90 mol%, both dielectric loss and magnetic loss of the products are small or even negligible in the frequency range of 2–7 GHz. It is worthwhile to note that μ' and the corresponding tan δ_μ of S3 show a complex variation and the curves exhibit a broad peak in the 5.5–6.5 range, suggesting a resonant behavior.

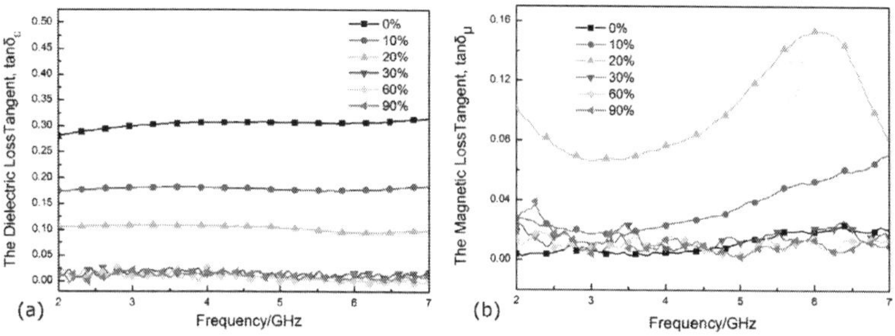

Figure 2.21 (a) Electric loss tangent and (b) magnetic loss tangent of the samples synthesized with different amounts of iron: 0, 10, 20, 30, 60, and 90 mol %.

As stated previously, in the cases of S2 and S3, relatively smaller amounts of Fe doping can effectively cause an increase in magnetic loss but a decrease in dielectric loss. Accordingly, the great dielectric loss exhibited by MnO_2 can be dominantly attributed to relaxation polarization. Fe doping brings out various crystal defects, such as lattice distortions mentioned above, and these defects would act as obstacles, stumbling the polarization process, which can cause a decrease in dielectric loss. It is found by Zhang et al.[33] that there is an inherent relationship between lattice distortions and dielectric properties in the tetragonal phase region that a smaller c/a value exhibits higher permittivity. The crystal structure of α-MnO_2 exactly fits into this region. According to the calculated results in Table 2.1, the c/a values of samples S1 and S3 show a slight increase, which might be a probable cause for decreasing permittivity in MnO_2 composites. The crystal structure of $(Fe_{0.67}Mn_{0.33})OOH$ is the hexagonal phase and not suitable for the above relationship. The negligible values of both dielectric loss and magnetic loss indicate that the compound $(Fe_{0.67}Mn_{0.33})OOH$ intrinsically belongs to a wave-transmitting material. To our best knowledge, reports on $(Fe_{0.67}Mn_{0.33})OOH$ are very scarce[34], and the corresponding studies need to be carried out in the future.

As the speculation of the structure and composition mentioned above, the mixture of Mn^{3+} and Mn^{4+} could be taken into account to compensate the charge of the introduced Fe^{3+}, which is likely to influence the magnetic coupling between the Mn^{3+} and Mn^{4+} ions. Meanwhile, the charge separation of the Mn^{3+} and Mn^{4+} ions should closely relate to the distribution of the intercalated ions located at the tunnel cavity. To relax the electrostatic repulsion, the Mn^{3+} and Mn^{4+} ions are expected to occupy the Mn sites close to the Fe ions and vacancies, respectively. As a result, a spin-glass behavior could be caused by the geometrical frustration and the mixture of Mn^{3+} and Mn^{4+}[35] in Fe-containing MnO_2 samples, which may bring about an increase of magnetic loss, according to a potential mechanism of spin-wave resonance. Besides, the formation of a substitutional solid solution by appropriate Fe doping may have good magnetism and result in a certain extent of increase in magnetic loss. In addition, the hysteresis loop measurements have been employed to examine the magnetic hysteresis loss properties of the samples. But according to the test results, the as-prepared samples exhibit no magnetic properties regardless of the iron content, which indicates that magnetic hysteresis loss could be excluded. In S4, S5, and S6, changes of the crystal structure and chemical composition caused by further Fe doping should play a decisive role in poor EM parameters. Not only that, the morphological change from the hollow structure to solid spheres should also be considered as another important reason for the decrease of the dielectric loss and magnetic loss. Considering a similar-sized particle, the hollow spheres have a higher volume ratio and surface area and thus can induce more EM energy attenuation than solid spheres. It can be also demonstrated in an investigation of Wang et al.[36], in which the as-obtained hollow magnetite nanospheres provided enhanced EM wave absorption performance. Hence, the solid structures exhibited by S4, S5, and S6 might be another factor for the poor EM parameters.

2.3.2 Ni-/Co-Doping Manganese Oxides

2.3.2.1 Synthesis

All the chemicals contained in the synthesis process were used directly without further purification. In a typical procedure, 30 mmol potassium permanganate and 60 mmol manganese sulfate were each dissolved in 120 mL distilled water. Subsequently, 10 mmol nickel sulfate was added to the manganese sulfate solution with stirring until the blended solution was translucent. Then the two separate solutions were mixed and stirred for 3 min before heating. The reaction system was placed in a water bath at 100°C for 3 h to ensure uniform temperature distribution. After the reaction was completed, the obtained precipitate was filtered, washed with distilled water and absolute ethyl alcohol several times until most of the soluble ions were removed from the final product, and finally dried at 150°C for 24 h. The obtained sample was marked as Ni-MnO_2, and the related chemical reaction is shown as follows:

$$2KMnO_4 + 3MnSO_4 + 2H_2O \rightarrow 5MnO_2 + 2H_2SO_4 + K_2SO_4 \quad (2.6)$$

Co-MnO_2 was synthesized by replacement of nickel sulfate with cobalt sulfate. For comparison, pure MnO_2 was synthesized in the absence of nickel sulfate or cobalt sulfate, while the other parameters were kept unchanged. The resulting sample was labeled p-MnO_2.

2.3.2.2 Structure analysis

The XRD patterns of the samples are shown in Fig. 2.22a–c. All the diffraction peaks can be readily indexed to the tetragonal phase of α-MnO_2 (JCPDS 44-0141). No peaks for other impurities are observed, indicating the purity of the as-synthesized MnO_2 samples. The similarity of the patterns indicates that the crystalline structure is not changed by adding Ni^{2+}/Co^{2+} to the reaction system. Further evidence for incorporation of Ni^{2+}/Co^{2+} is obtained by X-ray fluorescence (XRF) measurement, as shown in Table 2.2. It reveals that the main elements in the three samples are Mn and O, and only trace amounts of doping elements Ni/Co are detected. The existence of K is thought to balance the valence variation and hence the stability of the crystal structures.

The mathematical computation based on Table 2.2 shows that the concentration of Ni^{2+} in the final product is only 2.04% in the mole fraction, which is much smaller than the theoretical ratio of 13.33%. Analogously, the mole fraction of Co^{2+} in the obtained MnO_2 is 7.09%, and it is also far away from the theoretical value. So it is concluded that doping with Ni^{2+}/Co^{2+} is realized, though the concentration of dopants is much smaller than the theoretical prediction.

The basic structure of α-MnO_2 is formed by primary [MnO_6] octahedra: double chains of the octahedral form [2 × 2] tunnels by sharing the corners and edges. Typically, and in nature, the tunnels contain some foreign cations such as K^+ (in the cryptomelane structure) to stabilize the tunnel structures[37]. The crystal structure of the fabricated samples is schematically shown in Fig. 2.22d, and in this paper, Ni^{2+}/Co^{2+} are supposed to occur in the octahedral frameworks with Mn^{4+} being substituted. The substitution of dopants would inevitably destroy the original charge balance due to different valence states of Ni^{2+}/Co^{2+} and Mn^{4+}. But this process can be compensated by the modification of K^+ and H_3O^+, which is proved by the changed concentrations of K and H after doping, as shown in Table 2.2 (the content of H is not listed). The other compensation process is assisted by the mixture of Mn^{4+}/Mn^{3+}. Studies show that the major valence state of Mn in MnO_2 is +4, while there is also a small amount of Mn^{3+} in the structure, especially in wet chemically synthesized MnO_2[38,39].

Table 2.2 XRF analysis result of the synthesized p-/Ni-/Co-MnO_2, which confirms the existence of doping elements. The major elements are Mn and O, and the existence of K is claimed to keep the stability of the crystal structures

Samples	Mn[a]	O	K	Ni	Co
p-MnO_2	62.7991[b]	31.4993	5.4800	0	0
Ni-MnO_2	61.0643	32.1166	5.1957	1.3311	0
Co-MnO_2	58.7855	30.8947	5.4816	0	4.4729

[a]Only the main analytes are listed in the table.
[b]The unit of the data is wt%.

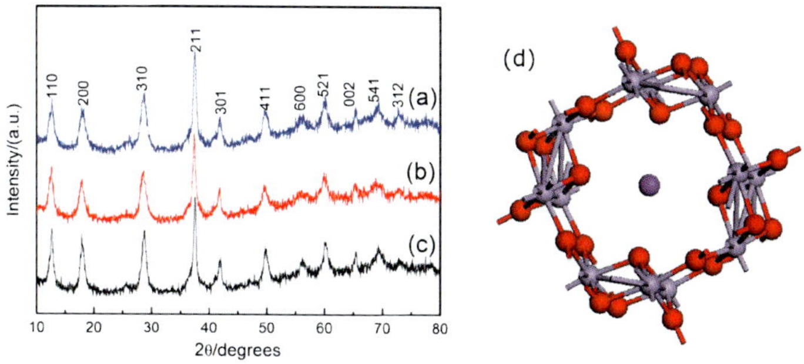

Figure 2.22 XRD patterns of the samples: (a) p-MnO_2, (b) Ni-MnO_2, and (c) Co-MnO_2. (d) Schematic illustration of the crystal structure of α-MnO_2 with K^+ locating at [2 × 2] tunnels to stabilize the tunnel structures, where Mn^{4+} are shown in large pink circles (in the center of octahedral), K^+ in large purple circles (in the center of the tunnel), and O^{2-} in small red circles. For Ni-/Co-MnO_2, Ni^{2+}/Co^{2+} ions are supposed to be located in the octahedral frameworks with Mn^{4+} being substituted, and the primary crystal structure keeps the same after doping.

2.3.2.3 Microstructure and morphologies

As shown in Fig. 2.23, SEM analysis indicates that all the three samples are totally composed of nanowires. Figure 2.23a reveals the morphology of p-MnO_2, and the bunched nanowires are clearly observed with several primary nanowires aggregating along the lateral direction. The diameter of primary nanowires ranges from 15 nm to 20 nm, while the length ranges from 100 nm to 300 nm. Aggregation in the lateral direction seems to occur via the oriented attachment mechanism. Oriented attachment involves spontaneous self-assembling of adjacent primary nanowires, the bonding among which reduces the surface energy. This mechanism is relevant in the cases where primary particles are free to move, for example, in the solution. SEM images of Ni-/Co-MnO_2 are shown in Fig. 2.23b,c. Both the samples share the same morphology: separately distributed nanowires are observed with a diameter of 15–20 nm and a length of 300–800 nm. Doping with Ni^{2+}/Co^{2+} does not change the one-dimensional nanostructure of α-MnO_2 but reduces aggregation in the lateral direction. In this report, the possible growth mechanism of MnO_2 nanowires is proposed as the processes of dissolution-recrystallization and oriented attachment. At the first stage of preparation, small particles are formed in the amorphous state after the two separate solutions are mixed, and these particles are unstable with high surface energy. Then these particles congregate to recrystallize through oriented attachment. During this process, the one-dimensional growth habit along the *c* axis of α-MnO_2 plays the dominant role[40]. The stabilities of different crystal planes are varied, so the oriented attachment tends to occur in some specific directions to reduce the surface energy. However, the suggested growth mechanism of the nanostructure in this report needs to be further studied by more experiments.

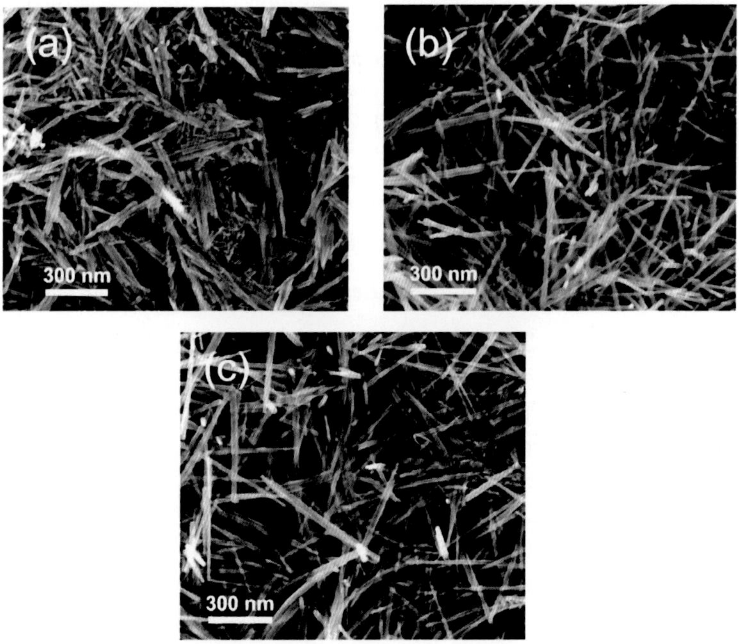

Figure 2.23 Typical SEM micrographs of the synthesized products: (a) p-MnO_2, (b) Ni-MnO_2, and (c) Co-MnO_2. One-dimensional morphology of α-MnO_2 is not changed by doping of Ni/Co, while the aggregation in lateral direction is reduced after doping.

2.3.2.4 Microwave dielectric response of Ni-/Co-doped MnO_2

From Fig. 2.24a, it can be seen that the real parts ε' of the three samples decrease gradually with the increase in the frequency, which is demonstrated as dielectric dispersion. The real parts ε' of Ni-/Co-MnO_2 show more distinct dielectric response characteristics than p-MnO_2 (declines from 11 to 7.77), as the former exhibit an abrupt decrease from 17.22/14.66 to 6.97/7.47, respectively. Manganese dioxide is a kind of dielectric loss material, and its permittivity is related to several electric polarization processes, such as atomic polarization, electronic polarization, dipole relaxation polarization, and ionic polarization. Different polarization processes work at different frequency ranges. The permittivity at microwave frequency is mainly contributed by dipole relaxation polarization, while the minor part is thought to originate from the contributions of electronic and atomic polarization[41]. When an external EM field is applied to the composite, the orientation of the dipoles will be adjusted from random to parallel to the external field, resulting in dipole polarization. However, in actual materials, this process is often associated with energy loss to overcome the resistance. When the EM frequency reaches a certain point, the rearrangement of the dipoles cannot catch up with the external field and makes virtually no contribution to dielectric polarization. So the dielectric response tends to gradually decrease when the frequency is further increased, as shown in Fig. 2.24a. Compared with the imaginary part ε'' of p-MnO_2, the values ε'' of Ni-MnO_2 and Co-MnO_2 exhibit significantly enhanced tendency throughout the whole testing frequency range, which indicate excellent dielectric loss behavior. The incorporation of Ni^{2+}/Co^{2+} destroys the original symmetry of the crystal lattice and results in some lattice defects due to the difference of ionic radius between Ni^{2+}/Co^{2+} and Mn^{4+}, as shown in Fig. 2.22d. As a result, (i) some lattice defects act as polarized centers[42], which are reflected in the enhanced real parts of permittivity for doped MnO_2 at some certain frequency range; and (ii) the destroyed crystal structure makes it more difficult for dipoles to alternate their directions, which is observed from the increased imaginary parts of permittivity after doping. Thus, more energy is attenuated to compensate the damped vibration, which is reflected in the increased ε'' of Ni- and Co-MnO_2.

Figure 2.24b shows the behavior of relative complex permeability as a function of frequency. Both the real and imaginary parts of the relative complex permeability are constant around 1.00 and 0.00 with changing frequency. The main reason is that manganese dioxide is a kind of nonmagnetic material, and the response of its magnetic moments to the applied magnetic field is negligible. Compared with complex permittivity, complex permeability contributes little to the energy attenuation. So the microwave attenuation mainly comes from dielectric loss, in which the dipole relaxation polarization plays an effective role. Generally, the loss tangent (tan δ) is applied to characterize the loss properties of the incident EM wave in an absorber. The dielectric loss tangent (tan $\delta_\varepsilon = \varepsilon'/\varepsilon''$) based on the measured complex permittivity of the composites was researched and is shown in Fig. 2.25. The dielectric loss tangent of the three samples are increased with the increasing frequency, and the values of tan δ_ε for Ni- and Co-MnO_2 are much larger than that of p-MnO_2. It can be deduced that the dielectric loss in the composite is markedly enhanced by doping with Ni^{2+}/Co^{2+}. The values of Ni-MnO_2 are almost the largest throughout the whole frequency range and increase from 0.29 to 0.75. However, the magnetic loss tangent (tan $\delta_\mu = \mu''/\mu'$) for all the composites fluctuates in the range of 0.00–0.07 (not shown) at 2–18 GHz, which differs from 0 by a negligible fraction. Thus the contribution of dielectric loss to the wave absorbing property is validated again, which is superior compared with that of magnetic loss. This is consistent with the analysis of EM parameters.

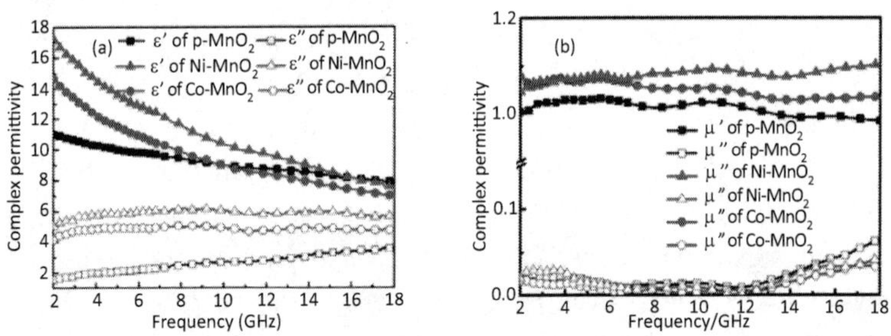

Figure 2.24 Microwave dielectric responses of the sample/paraffin composites: (a) relative complex permittivity and (b) permeability of the composites. Increased imaginary part of permittivity and distinct dielectric response are achieved after doping of Ni/Co.

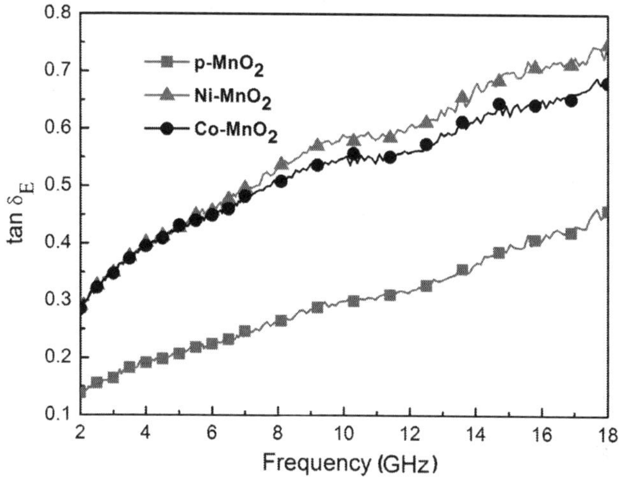

Figure 2.25 Dielectric loss tangent (tan δ_ε) versus frequency for as-synthesized composites of p-/Ni-/Co-MnO_2 and paraffin matrix.

2.4 Temperature-Dependent Dielectric Characterization

Dielectric permittivity implies polarization degrees of MnO_2 materials under an external EM field. The more dipoles exist in the field, the larger is the permittivity. With an increase in temperature, more of the localized dipoles are liberated and align their direction with the field, thus resulting in larger polarizability as well as an increase in ε[43].

2.4.1 Synthesis of MnO_2 Nanostructures

The synthesis of MnO_2 nanostructures can be expressed as follows:

$$3MnSO_4{\cdot}H_2O + 2MnO_4^- \rightarrow 3SO_4^{2-} + 5MnO_2\,(s) + H_2O + 4H^+ \quad (2.7)$$

Three different $KMnO_4$ concentrations (0.025 M, 0.0625 M, and 0.125 M) were used, while maintaining the molar ratio of $KMnO_4$ to $MnSO_4{\cdot}H_2O$ as 2:3, named sample A1, A2, and A3, respectively.

2.4.2 Effect of Temperature on Dielectric Properties of MnO_2

Figure 2.26 shows temperature-dependent dielectric permittivity of samples A1, A2, and A3 at different frequencies of 1 kHz, 10 kHz, 100 kHz, and 1000 kHz and temperatures ranging from 223 K to 393 K. Taking an overview of Fig. 2.27, it is clear that all samples exhibit the general trend of an increase in ε' with increasing temperature and decreasing ε' with increasing frequency. Sample A3 prepared with the largest reactant concentration has the maximum values of dielectric constants at the same temperature, and A1 has the least values of permittivity in the whole temperature range at all frequencies. Take the frequency at 1 kHz, for example. When the temperature changes from 223 K to 393 K, the permittivity of sample A3 increases from 35.3 to 133.6; however, the permittivity of sample A1 only changes from 13.3 to 36.1. At the same temperature of 293 K, when the frequency changes from 1 kHz to 1000 kHz, the permittivity of samples A3 turns from 60.2 to 7.1, and sample A1 changes from 23.2 to only 3.1.

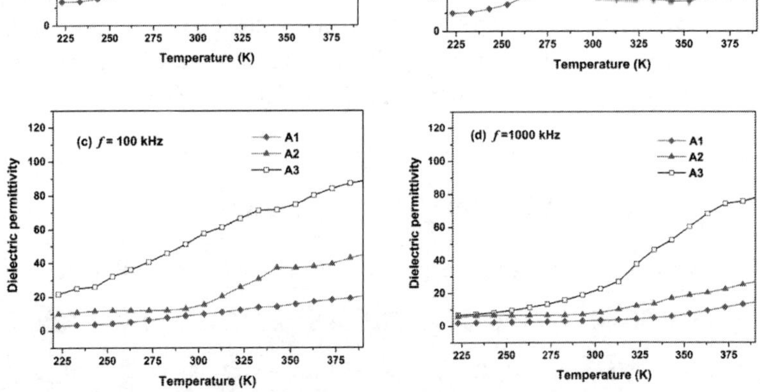

Figure 2.26 Temperature dependence of real permittivity of samples A1, A2, and A3 at different frequencies: (a) 1 kHz, (b) 10 kHz, (c) 100 kHz, and (d) 1000 kHz.

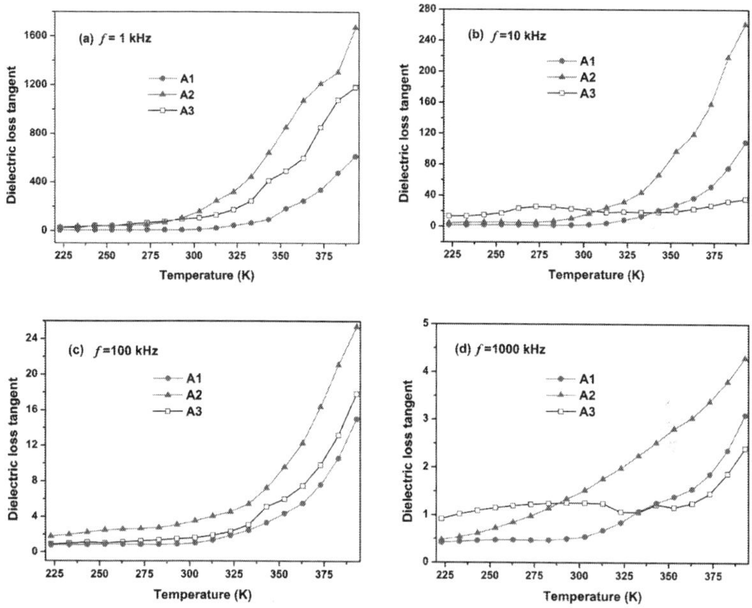

Figure 2.27 Temperature dependence of dielectric loss tangents of samples A1, A2, and A3 at different frequencies: (a) 1 kHz, (b) 10 kHz, (c) 100 kHz, and (d) 1000 kHz.

Figure 2.27 shows the relation between dielectric loss tangents ($\varepsilon''/\varepsilon'$) and the absolute temperature (K) at frequencies of 1 kHz, 10 kHz, 100 kHz, and 1000 kHz for samples A1, A2, and A3 in the temperatures ranging from 223 K to 390 K. It reveals that the dielectric loss tangents increase with the temperature and decrease with the testing frequency, which is the same trend as dielectric permittivity shown in Fig. 2.26. With a closer look at Fig. 2.27, one can see that sample A2 has the greatest dielectric loss values within the three samples, and with increasing temperature and frequency, the dielectric loss differences between sample A2 and the other two samples turn out to be larger.

The differences of dielectric loss tangents can be ascribed to the particle morphology and crystallinity variations of the three samples. As is known, the dielectric loss tangent implies the polarization ability of a material under an external EM field. The variations of crystalloid states of MnO_2 materials can result in a change of the crystal lattice, the differences of crystal defects and holes, which all play key roles in the polarization of materials[44]. The morphological structure variations also induce different extinction cross sections and so result in various EM polarization abilities. The stronger the polarization, the greater the dielectric loss. For the as-prepared MnO_2 materials, particles with different reactant concentrations have different crystalline phases and morphological shapes, so they have various dielectric loss tangents[45].

2.5 Theoretical Study of MnO_2

Density functional theory (DFT) has proved to be a powerful method to investigate the electronic, magnetic, or optical performances of materials containing defects. The first-principles calculations were performed with the approximation of Perdew, Burke, and Ernzerhof (PBE), utilizing the generalized gradient approximation (GGA) scheme, using the CASTEP (Cambridge Serial Total Energy Package) module of Materials Studio[46].

2.5.1 Method and Computational Details

The MnO_2 models for calculations are constructed as follows. The undoped MnO_2 model is an ideal α-MnO_2 supercell ($2 \times 1 \times 1$, $Mn_{16}O_{32}$), as shown in Fig. 2.28a. The Fe-doped MnO_2 model ($Mn_{15}FeO_{32}$) is built by replacing one of the Mn atoms with an Fe atom, on the basis of an undoped MnO_2 model, as shown in Fig. 2.28b. The corresponding doping concentration by molar ratio is 6.25%. It should be noted that the outer electronic configurations of Mn, Fe, and O atoms are $3d^54s^2$, $3d^64s^2$, and $2s^22p^4$, respectively. Both Mn and Fe are magnetic atoms. Therefore the magnetic arrangement of Mn atoms is considered, and the ground-state magnetic structure of α-MnO_2 is built as follows: the Mn–Mn coupling between corner-sharing [MnO_6] is antiferromagnetic, while the coupling between edge-sharing [MnO_6] is ferromagnetic, as shown in Fig. 2.28. The calculation process is conducted in two steps: geometry optimization and energy calculation.

The first-principles calculations are carried out using the CASTEP code[47] using the DFT framework on the basis of the plane-wave basis set and the ultrasoft pseudo-potential[48,49]. To overcome the limitations of the conventional GGA, for example, the underestimation of the band gap, a Coulomb correction of U = 2.5 eV is induced. Considering both the calculation efficiency and accuracy, structural geometries and forces are well converged for a 1 × 3 × 9 Monkhorst–Pack grid with a cutoff energy of 340 eV. The total energy is converged to 1.0×10^6 eV per atom. A Gaussian smearing of 0.1 eV was used for Fermi surface broadening. Relaxations of atomic positions and lattice vectors were performed until residual forces were 0.03 eV·Å^{-1} or less. The whole calculation process is carried out in the reciprocal space.

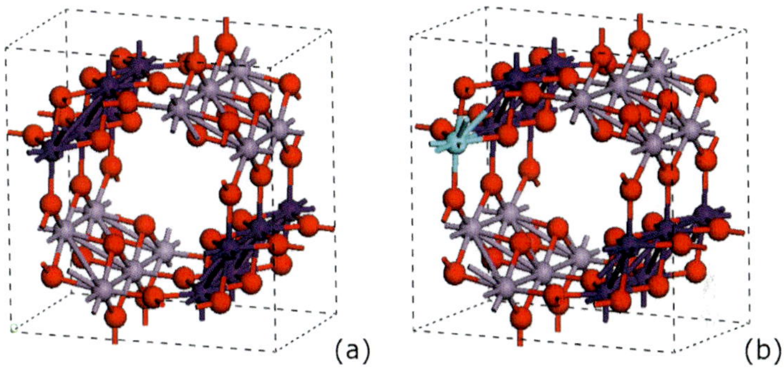

Figure 2.28 Constructed models of (a) pure and (b) Fe-doped α-MnO_2. Mn atoms in spin-up states are sketched in dark purple spheres, Mn atoms in spin-down states in light purple spheres, O atoms in red spheres, and the Fe atom in the blue sphere.

2.5.2 Magnetism Properties

The magnetic properties of a material are mainly determined by magnetic moments resulting from the orbiting and spinning of electrons around the nucleus. Therefore, the magnetism of doped α-MnO_2 can be analyzed by calculating the spin density. Figure 2.29a,b shows the density of states (DOS) of pure and Fe-doped MnO_2. The Fermi energy is expressed by the perpendicular dotted line. It can be seen from Fig. 2.29 that the valence bands of pure and Fe-doped MnO_2 consist mainly of two parts. The upper valence band from 7 eV to 0 eV is contributed by the Mn 3d, O 2p, and Fe 3d states, while the lower valence band from 19.5 eV to 16.5 eV is dominated by the O 2s states. The localized density state indicates strong hybridization among the Mn 3d, O 2p, and Fe 3d orbitals. Owing to the hybrid orbitals, stabilized chemical bonds are formed between the metal atoms and oxygen atoms. The spin-up and spin-down states in the DOS of pure MnO_2 exhibit a symmetrical distribution, as shown in Fig. 2.29a. As for the Fe-doped MnO_2, however, Fe doping leads to asymmetrical distribution in the total spin electronic partial density of states (PDOS) at the conduction band near the Fermi energy shown in Fig. 2.29b, which is absent in that of Ni-/Co-doped MnO_2, as reported in Ref. [34]. This specific distribution of the spin state indicates that the doped α-MnO_2 possesses enhanced magnetic characteristics.

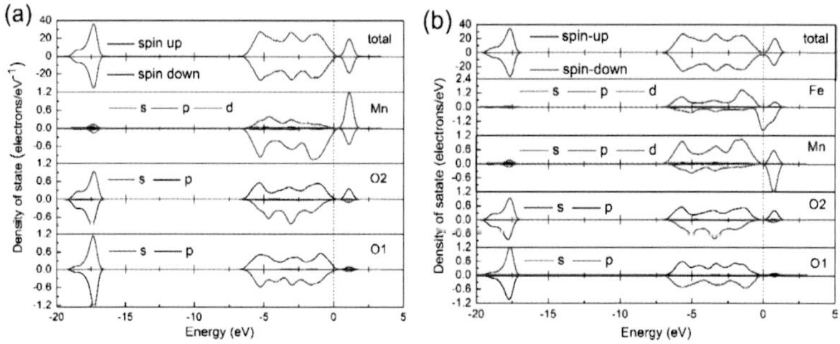

Figure 2.29 Total and partial density of the states of (a) pure and (b) Fe-doped α-MnO_2. The Fermi energy is expressed by the perpendicular dotted line.

As a result, in its calculation, it is supposed that MnO_2 has weak magnetic performance, because the spin-up and spin-down PDOSs of Mn in MnO_2 are not equal. However, it should be noted that this is just the theoretical calculation result. While considering the experimental situation, this magnetic ordering at room temperature is expected to be only short range, and the thermal fluctuation above a particular temperature would readily randomize the magnetic interactions of Mn–Mn to some degree. So the practically measured complex permeability of α-MnO_2 exhibited nonmagnetic character.

In the exported calculation results, the value of integrated spin density (2) for pure MnO_2 is 1.2443×10^{-6}, while the absolute value of the integrated spin density (2) is 49.8344. As for Fe-doped MnO_2, the former value is −0.9995 and the latter value is 48.5909. The vector sum of the spin magnetic moments for each type of ion in pure α-MnO_2 is listed in Table 2.3, and the net magnetic moment of the whole cell is approximated to 0. The net magnetic moment of Fe-doped MnO_2 is calculated to be 0.9995, as shown in Table 2.3, which further supports appearance of magnetism.

Table 2.3 Net spin magnetic moment of pure and Fe-doped MnO_2

System	Mn	O1	O2	Fe	Total
pure MnO_2	0	0	0	–	1.244×10^{-6}
Fe-doped MnO_2	−1.49	−0.11	0.01	0.85	−0.9995

It should be noted that the increased asymmetrical distribution of the DOS for doped MnO_2 is mainly attributed to the contribution of the Fe 3d state. Fe doping introduces spin polarization around the Fermi level; and hence, the net magnetic moment in MnO_2, which results in the magnetic configuration of Fe-doped MnO_2. It is in line with the experimental observations in our previous work[50]. It is supposed that the specific electronic configuration of Fe brings this intriguing magnetic enhancement, which is not observed in Ni-/Co-doped MnO_2[51].

References

1. Cheng F, Su Y, Liang J (2009). MnO_2-based nanostructures as catalysts for electrochemical oxygen reduction in alkaline media, *Chem Mater*, **22**(3), 898–905.
2. Luo Y, Jiang J, Zhou W (2012). Self-assembly of well-ordered whisker-like manganese oxide arrays on carbon fiber paper and its application as electrode material for supercapacitors, *J Mater Chem*, **22**(17), 8634–8640.

3. Cao J, Mao Q, Shi L (2011). Fabrication of γ-MnO_2/α-MnO_2 hollow core/shell structures and their application to water treatment, *J Mater Chem*, **21**(40), 16210–16215.
4. Bach S, Henry M, Baffier N (1990). Sol-gel synthesis of manganese oxides, *J Solid State Chem*, **88**(2), 325–333.
5. Li WN, Yuan J, Shen XF (2006). Hydrothermal synthesis of structureand shape controlled manganese oxide octahedral molecular sieve nanomaterials, *Adv Funct Mater*, **16**(9), 1247–1253.
6. Machefaux E, Verbaere A, Guyomard D (2006). Preparation of nanowires of M substituted manganese dioxides (M = Al, Co, Ni) by the electrochemical–hydrothermal method, *J Phys Chem Solids*, **67**(5), 1315–1319.
7. Calvert C, Joesten R, Ngala K (2008). Synthesis, characterization, and rietveld refinement of tungsten-framework-doped porous manganese oxide (K-OMS-2) material, *Chem Mater*, **20**(20), 6382–6388.
8. Cai J, Liu J, Willis WS (2001). Framework doping of iron in tunnel structure cryptomelane, *Chem Mater,* **13**(7), 2413–2422.
9. Qu D, Diehl D, Conway BE (2005). Development of high-capacity primary alkaline manganese dioxide/zinc cells consisting of Bi-doping of MnO_2, *J Appl Electrochem*, **35**(11), 1111–1120.
10. Jana S, Pande S, Sinha AK (2009). A green chemistry approach for the synthesis of flower-like Ag-doped MnO_2 nanostructures probed by surface-enhanced Raman spectroscopy, *J Phys Chem C*, **113**(4), 1386–1392.
11. Yu P, Zhang X, Wang D (2008). Shape-controlled synthesis of 3D hierarchical MnO_2 nanostructures for electrochemical supercapacitors, *Cryst Growth Des*, **9**(1), 528–533.
12. Shen YF, Suib SL, O'Young CL (1994). Effects of inorganic cation templates on octahedral molecular sieves of manganese oxide, *J Am Chem Soc*, **116**(24), 11020–11029.
13. Ragot F, Badot JC, Baffier N (1995). Influence of the microstructure on dielectric and conducting properties of vanadium pentoxide, *J Mater Chem*, **5**(8), 1155–1161.
14. Wang G, Zhang L, Zhang J (2012). A review of electrode materials for electrochemical supercapacitors, *Chem Soc Rev*, **41**(2), 797–828.
15. Donne SW, Hollenkamp AF, Jones BC (2010). Structure, morphology and electrochemical behaviour of manganese oxides prepared by controlled decomposition of permanganate, *J Power Sources*, **195**(1), 367–373.

16. Wang HE, Qian D (2008). Synthesis and electrochemical properties of α-MnO_2 microspheres, *Mater Chem Phys*, **109**(2), 399–403.

17. Du JH, Sun C, Bai S (2002). Microwave electromagnetic characteristics of a microcoiled carbon fibers/paraffin wax composite in Ku band, *J Mater Res*, **17**(05), 1232–1236.

18. Guan H, Liu S, Zhao Y (2006). Electromagnetic characteristics of nanometer manganese dioxide composite materials, *J Electron Mater*, **35**(5), 892–896.

19. Wu M, He H, Zhao Z (2000). Preparation of magnetic cobalt fibres and their microwave properties, *J Phys D: Appl Phys*, **33**(22), 2927.

20. Wen B, Zhao J, Duan Y (2006). Electromagnetic wave absorption properties of carbon powder from catalysed carbon black in X and Ku bands, *J Phys D: Appl Phys*, **39**(9), 1960.

21. Tang X, Tian Q, Zhao B (2007). The microwave electromagnetic and absorption properties of some porous iron powders, *Mater Sci Eng A*, **445**, 135–140.

22. Qin S, Wang CQ (2006). *Mineralogy Foundation* (in Chinese), Peking University Press, Beijing.

23. Ge FD, Jing JZ, Chen L (1996). Effects of inclusion shape on absorbing ability of microwave absorbing materials, *Acta Electron Sinica*, **24**(6), 82–86.

24. Naito Y, Suetake K (1971). Application of ferrite to electromagnetic waver absorber and its characteristics, *IEEE Trans Microw Theory Techn*, **19**(1), 65–72.

25. Zhang J, Duan Y (2010). The effects of high magnetic field on the morphology and microwave electromagnetic properties of MnO_2 powder, *J Solid State Chem*, **183**(7), 1490–1495.

26. Chen N, Mu GH, Pan XF, Gan KK (2007). Microwave absorption properties of $SrFe_{12}O_{19}/ZnFe_2O_4$ composite powders, *Mater Sci Eng*, **139**, 256–260.

27. Muraoka Y, Chiba H, Atou T (1999). Preparation of α-MnO_2 with an open tunnel, *J Solid State Chem*, **144**(1), 136–142.

28. Li B, Rong G, Xie Y (2006). Low-temperature synthesis of α-MnO_2 hollow urchins and their application in rechargeable Li^+ batteries, *Inorg Chem*, **45**(16), 6404–6410.

29. He X, Yang M, Ni P (2010). Rapid synthesis of hollow structured MnO_2 microspheres and their capacitance, *Colloid Surface A*, **363**(1), 64–70.

30. Fu X, Feng J, Wang H (2010). Fast synthesis and formation mechanism of γ-MnO_2 hollow nanospheres for aerobic oxidation of alcohols, *Mater Res Bull*, **45**(9), 1218–1223.

31. Wang X, Li Y (2002). Selected-control hydrothermal synthesis of α-and β-MnO_2 single crystal nanowires, *J Am Chem Soc*, **124**(12), 2880–2881.

32. Portehault D, Cassaignon S, Baudrin E (2007). Morphology control of cryptomelane type MnO_2 nanowires by soft chemistry. Growth mechanisms in aqueous medium, *Chem Mater*, **19**(22), 5410–5417.

33. Zhang DM, Yan WS, Zhong ZC (2004). Study on the relation between the dielectric properties and lattice distortions in PZT ferroelectric tetragonal phase region, *Acta Phys Sin*, **53**(5), 1316–1320.

34. Scano A, Ennas G, Frongia F (2011). Mn–ferrite nanoparticles via reverse microemulsions: synthesis and characterization, *J Nanopart Res*, **13**(7), 3063–3073.

35. Luo J, Zhu HT, Liang JK (2010). Tuning magnetic properties of α-MnO_2 nanotubes by K^+ doping, *J Phys Chem C*, **114**(19), 8782–8786.

36. Wang F, Liu J, Kong J (2011). Template free synthesis and electromagnetic wave absorption properties of monodispersed hollow magnetite nano-spheres, *J Mater Chem*, **21**(12), 4314–4320.

37. Johnson CS (2007). Development and utility of manganese oxides as cathodes in lithium batteries, *J Power Sources*, **165**(2), 559–565.

38. Hui J, Yuping D, Zhuo L (2012). Influence of Fe-doping on the microstructure and electromagnetic performance of manganese oxides, *Physica B*, **407**(6), 971–977.

39. Luo J, Zhu HT, Liang JK (2010). Tuning magnetic properties of α-MnO_2 nanotubes by K^+ doping, *J Phys Chem C*, **114**(19), 8782–8786.

40. Portehault D, Cassaignon S, Baudrin E (2007). Morphology control of cryptomelane type MnO_2 nanowires by soft chemistry. Growth mechanisms in aqueous medium, *Chem Mater*, **19**(22), 5410–5417.

41. Chen LF, Ong CK, Neo CP, Varadan VV, Varadan VK (2004). *Microwave Electronics: Measurement and Materials Characterization*, John Wiley & Sons.

42. Watts PCP, Ponnampalam DR, Hsu WK (2003). The complex permittivity of multi-walled carbon nanotube–polystyrene composite films in X-band, *Chem Phys Lett*, **378**(5), 609–614.

43. Guan H, Wang Y, Chen G, Zhu J (2012). Frequency and temperature effects on dielectric and electrical characteristics of α-MnO_2 nanorods, *Powder Technol*, **224**, 356–359.

44. Duan Y, Ma H, Li XG (2010). The microwave electromagnetic characteristics of manganese dioxide with different crystallographic structures, *Phys Rev B*, **405**(7), 1826–1831.

45. Guan H, Chen G, Zhu J (2010). Temperature dependent dielectric characterization of manganese dioxide nanostructures with different morphologies at low frequency, *J Alloy Compd*, **507**(1), 126–132.

46. Duan Y, Liu Z, Zhang Y (2013). A theoretical study of the dielectric and magnetic responses of Fe-doped α-MnO_2 based on quantum mechanical calculations, *J Mater Chem C*, **1**(10), 1990–1994.

47. Clark SJ, Segall MD, Pickard CJ (2005). First principles methods using CASTEP, Z *Kristallogr*, **220**, 567–570.

48. Segall MD, Lindan PJD, Probert MJ (2002). First-principles simulation: ideas, illustrations and the CASTEP code, *J Phys: Condens Mat*, **14**(11), 2717.

49. Wright AF, Nelson JS (1994). Explicit treatment of the gallium 3d electrons in GaN using the plane-wave pseudopotential method, *Phys Rev B*, **50**(4), 2159.

50. Duan YP, Jing H, Liu Z (2012). Controlled synthesis and electromagnetic performance of hollow microstructures assembled of tetragonal MnO_2 nano-columns, *J Appl Phys*, **111**(8), 084109.

51. Duan YP, Liu Z, Jing H (2012). Novel microwave dielectric response of Ni/Co-doped manganese dioxides and their microwave absorbing properties, *J Mater Chem*, **22**(35), 18291–18299.

Chapter 3

Fe-Based Composite Absorbers

Among multitudinous magnetic materials, iron-based alloys have received appreciable attention because of their excellent soft magnetic properties, such as high saturation magnetization, high permeability, and good absorbing properties[1–3]. Extensive studies have been carried out on alloying of Fe-based alloys in recent years. The influencing factors on electromagnetic performance and working mechanisms have been investigated in different emphasis. Kim[4] produced nanocrystalline Fe-Co powders and systematically analyzed the influences of grain size on internal strain, M_S, and H_C. Murakami[5] studied the effects of the magnetic domain structure in Co-Ni-Al ferromagnetic-shaped memory alloys and found that it exhibited the B2 to $L1_0$ martensitic transformation upon cooling. Koohkan[6] focused on the relations between milling time and the morphology, H_C, and saturation intrinsic flux density (J_s) of the Fe-Ni alloy. In the whole process of milling (0–100 h), the lattice strain, H_C, and J_s of the Fe-Ni alloy all increased with milling. Wang[7] placed emphasis on the aspect ratio, electromagnetic parameter ($\mu = \mu' - j\mu''$, $\varepsilon = \varepsilon' - j\varepsilon''$), and reflection loss (RL) of the Fe-Cr-Si-Al alloy and pointed out that the aspect ratio played a key role in the increase of RL. $Fe_{1.5}M_{0.5}CoSi$ (M = V, Cr, Mn, Fe) alloys were synthesized, and their electronic and magnetic properties were studied theoretically using the simulated computation method[8]. Theoretical calculations

Microwave Absorbing Materials
Yuping Duan and Hongtao Guan

ISBN 978-981-4745-10-9 (Hardcover), 978-981-4745-11-6 (eBook)
www.panstanford.com

suggested that doping of low-valent atoms helped to change the band structures, leading to 100% spin polarization in $Fe_{1.5}Mn_{0.5}CoSi$ and $Fe_{1.5}Cr_{0.5}CoSi$ and ultimately resulted in magnetic evolution.

3.1 FeSi Alloy

Compared to other Fe-based powders, FeSi alloy powders have higher resistivity, which is beneficial to suppress the unfavorable eddy current effect in the GHz frequency region.

Figure 3.1 shows the scanning electron microscopy (SEM) micrographs of sieved FeSi powders with a particle size of less than 53, 53–75, 75–106, and 106–150 μm. Most of the particles are spherical, and they could be dispersed uniformly into paraffin due to good sphericity. What's more, the spherical powders have no shape anisotropy, so when we investigate the influence of particle size on electromagnetic (EM) behavior and microwave absorption properties, the shape anisotropy effect could be neglected.

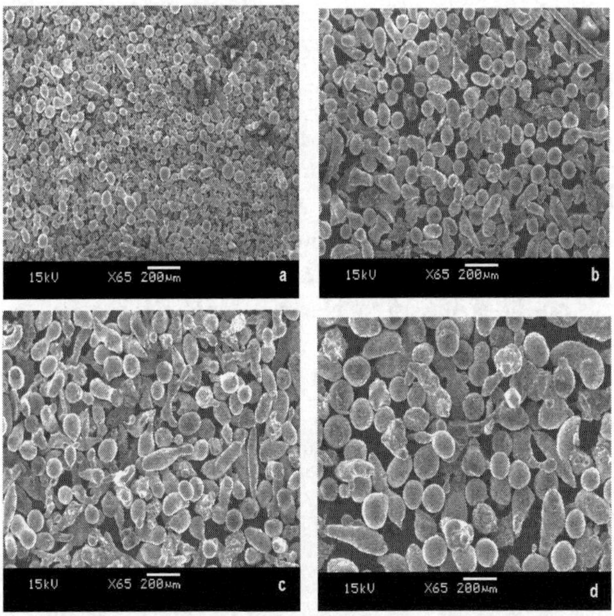

Figure 3.1 SEM micrographs of FeSi powders with a particle size of (a) less than 53 μm, (b) 53–75 μm, (c) 75–106 μm, and (d) 106–150 μm.

3.1.1 Electromagnetic Properties

Figure 3.2a,b shows complex permittivity ($\varepsilon = \varepsilon' - j\varepsilon''$) of the composites containing FeSi powders with different particle sizes in the frequency range of 2–7 GHz. As shown in the figure, the real part, ε', decreases with increasing frequency, namely ε', which obtains the largest value at 2 GHz in the frequency range of 2–7 GHz. However, the imaginary part, ε'', ascends with frequency. Besides, both ε' and ε'' increase with a decrease in particle size; the sample with a particle size of less than 53 μm exhibits better dielectric properties compared to others.

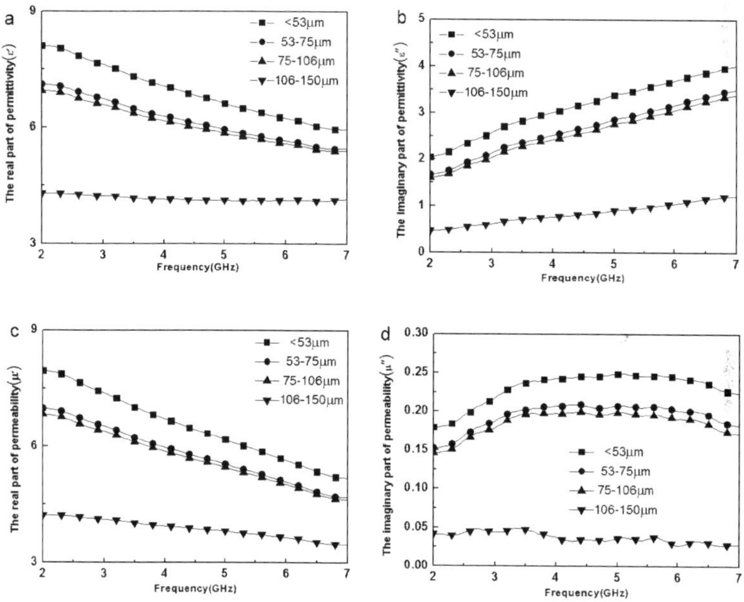

Figure 3.2 Complex permittivity (a, b) and permeability (c, d) of FeSi/paraffin composites with different particle sizes in the 2–7 GHz range.

The results may be interpreted as follows: First, for metal particles dispersed in an insulating matrix, space-charge polarization between metal/insulator interfaces is the dominant mechanism that determines the dielectric properties of composites. For smaller-particle-size samples, large surface areas are obtained, so the space-charge polarization is obviously enhanced due to the increase in interface areas[9,10]. Second, the metal particle volume fraction in composites increases due to the decrease in particle sizes; according to effective medium theory, the dielectric properties of composites with small particle sizes are improved[11,12]. Third, for samples with small particle sizes, more free electrons from the enhanced metal fractions promote charge accumulation at the interface, and high permittivity is attained owing to improved electrical conductivity[13].

Frequency dependencies of complex permeability ($\mu = \mu' - j\mu''$) of the composites employing FeSi powders with different particle sizes in the range of 2–7 GHz are shown in Fig. 3.2c,d. As seen from Fig. 3.2c, the values of μ' exhibit a declining trend over the range of 2–7 GHz. However, as shown in Fig. 3.2d, the values of μ'' slightly ascend with frequency in the range of 2.0–3.4 GHz and remain nearly invariable in the range of 3.4–7.0 GHz. In addition, both μ' and μ'' of the samples with small-particle-size FeSi powders are higher than those of the larger ones, which indicates that a small particle size is good for enhancing the permeability of composites.

Generally speaking, the main mechanisms determining the permeability of metal particles include hysteresis loss, domain wall motion, gyromagnetic spin rotation, and the eddy current effect. Hysteresis loss is caused by irreversible magnetization, and the energy loss is proportional to the area encircled by the hysteresis loop, which is mainly determined by saturation magnetization (M_S) and coercivity (H_c)[14]. The static magnetic properties of FeSi powders with different sizes are measured by vibrating sample magnetometer (VSM) at room temperature, and the measured results of M_S and H_c are listed in Table 3.1. As shown in the table, both M_s and H_c vary mildly for different-particle-size powders, so it can be deduced that particle size has no obvious influence on hysteresis loss. In general, domain wall motion and gyromagnetic spin rotation are usually coupled with resonance phenomena in the permeability spectrum[15]. As shown in Fig. 3.2c,d, no explicit resonance peak corresponding to the domain wall motion or gyromagnetic spin rotation is found in the measured frequency region, so the difference in permeability of FeSi powders with different particle sizes is related to the eddy current effect. As we know, the decrease in the particle size is beneficial to suppress the unfavorable eddy current effect[16]. Thus the permeability of FeSi powders with a small size remains higher than that of larger particles. In addition, along with the decrease in the particle size, the interface area and particle volume fraction in composites are increased, so the exchange coupling reaction of the magnetic moment of neighboring particles is enhanced[17].

Table 3.1 Saturation magnetization (M_S) and coercivity (H_c) of FeSi powders with different particle sizes

Sample (μm)	M_s (emu/g)	H_c (Oe)
<53	146	46
53–75	147	44
75–106	147	43
106–150	145	42

3.1.2 Microwave Absorption Properties

On the basis of measured complex permittivity and permeability, we assume that a single layer of FeSi/paraffin composites is attached to a metal plate. Then the microwave absorbing performances are investigated by calculating the attenuation constant (α) and RL using the following equations:[16,18]

$$Z_{in} = Z_0\sqrt{\mu/\varepsilon}\cdot\tanh(\gamma d) \tag{3.1}$$

$$\gamma = \alpha + j\beta = j(2\pi f/c)\sqrt{\mu\varepsilon} \tag{3.2}$$

$$\mathrm{RL\,(dB)} = 20\log|(Z_{in} - Z_0)/(Z_{in} + Z_0)| \tag{3.3}$$

where Z_{in} is the normalized input impedance at the free space and material interface; γ is the EM wave propagation constant in the material; α is the attenuation constant, which is defined as the real part of γ; β is the phase constant and is equal to the imaginary part of γ; c is the velocity of light; d is the thickness of the single absorber; f is the frequency; and μ and ε are the complex permeability and permittivity, respectively.

Figure 3.3 shows the frequency dependencies of the attenuation constant, α, for FeSi/paraffin composites with different particle sizes. As is seen from the figure, the values of α ascend with frequency in the 2–7 GHz range; the result indicates that FeSi/paraffin composites show better microwave absorption properties in a higher frequency range. Besides, the sample containing smaller-particle-size FeSi powders has a bigger α in the whole frequency range, indicating excellent attenuation or EM wave absorption.

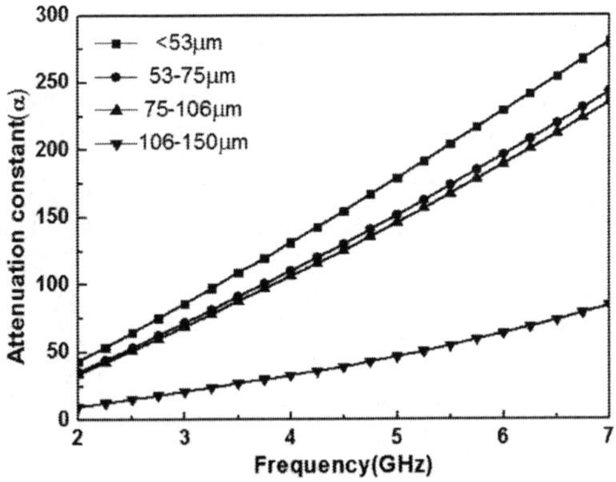

Figure 3.3 Attenuation constant of FeSi/paraffin composites with different particle sizes versus frequency.

Figure 3.4 shows effects of particle size on the calculated RL curves of FeSi/paraffin composites (thickness = 4 mm), and the detailed data for the microwave absorption properties of the four samples are listed in Table 3.2. As we can see, the matching frequency, f_m, which describes the position of the absorption peak, shifts toward the lower frequency range by decreasing the particle size of FeSi powders. Meanwhile, the effective absorption bandwidth (Δf), which is defined as the frequency range of RL < –10 dB, is enhanced with the reduction of particle size. Usually, Δf is an important parameter to estimate microwave absorption properties, so the sample with a particle size of less than 53 μm exhibits the best EM wave absorption properties, and Δf achieves 3.47 GHz.

The shift trend of f_m as the variation of particle size may be explained by the following equation[19]. When the EM wave enters the absorber backed up with a metal plate, a part of it is reflected at the air–absorber interface, while some part is reflected from the absorber–metal interface. The two reflected waves are out of phase by 180° and cancel each other at the air–absorber interface when the quarter-wave thickness criterion is satisfied:[20]

$$f_m = \frac{c}{4d_m}\frac{1}{\sqrt{\varepsilon'\mu'}}\left(1+\frac{1}{8}\tan^2\delta_\mu\right)^{-1} \tag{3.4}$$

where f_m, d_m, and $\tan\delta_\mu$ are the matching frequency, matching thickness, and magnetic loss tangent, respectively. In our study, we can clearly observe tan $\delta_\mu = (\mu''/\mu') << 1$ from Fig. 3.2c,d, so f_m is inversely proportional to $\sqrt{\varepsilon'\mu'}$ at a given thickness. Hence, for FeSi/paraffin composites, higher ε' and μ' values of samples with smaller particle sizes result in a lower f_m.

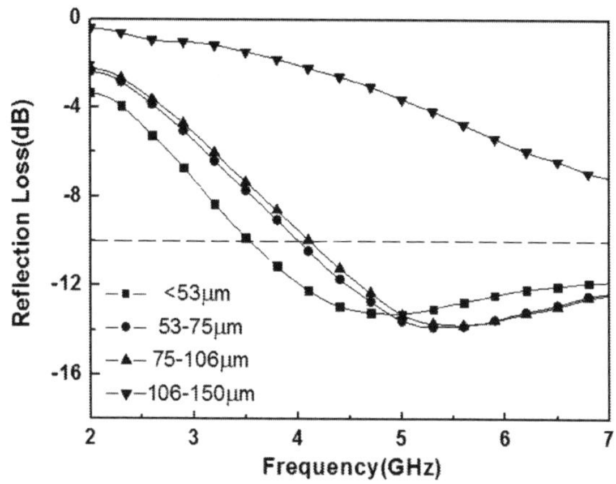

Figure 3.4 Reflection loss of FeSi/paraffin composites with different particle sizes.

Better microwave absorption properties are obtained from a sample containing smaller-particle-size FeSi. The result may be interpreted by the following two factors. The excellent complex permittivity and permeability of the composites could improve microwave absorption properties; especially, the higher ε'' and μ'' signify larger dielectric and magnetic loss, respectively, to increase the microwave absorption[21]. Furthermore, the enhanced absorption properties as particle size reduces are attributed to the increase of scattering effects of FeSi powders. When an EM wave enters the absorber along the x axis direction, the interface between absorber and air is defined as $x = 0$, and the energy of the EM wave at x point $I_{(x)}$ is given by[22]

$$I_{(x)} = I_0 \exp(-n\sigma_{sc}x) \tag{3.5}$$

where I_0 is the energy of the incident EM wave, n is the number of absorbent particles per unit volume in absorber, and σ_{sc} is the scattering cross section of absorbent particles. By reducing the particle size of FeSi powders, the enhanced surface area of particles leads to the increase of $n\sigma_{sc}$. Therefore, according to Eq. 3.5, more energy of the EM wave is scattered for composites with smaller-particle-size FeSi.

Table 3.2 Microwave absorption properties of samples containing different-particle-size FeSi powders

Sample (μm)	Frequency of absorption peak (GHz)	Minimum RL value (dB)	Absorption band (GHz) RL < −10 dB
<53	4.95	−13.3	3.47(3.53–7)
53–75	5.25	−13.9	3.03(3.97–7)
75–106	5.6	−13.8	2.84(4.16–7)
106–150	7	-7.1	None

3.1.3 Carbonyl-Iron/FeSi Composites

FeSi as a kind of magnetic metallic material has relatively high resonance frequency, permeability, and electric resistivity[23]. Meanwhile, carbonyl-iron as a cheap and popular magnetic metallic material has superior electric conductivity and large saturation magnetization but weak frequency dependence of complex permeability due to the eddy current loss induced by EM waves[24]. Therefore, merging the characteristics of these two kinds of materials could be attractive for microwave absorption materials, which can produce the best-possible results for the furthest attenuation microwave.

3.1.3.1 EM characteristics

The frequency dependence on the relative permittivity of the carbonyl-iron powders (CIPs), $Fe_{91}Si_9$ powders (FSPs), and composite samples CF1, CF2, and CF3 are shown in Fig. 3.5. The weight ratios of CF1, CF2, and CF3 are 9:1, 2.3:1, and 1:1, respectively.

It is found that the values of complex permittivity are sensitive to the content of FSP addition. Both the values of ε' and ε'' decrease gradually with increasing weight percentage of FSPs in the composites.

For the CF1 composites, the values of ε' and ε'' decline from 34.6 and 13.1 to 28.5 and 6.2, respectively, gradually with increasing frequency in the 2–7 GHz range. For the CF2 composites, the values of ε' show less variation ($\varepsilon' = \sim 23$) and the values of ε'' decline from 4.3 to 1.7 between 2 and 7 GHz. The values of ε' and ε'' are almost constant over the whole measured frequency range in the composites of CF3 ($\varepsilon' = \sim 14.5$ and $\varepsilon'' = \sim 1.4$). According to free-electron theory[25], $\varepsilon'' = \sigma/2\pi f\varepsilon_0$), where σ is the conductivity. It can be found clearly that the sample CF3 has higher resistivity than the samples CF1 and CF2.

The real part, μ', and the imaginary part, μ'', of relative permeability of the CIPs, FSPs, and composite samples CF1, CF2, and CF3 are plotted as a function of frequency in Fig. 3.6.

It is obvious that μ' decreases gradually with increasing frequency, but μ'' shows an increasing trend with increasing frequency. It is found that μ' and μ'' exhibit a linear variation as a function of the increasing addition of FSPs. The changes in μ' and μ'' with the frequency for the composites obey the Lichitenecker's logarithm mixed law[26],

$$\log \mu = \delta_1 \log \mu_1 + \delta_2 \log \mu_2 \quad (3.6)$$

where δ_1 and δ_2 are the normalized volume ratios of CIPs and FSPs, respectively μ, μ_2, and μ_2 are the complex permeability for the composite samples, CIPs and FSPs, respectively. Compared to CIPs, FSPs have lower μ' and μ'', so the addition of FSPs will result in a decrease in relative permeability.

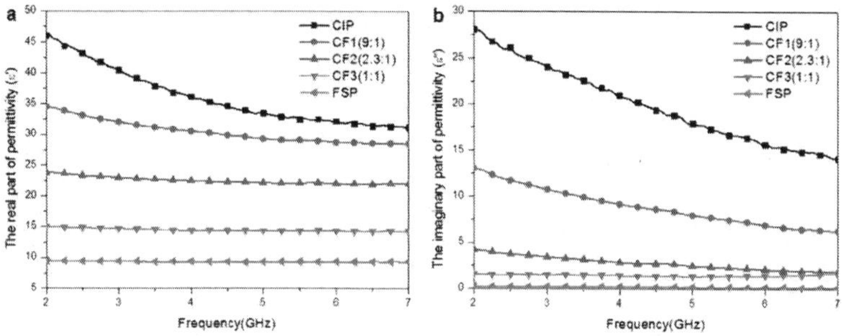

Figure 3.5 Frequency dependence on the relative permittivity, (a) the real part ε', and (b) the imaginary part ε'', of the carbonyl-iron, $Fe_{91}Si_9$, and carbonyl-iron/$Fe_{91}Si_9$ paraffin matrix composites with different weight ratios of carbonyl-iron to $Fe_{91}Si_9$.

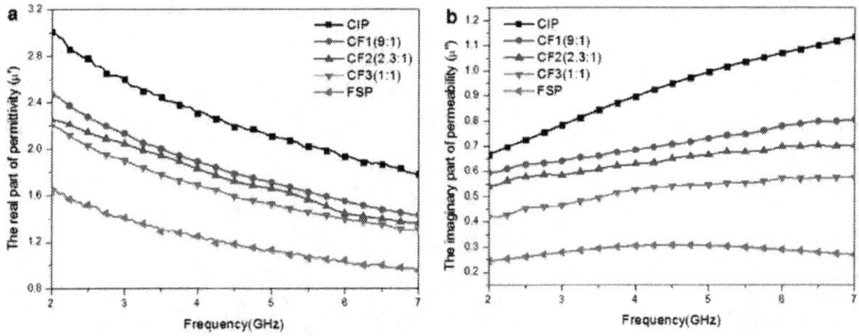

Figure 3.6 Frequency dependence on the relative permeability, (a) the real part μ' and (b) the imaginary part μ'', of the carbonyl-iron, $Fe_{91}Si_9$, and carbonyl-iron/$Fe_{91}Si_9$ paraffin matrix composites with different weight ratios of carbonyl-iron to $Fe_{91}Si_9$.

3.1.3.2 Microwave absorption properties

Figure 3.7 shows a typical relationship between RL and frequency for paraffin matrix composites with 80 wt% CIPs/FSPs. It is seen that the RLs of carbonyl-iron and FeSi are very low and the peak values are more than –20 dB in the frequency range of 2–7 GHz. The CF1 sample exhibits poor microwave absorption performance with RL values more than –20 dB from 2 to 7 GHz, which may be due to the low EM impedance match between the high permittivity and relative low permeability. However, with increasing FSP content, microwave absorption is evidently improved. The RL values of the CF2 sample are less than –20 dB in the range of 2.0–3.5 GHz over absorber thickness of 5.2–3.0 mm, and the optimal RL value is –33 dB at the frequency of 3 GHz with a matching thickness (d_m) of 3.6 mm. The CF3 sample provides the best microwave absorption performances. RL values of the CF3 sample of less than –20 dB are obtained in the 3.7–6.7 GHz frequency range, with a thickness of 4.0–2.4 mm, respectively. In particular, the optimal RL value of –45 dB is observed at 5.2 GHz with a d_m of 3.0 mm. To make the results more clear, the EM wave absorption properties of the composite samples CF1, CF2, and CF3 prepared under optimized conditions are summarized in Table 3.3.

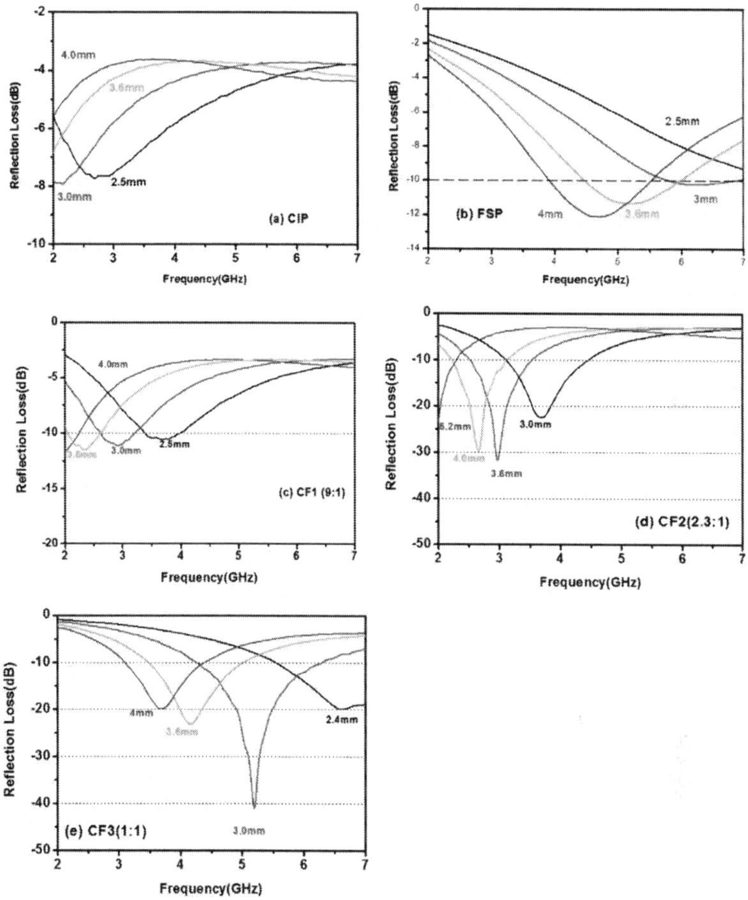

Figure 3.7 Frequency dependences of reflection loss values for carbonyl-iron, $Fe_{91}Si_9$, and carbonyl-iron/$Fe_{91}Si_9$ paraffin matrix composites with different weight ratios of carbonyl-iron to $Fe_{91}Si_9$.

It is clear that the frequency band of RL < –20 dB gets broader and the thickness gets thinner with increasing FSP content in the frequency range of 2–7 GHz. The improvement in microwave absorption of the composites with the addition of FSPs is suggested to originate from the efficient combination of CIPs and FSPs[27]. Generally, excellent EM wave absorption results from efficient complementarities between relative permittivity and permeability in materials. Either only the magnetic loss or only the dielectric loss may result in weak EM wave absorption properties due to the imbalance of the EM impedance match[28]. The introduction of FSPs to CIPs weakens the dielectric loss but does not weaken the magnetic loss too much. Thus, a better EM impedance match could be established due to the combination of the reduced dielectric loss and nearly invariable magnetic loss, resulting in enhanced microwave absorption[29].

Table 3.3 Electromagnetic wave absorption properties of carbonyl-iron/$Fe_{91}Si_9$ paraffin matrix composites with different weight ratios of carbonyl-iron to $Fe_{91}Si_9$

Sample (80 wt%)	Frequency (GHz) (RL < −20 dB)	Thickness (mm) (RL < −20 dB)	Optimal RL value (dB)	f_m (GHz) optimal RL	d_m (mm) optimal RL
CF1(9:1)	–	–	12	2.0	4.0
CF2(2.3:1)	2.0–3.5	3.0–5.2	33	3.0	3.6
CF3(1:1)	3.7–6.7	2.4–4.0	45	5.2	3.0

Figure 3.8 shows the distribution of FSPs and CIPs in a paraffin matrix. The FSP acts not only as a magnetic material, increasing the permeability of the composite powder, but also as an insulating matrix distributed among the gaps between carbonyl-iron particles, which could reduce the eddy current loss through increasing electric resistivity, which is an important reason for bringing about excellent microwave absorption[30].

3.2 Fe-Co-Ni Alloy

Ternary alloys of Fe-Co-Ni have been found to possess excellent magnetically soft properties, including relatively high values of saturation magnetization and extremely low coercivities[31]. All of these indicate that Fe-Co-Ni alloys process excellent absorption properties.

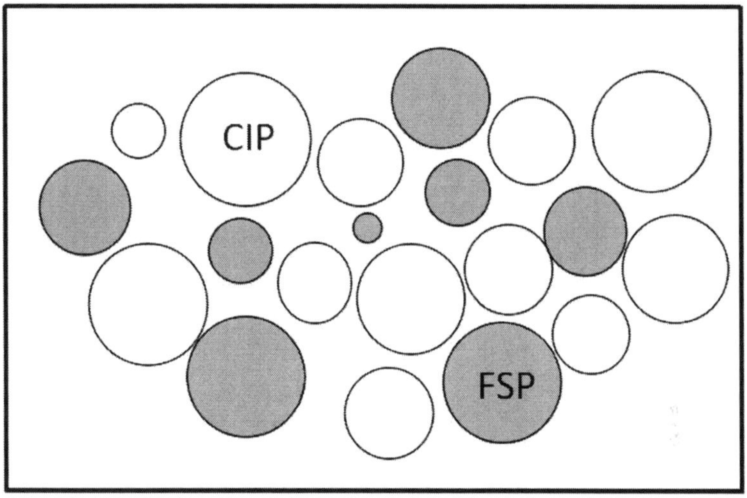

Figure 3.8 Distribution of $Fe_{91}Si_9$ and carbonyl-iron powders in a paraffin matrix.

Figure 3.9 shows SEM micrographs of original powder mixtures and milled powders after different milling times (25–90 h). It can be seen that different morphologies are presented during the mechanical alloying stages. For the unmilled powders, the existence of massive iron, spicate nickel, and fluffy cobalt is clearly observed (Fig. 3.9a). After milling for 25 h, the powders are crushed and changed into flake shapes of different sizes (Fig. 3.9b). With a further increase in the milling time (90 h), flaky particles with uniform size distribution are obtained after deformation and cold welding (Fig. 3.9e).

The changes of the microstructure, including phase transition, average grain size, and internal strain of the powders in the ball-milling process, can be observed in X-ray diffraction (XRD) patterns, as shown in Fig. 3.10. The XRD pattern of the initial powders mixture confirms the coexistence of the body-centered cubic (bcc) alpha iron (α-Fe), face-centered cubic (fcc) nickel (β-Ni), and fcc cobalt (β-Co). It is noticeable that these diffraction peaks are difficult to distinguish due to their similar structures. As the milling time increases, the characteristic peaks of β-Ni and β-Co at 51.5° and 76° (reflection angles, 2θ) vanish and are accompanied by line broadening of α-Fe at

64.2° and 81.5°, indicating refinement of the grain size as well as an increase in internal strain. Moreover, the peak of α-Fe corresponding to 44° shifts to higher angles with increasing milling time, as shown in the drawing of partial enlargement for Fig. 3.10. That indicates the formation of a bcc Fe-Co-Ni solid solution, which can be explained by the differences of α-Fe (Co, Ni) lattice parameters resulting from diffusion of Co and Ni into Fe, followed by the formation of a solid solution of α-Fe(Co, Ni)[32].

Figure 3.9 SEM micrographs of $Fe_{50}Co_{40}Ni_{10}$ (wt%) powders after milling for (a) 0 h, (b) 25 h, (c) 40 h, (d) 55 h, and (e) 90 h. The morphology of the powders was changed into flaky with milling.

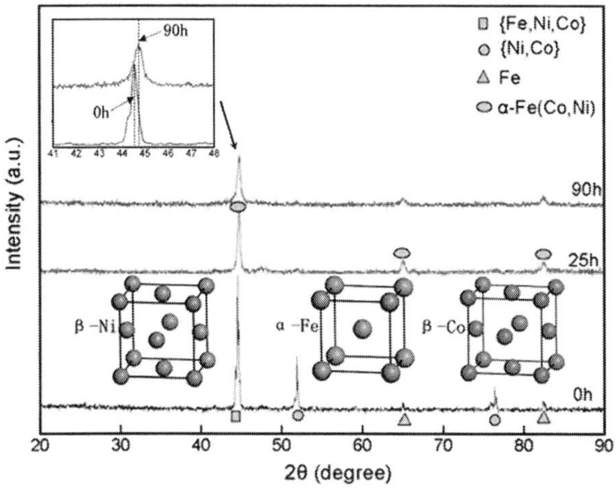

Figure 3.10 XRD patterns of $Fe_{50}Co_{40}Ni_{10}$ (wt%) powders milled for different times (0 h, 25 h, and 90 h).

Figure 3.11 presents transmission electron microscopy (TEM) and the corresponding high-resolution TEM (HRTEM) images of the milled powders. As shown in Fig. 3.11a–d, compared to the fusiform shape for the initial powder, the alloy particles are basically flaky with irregular edges, as indicated by the SEM pictures in Fig. 3.9a–e. Figure 3.11e–h shows the HRTEM images and corresponding fast Fourier transformation (FFT). It is clear that the corresponding selected area diffraction contains diffraction spots arising from the crystalline regions, along with the halo pattern attributed to the existence of the amorphous phase. According to Fig. 3.11e–h, the amount of the amorphous phase increases significantly after an excessive milling process and the maximum value is obtained at 90 h milling. On the other hand, even the formation of the α-Fe (Co, Ni) phase is inevitable, and it is difficult to distinguish the α-Fe and α-Fe (Co, Ni) phases from the FFT of the HRTEM and lattice fringe distance measurements in Fig. 3.11e–h. The grain size distribution histograms of the magnetic powders are shown in Fig. 3.12. It is clear that most of the grains are in the size range of 2–10 nm, and the occurrence frequency of the grains in the 2–4 nm range increases with extension of the milling time. Furthermore, appropriate ball milling makes the grain size present a normal distribution, as shown in Fig. 3.12e.

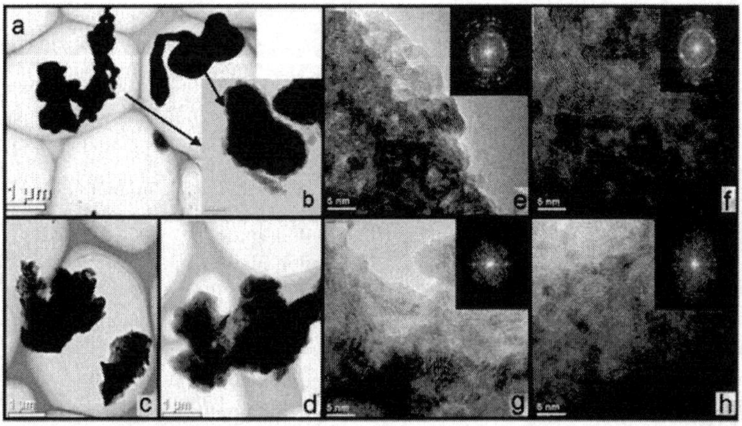

Figure 3.11 TEM micrographs (a–d) of powders for 0 h, 55 h, and 90 h milling and the HRTEM images (e–h) of powders after 25 h, 40 h, 55 h, and 90 h milling. The inset shows the corresponding FFT patterns.

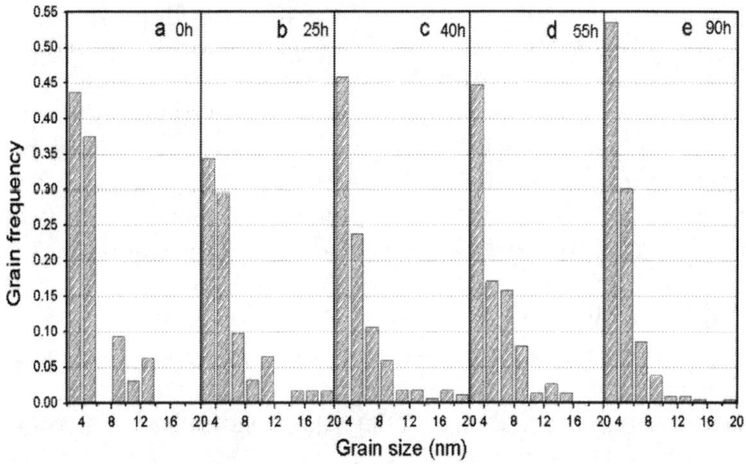

Figure 3.12 Grain size distribution of the powder alloy after different milling times.

3.2.1 Magnetic Response

Figure 3.13 shows how magnetization curves of powder samples and the related magnetic properties such as saturation magnetization

(M_S), coercivity (H_C), and remanent magnetization (M_r) can be obtained. As shown in Fig. 3.13, M_S reaches its maximum value at the milling time of 25 h and then decreases with increasing milling time after 25 h. The inset reflects the changes of H_C of the powders. In contrast with M_S, H_C shows a general increase tendency with milling. M_r and M_r/M_S show increased zigzag tendencies with increased milling time. The specific numerical results for M_S, H_C, M_r, and M_r/M_S are presented in Fig. 3.14.

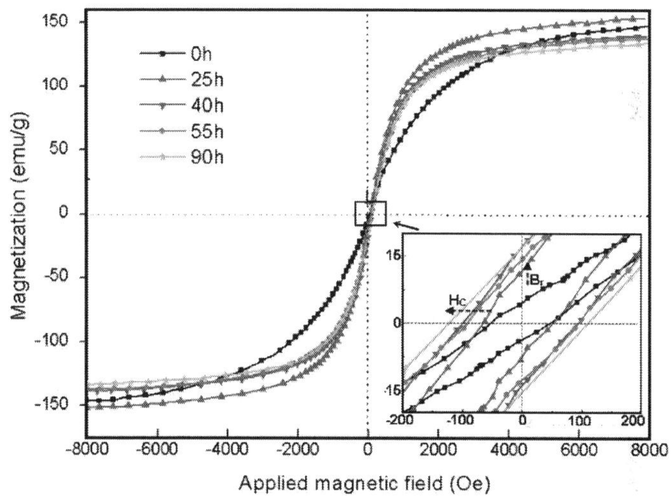

Figure 3.13 Hysteresis loops of Fe-Co-Ni powders milled for different times (0–90 h). The inset shows the corresponding magnification of hysteresis loops near the origin.

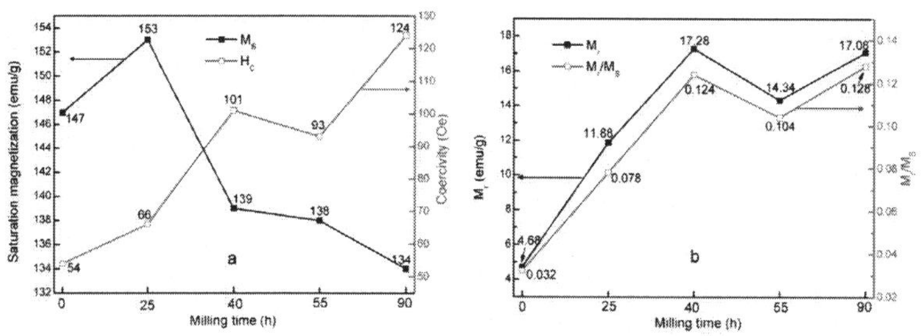

Figure 3.14 Change of M_S and H_C (a) as a function of milling time for Fe-Co-Ni powders and (b) the valuations of M_r and M_r /M_S with the milling process.

3.2.1.1 Saturation magnetization (M_S) and coercivity (H_C)

Saturation magnetization (M_S) strongly depends on the chemical composition of the local environment of atoms and their electronic structures. M_S increases slightly from 147 to 153 emu/g at the early stage, up to 25 h. This trend is attributed to the formation of a solid solution of α-Fe (Ni, Co) and charge transfer between Fe, Ni, and Co atoms[33]. To better explore the relationship between the variations of M_S and the electron distribution, models of α-Fe, α-Fe (Ni), and α-Fe (Co) are built and the corresponding properties, such as electron distribution and Bohr magneton, are calculated using first principles density functional theory (DFT) within the generalized gradient approximation (GGA).

Figure 3.15a shows models of a solid solution with different compositions, and Fe_7Co_1, Fe_6Co_2, Fe_9Ni_1, and Fe_8Ni_2 represent the atomic ratio. In the atom substitution process, the stability of alloys was considered on the basis of the law of the lowest energy. Figure 3.15b illustrates the variation of the Bohr magneton of a different solid solution. It is clear that a proper amount of Co or Ni solution in α-Fe gives the benefit of an increase in magnetism in α-Fe (Ni, Co), but excessive amount of Co or Ni reduces magnetism. One reason is the electron distribution of Ni, Co, and Fe atoms: a transfer of electrons from Fe atoms to Ni or Co atoms, as shown in Table 3.4, causes a spin polarizability increase and ultimately decides the value of the magnetic moment. Moreover, from Table 3.4 one can see that charge transfer is mainly attributed to Fe atoms of the 3d orbital, which is in good agreement with the theory that ferromagnetism (Fe, Co, and Ni) essentially originates from the 3d orbital[34].

M_S keeps decreasing with ball-milling time extension. The explanations are as follows:[35]

- Too much of Co and Ni solution in α-Fe is adverse for M_S, as shown in Fig. 3.15b.
- The internal strain of powders increases with further milling, and large quantities of disordered regions (the grain boundaries) and extended imperfections (especially point defects and dislocations) are induced at this period, which impedes the domain wall motion and reduces M_S. For

coercivity (H_C), it is highly structure sensitive and originates from displacement of the domain wall. In the milling stage, the continuous increase in interfaces hinders the movement of magnetic domain walls, causing a rapid increase in H_C. In addition, the remanent magnetization (M_r) and remanence ratio (M_r/M_S) are usually discussed in permanent magnet alloys.

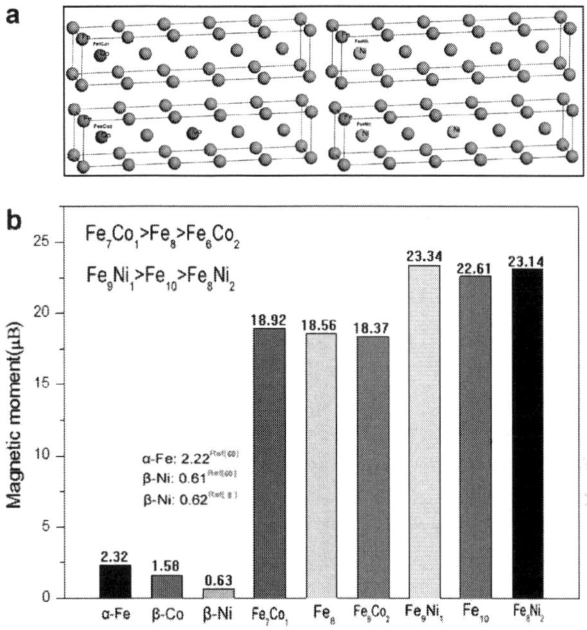

Figure 3.15 Models of a solid solution with different compositions of (a) Fe_7Co_1, Fe_6Co_2, Fe_9Ni_1, and Fe_8Ni_2. (b) Variations of the Bohr magneton of α-Fe, β-Ni, and different solid solutions. The other theoretical results are also listed for comparison.

In this work, the values of M_r and M_r/M_S for the Fe-Co-Ni alloy increase with ball milling, as shown in Fig. 3.14b. This is supposedly caused by induced magnetic anisotropy, which is attributed to the formation of a solid solution during the milling process. Previous studies said that compounds such as α-Fe (Co) and α-Fe (Ni) had a higher Curie temperature, leading to higher possibility of orientational crystallization, and caused higher M_r and M_r/M_S[36].

Table 3.4 Calculated charge transfer between Fe, Co, and Ni atoms for the alloys of Fe-Co and Fe-Ni

Species	Ion	Number	Charge	3d orbital
Fe_7Co_1	Fe	7	−0.06	−0.08
	Co	1	−0.06	–
Fe_6Co_2	Fe	6	−0.12	−0.10
	Co	2	+0.12	–
Fe_9Ni_1	Fe	9	−0.19	−0.12
	Ni	1	+0.19	–
Fe_8Ni_2	Fe	8	−0.36	0.20
	Ni	2	+0.36	–

3.2.1.2 Electromagnetic parameter

The changes of EM parameters, including complex permeability ($\mu = \mu' - j\mu''$) and complex permittivity ($\varepsilon = \varepsilon' - j\varepsilon''$) of Fe-Co-Ni powders milled for different times at 2–18 GHz, are shown in Fig. 3.16. It is observed that the real part and the imaginary part of complex permeability increase with the increased milling time (Fig. 3.16a). The influence of Co and Ni on the complex permittivity of alloys is adverse[37,38], in which case, the results are attributed to the flake-like alloy powders after ball milling. More specifically, (i) the electrical charge of flaky powders can be more easily polarized, (ii) space-change polarization and the exchange coupling reaction of the magnetic moment between particles enhance with an increase in the surface area, and (iii) the eddy current loss reduces with the particle shape changing from spherical to flaky[7]. In addition, the μ' values for all samples decline with frequency in the test frequency range (2–18 GHz). For μ'', the line presents an open downward parabolic shape and the resonance peak for all samples shifts to a higher frequency after milling. This is due to the decrease in demagnetization resulting from the eddy current loss with the formation of flaky alloy powders[39]. In particular, μ'' for the milled powders at 2–18 GHz increases with the alloying process from 0 to 90 h, which is different from the situation that μ'' increases only for a short time, as in our previous research. That gives another piece of evidence of the alloying success and impacts of α-Fe (Ni, Co) on the magnetic properties.

Figure 3.16b shows frequency dependence of the real part and the imaginary part of complex permittivity of the Fe-Co-Ni alloy at different milling times. It can be seen that the values of ε'' for both samples increase with frequency, while ε' remains constant in the range of 2–18 GHz. Moreover, the values of ε' and ε'' for milled powders are higher than those of simply mixed powders (0 h). The difference is that the variation of ε' and ε'' shows an increase-decrease-increase trend with extension of the milling time. The alloy powders after 40 h and 55 h have the lowest values of ε' and ε'', respectively, as shown in Fig. 3.16b. Besides, the ε'' resonance peak is observed, which is due to the resonance of interfacial polarization resulting from the heterogeneous system of conductivity particles separated by higher-resistivity particle boundaries.

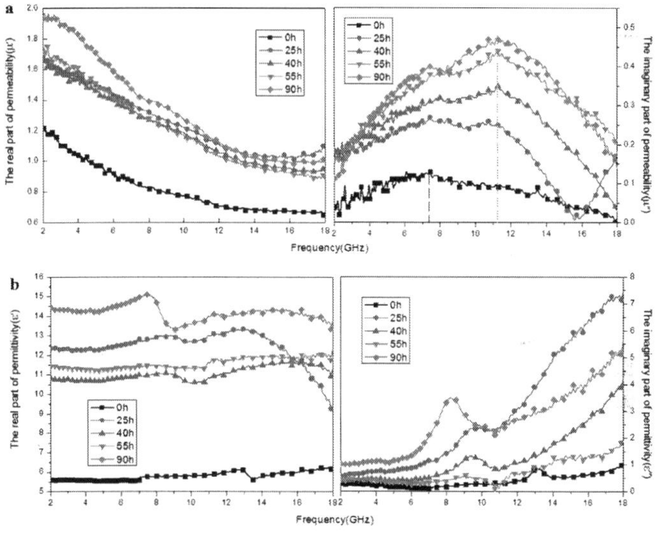

Figure 3.16 Frequency dependence of (a) complex permeability and (b) complex permittivity of powders at different milling times.

3.2.2 Microwave Absorption Property

The changes of morphology, complex permeability, and permittivity indicate that the Fe-Co-Ni alloy particles can be considered a good candidate for EM wave absorbers in the specific frequency band.

Figure 3.17a shows frequency dependence of the calculated RL for the powders after milling with a thickness of 2 mm. It is evident that the absorption property of powders after milling enhances, and the peak frequency (f_m) for all milled powders located at the X-band (8–12 GHz). Especially, the powders after 90 h milling show adequate microwave absorption (RL < −10 dB) at a frequency from 6.8 to 11.4 GHz and the minimum value of −32.4 dB is obtained at about 9 GHz. The other samples milled for 25 h, 40 h, and 55 h show a lower minimum value than that of unmilled powders. The enhancement of absorbing performance is partly because of the improved impedance matching as a result of the proper combination between modified complex permittivity and permeability after ball milling and partly because powders with the flake shape as a layer utilize a combination of loss and resonant cancelation for microwave absorption.

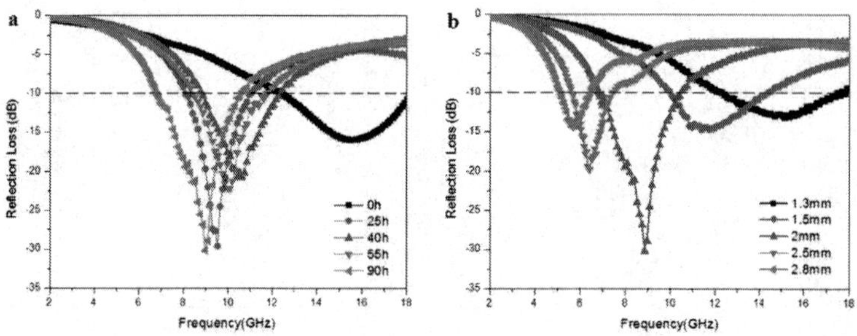

Figure 3.17 (a) Frequency dependence of the calculated reflection loss for powders at different milling times with a thickness of 2 mm and (b) reflection loss for the samples after 90 h milling with different single-layer thicknesses (2.8 mm, 2.5 mm, 2 mm, 1.5 mm, 1.3 mm).

3.2.2.1 Loss tangent of magnetic/dielectric

To further investigate the intrinsic reasons for enhanced microwave absorption for milled Fe-Co-Ni alloy powders, the loss tangent of the magnetic/dielectric are studied and can be expressed as tan $\delta_\mu = \mu''/\mu'$ and tan $\delta_\varepsilon = \varepsilon''/\varepsilon'$, respectively. Figure 3.18 shows the frequency dependence of tan δ_μ and tan δ_ε of powders after different milling times. The values of tan δ_μ increase with the extension of milling time and present a parabola-like curve in the entire frequency. The increase of tan δ_μ denotes that the process of mechanical alloying promotes magnetic loss after ball milling, and results in the enhancement of the absorbing property. Furthermore, the tan δ_ε distributions for all samples show a slight resonance peak in the range of 8–14 GHz and then markedly increase between 12 and 18 GHz. It is ascribed to the higher resistivity and polarization ability due to the variations of particle morphology and crystallinity and ultimately results in the improvement of absorbing ability.

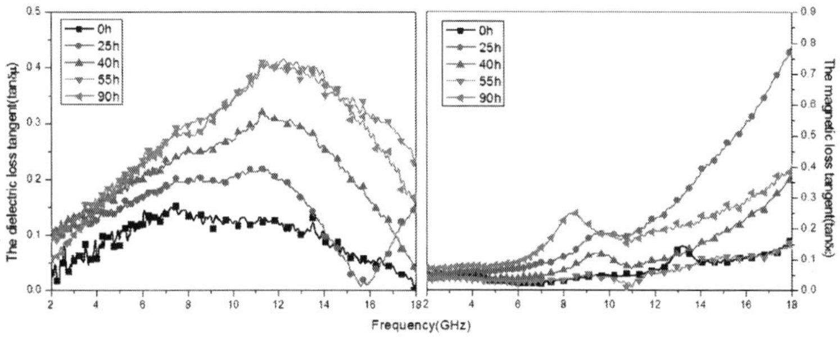

Figure 3.18 (a) Magnetic loss tangent (tan $\delta_\mu = \mu''/\mu'$) and (b) dielectric loss tangent (tan $\delta_\varepsilon = \varepsilon''/\varepsilon'$) of samples within 2–18 GHz.

3.2.2.2 Coefficient of electromagnetic matching (δ)

The optimal EM matching is illustrated by the following equation to meet the wave absorption property: $\mu'' = \varepsilon''$, $\mu' = \varepsilon'$. However, μ and ε cannot meet the equation in different frequencies in the same medium because μ and ε are functions of frequency and most of the magnetic materials have a higher μ than ε. Under this circumstance, the coefficient of EM matching δ is used as a criterion to understand the absorbing mechanism and defined as the following equation:[40]

$$\delta = \frac{\mu''/\varepsilon''}{\mu'/\varepsilon'} = \frac{\mu''\varepsilon'}{\mu'\varepsilon''} \tag{3.7}$$

According to the generalized EM matching theory, the EM matching performs much better with the δ close to 1. Figure 3.19 shows the δ values of Fe-Co-Ni powder alloys at different milling times as a function of frequency. It can be found clearly that the values of δ of the powders after milling for 25 and 90 h are mostly closed to constant one, and the bandwidth of δ = 1–2 are 7.7–11.8 GHz and 6.8–10.2 GHz, respectively. This is consistent with the situation that the powders of 25 and 90 h milling had a stronger absorbing ability (RL < −10 dB) in the frequency range of 8.1–11 GHz and 6.8–11.4 GHz, respectively, as shown in Fig. 3.16a. Thus, it explains the enhanced microwave absorption mechanism in another way.

3.2.2.3 Impact factor of simulation thickness (*d*)

Figure 3.17b shows the frequency dependence of RL for the samples after 90 h milling with different single-layer thicknesses (2.8 mm, 2.5 mm, 2 mm, 1.5 mm, 1.3 mm). It is shown that both the minimum RL f_m and the bandwidth (RL<−10 dB) are dependent on the absorber thickness. For all milled powders, f_m decreases with an increase in the calculated thicknesses. For samples with thicknesses of 1.3 mm and 1.5 mm, f_m locates at the Ku-band and the f_m for other samples with bigger thicknesses are all in the X-band and the C-band, respectively. Here we introduce the matching thickness d_m:[41]

$$d_m = \frac{\lambda_m}{4} = \frac{c}{4 f_m \sqrt{|\varepsilon\mu|}} \tag{3.8}$$

where $\lambda_m = \left(c/4 f_m \sqrt{|\varepsilon\mu|}\right)$ s the wavelength in the absorber and c is the speed of light. According to Eq. 3.8, it is deduced that the peak frequency shifts toward a lower frequency with increasing thickness. Moreover, it is noted that the bandwidth (RL < −10 dB) decreases with increasing thickness, as shown in Fig. 3.17b.

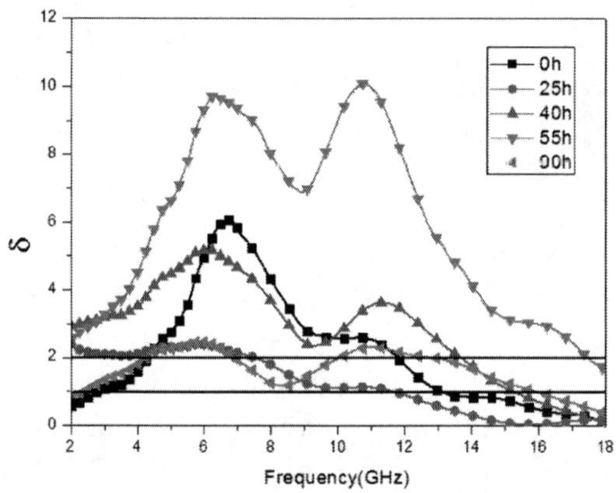

Figure 3.19 Coefficient of electromagnetic matching (δ) of samples after different milling times as a function of frequency.

3.3 Fe-Ni Alloy

Among iron-based soft magnetic materials, Fe-Ni alloys are widely studied and applied in the field of electronic devices due to their excellent performance[42,43]. Many studies have suggested that the magnetic properties of Fe-Ni alloys are dependent on the composition and microstructures, such as lattice distortion[44], domain wall motion[45], magnetic anisotropy,[46] and composition[47]. The experimental results are important to the analysis. However, different experimental conditions and preparation methods also have significant influence on their EM properties, and it is hard to control the experimental environment to understand their internal mechanisms. Theoretical calculations can overcome these insufficiencies and help us to analyze the microscopic electronic structures of Fe-Ni alloys[48] and understand the effect mechanism of addition of Ni on the structures and magnetic properties of pure iron.

3.3.1 Computational Methods and Crystal Structure

The geometry optimization and related calculations are implemented with Cambridge Serial Total Energy Package (CASTEP) code in the Material Studio 4.3 package, which is based on DFT using a plane-wave pseudopotential method[49]. The GGA following the Perdew–Burke–Ernzerhof (PBE) as well as Hubbard+U functional form is used for the exchange correlation potential[50]. The electron interaction is described by using Vanderbilt ultrasoft pseudopotentials[51]. The numerical integration of the Brillouin zone is performed using a 2 × 7 × 10 Monkhorst–Pack *k*-point sampling, and the cutoff energy of the plane wave has been assumed to be 300 eV. The valence electron configurations for Fe and Ni are chosen as Fe ($3d^6 4s^2$) and Ni ($3d^8 4s^2$), respectively.

The 5 × 1 × 1 supercell containing 10 atoms is adopted for pure Fe, as shown in Fig. 3.20. Two different configurations based on this supercell are considered to study the effect of doping concentration of Ni. One model is obtained by replacing one Fe with a Ni atom in body-centered *d* position (shown in Fig. 3.20), which approximately corresponds to the composition of Fe-10wt%-Ni alloy. For the Fe-20wt%-Ni alloy, two Fe atoms are substituted by Ni.

As a result, the chemical formulas of the above supercell are Fe_9Ni_1 and Fe_8Ni_2, respectively. Considering the stability of the alloy on the basis of the law of the lowest energy, the energies of Fe_8Ni_2 alloy for different substituted positions (site 2, site 3, . . .) are compared and the optimized position (site 2) is chosen to place the Ni atoms (shown in Fig. 3.21). In the process of optimization of structure, the Broyden–Fletcher–Goldfarb–Shanno (BFGS) algorithm is used. The energy change and the maximum tolerances of the force, stress, and displacement are set as 1×10^{-5} eV/atom, 0.03 eV/Å, 0.05 GPa, and 0.001 Å, respectively.

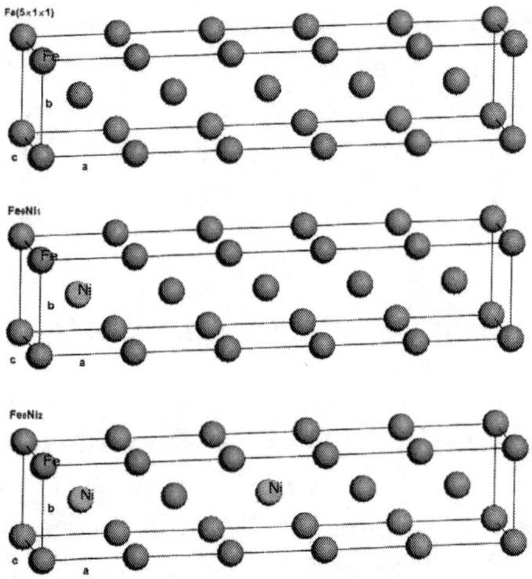

Figure 3.20 Models of (a) pure α-Fe (5 × 1 × 1), (b) Fe_9Ni_1, and (c) Fe_8Ni_2. The gray and red balls represent the Fe and Ni atoms, respectively. Supercell dimensions were measured with a, b, and c.

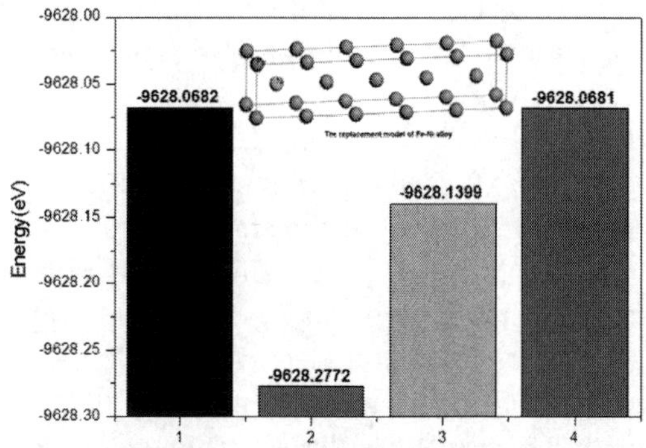

Figure 3.21 Variations of energy of Fe_8Ni_2 for different substituted positions (site 2, site 3, . . .), and model 2 with the lowest energy was chosen as the most stable structure.

3.3.2 Crystal Structures and Phase Stability of the Fe-Ni Alloy

After calculation, we find that the lattice constant of pure iron with a bcc structure is 2.866 Å, which is in good agreement with the previous value of 2.87 Å in Ref. [52]. The energies of Fe_8Ni_2 after optimization for different models are listed in Fig. 3.21, and model 2 with the lowest energy is taken as the substitution position. After optimization, the lattice parameters of Fe_9Ni_1 and Fe_8Ni_2 are changed as a result of the Ni substitutions, as shown in Table 3.5. It can be seen that the Bravais lattice of the supercell slightly changes from bcc to body-centered tetragonal. To study the stability of the Fe-Ni alloy, the defect formation energy (E_{form}) is discussed, which can be calculated according to the following formula:[53,54]

$$E_{form} = E_{doped} - E_{pure} + n(\mu_{Fe} - \mu_{Ni}) \tag{3.9}$$

where E_{doped} and E_{pure} are the energy of Fe-Ni alloy and pure Fe supercell, respectively; n is the number of Fe atoms replaced by Ni; and μ_{Fe} and μ_{Ni} are the chemical potentials of Fe and Ni, respectively. The energies of an isolated Fe and Ni atom under a periodic boundary condition are calculated to be represented by their chemical potential[53], and the calculated μ_{Fe} and μ_{Ni} were −864.668 eV and −1354.195 eV, respectively, which are very similar to the calculated values in Ref. [55]. The calculated formation energy is −4.177 eV for Fe_9Ni_1 and −3.817 eV for Fe_8Ni_2 (see Table 3.5). This negative formation energy shows that Fe_9Ni_1 and Fe_8Ni_2 alloys can be fabricated more easily through experiments.

Table 3.5 Lattice constant and defect formation energies (E_{form}) of Fe_9Ni_1 and Fe_8Ni_2 after Ni substitutions

Species	Structure (Å)	E_{doped}(eV)	E_{pure}(eV)	E_{from}(eV)
Fe (5 × 1 × 1)	a, b, c, v a = 14.332 $b = c$ = 2.866, v = 117.722	–	–	
Fe_9Ni_1	a = 12.318, c = 2.467 b = 3.496, v = 106.238	-9139.110	-8645.406	-4.177
Fe_8Ni_2	a = 17.684 $b = c$ = 2.465, v = 107.452	-9628.277	-8645.406	-3.817

3.3.3 Electronic Properties

To investigate the effect of Ni substitution on the electronic structures of pure Fe, their band structures, total density of states (TDOS), local density of states (LDOS), and partial density of states (PDOS) are calculated. Figure 3.22 shows the band structures of Fe and Fe-Ni alloys along the high-symmetry lines of the Brillouin zone. As shown in Fig. 3.22a, α-Fe, Fe_9Ni_1, and Fe_8Ni_2 have nearly the same band structures. A lot similar to pure Fe, the spin-up and spin-down bands of Fe-Ni obviously spill and the Fermi energy (E_f) goes across the valence band (VB). This indicates that Fe_9Ni_1 and Fe_8Ni_2 exhibit metallic properties as pure Fe. In addition, it is found that the band structures are to be separated after the doping of Ni, as shown in the elliptical area. Also, the Fe-Ni alloys have lower energy in the band structure, as shown in Fig. 3.22b. That would cause changes in the DOS of the Fe-Ni alloy.

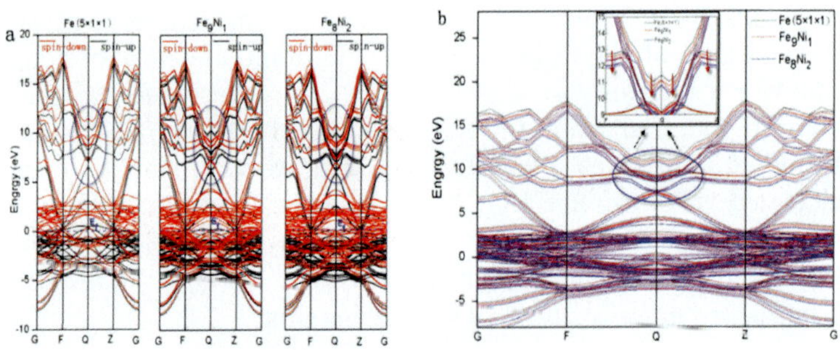

Figure 3.22 (a) Band structures of Fe, Fe_9Ni_1, and Fe_8Ni_2. (b) Comparison of the energy of spin-up electrons of Fe, Fe_9Ni_1, and Fe_8Ni_2. The inset is the enlargement of the elliptical area at point Q. The Fermi energy is marked by the dashed blue line.

Figure 3.23 shows the TDOS of pure Fe (5 × 1 × 1), Fe_9Ni_1, and Fe_8Ni_2. Compared to α-Fe, the obvious differences in the DOS of Fe-Ni alloys are that the Fermi level shifts upward to high energy, and the DOS shifts to lower energy with an increase in Ni contents. This results in a decline of the electronic number near the Fermi level, as the inset shows in Fig. 3.23. Deng et al.[56] pointed out that this variation would lead to an increase of the band gap. According to this theory, the Fe-Ni alloy would have higher resistivity than pure Fe, in agreement with the view mentioned in Ref. [57]. Figures 3.24a and 3.24b show the PDOS of Fe_9Ni_1 and Fe_8Ni_2, respectively. It could be seen that the DOS mainly originates from the Fe 3d and Ni 3d electrons. The VB of Fe_9Ni_1 and Fe_8Ni_2, located at a value range from −8 to 1 eV, comes from the contribution of 3d electrons of Fe and Ni. The conduction bands (CBs) from 1 to 3.5 eV mainly come from the Fe 3d orbitals and others are contributed by the s and p electrons of Fe.

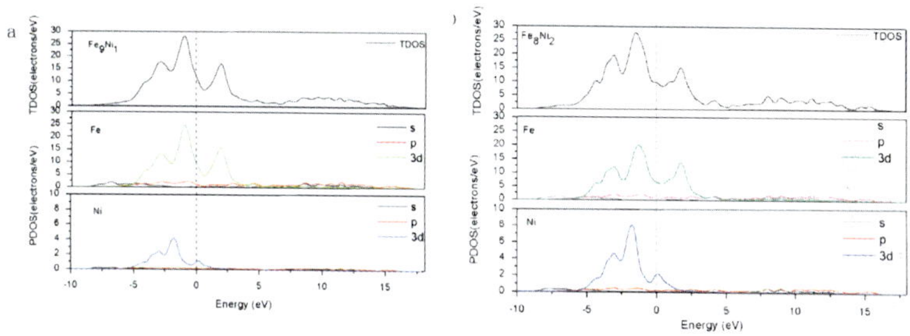

Figure 3.23 PDOS of (a) Fe_9Ni_1 and (b) Fe_8Ni_2, and the Fermi energy is marked by the dashed vertical line.

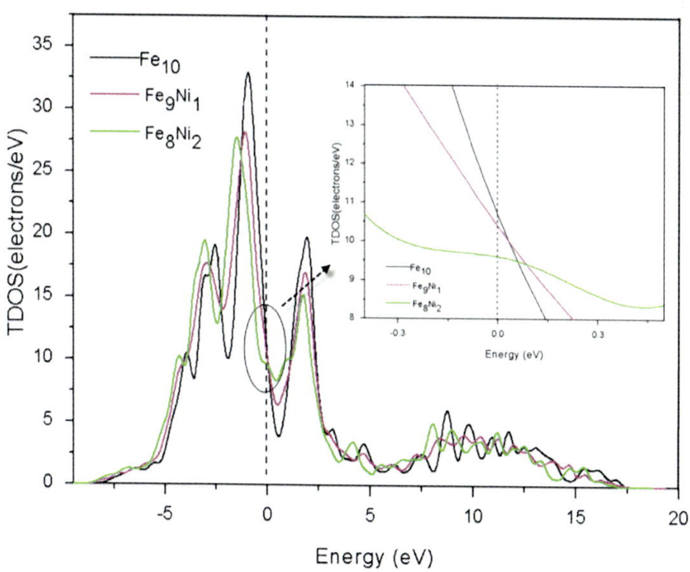

Figure 3.24 TDOS of pure α-Fe (5 × 1 × 1), Fe_9Ni_1, and Fe_8Ni_2. The inset is the enlargement of the elliptic area at the Fermi energy, and the Fermi energy is marked by the dashed vertical line.

3.3.4 Mulliken Population and Charge Density

To investigate the effects of the addition of Ni on the electronic configurations of Fe, it is essential to have an insight into the Mulliken population and the charge transfer situation. Figure 3.25 shows the electron density difference of Fe_9Ni_1 and Fe_8Ni_2. It is noted that the Fe atoms near Ni have fewer electrons than Fe atoms at a longer distance from Ni. To clearly identify the Fe atoms at different coordinates, they are numbered separately in Arabic numerals from 1 to 9, as shown in Fig. 3.25. Figure 3.26 shows the DOS of the Fe atoms at different positions, from which one can clearly see that the DOS of Fe atoms located at site 1 (Fig. 3.26a) near Ni is generally lower than that of Fe atoms located at site 5. And the same situation appears in Fig. 3.26b for Fe_8Ni_2. In Table 3.6, the average electronic number occupancies over each orbit (s, p, and d) and the charge transfer situation are analyzed quantitatively and systematically. Considering the crystal periodic structure of the models, some coordinate positions of Fe atoms in the Fe-Ni alloy are equivalent, for example, site 1 and site 2 in the Fe_9Ni_1 model and site 2 and site 4 in Fe_8Ni_2 model. Specifically, these equivalent relations are listed in Table 3.6.

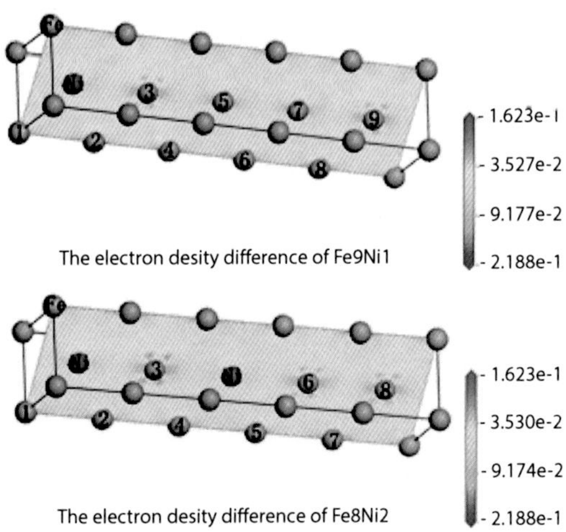

Figure 3.25 Calculated electron density differences of Fe_9Ni_1 and Fe_8Ni_2.

Table 3.6 Calculated atom population for each orbit, corresponding charge transfer situation, as well as the local magnetic moment for each atom

Species	Ion	Site	s	p	d	Total	Charge (e)	Magnetic moment (μ_B)
α-Fe	Fe	-	0.62	0.77	6.61	8.00	0.00	2.32
β-Ni	Ni	-	0.53	0.80	8.66	10.00	0.00	0.63
Fe_9Ni_1	Fe(9)	1.2	0.58	0.84	6.55	7.96	-0.04	2.83
		3.9	0.61	0.74	6.62	7.97	-0.03	2.48
		4.8	0.62	0.75	6.61	7.98	-0.02	2.46
		5.7	0.62	0.76	6.61	7.99	-0.01	2.40
		6	0.62	0.76	6.61	8	0.00	2.38
	Ni(1)	-	0.63	0.92	8.64	10.19	0.19	0.60
Fe_8Ni_2	Fe(8)	1.5	0.57	0.83	6.55	7.96	-0.04	2.86
		2.4	0.57	0.82	6.55	7.94	-0.06	2.86
		6.8	0.61	0.73	6.62	7.96	-0.04	2.50
		3	0.59	0.71	6.64	7.94	-0.06	2.48
		7	0.62	0.74	6.61	7.97	-0.03	2.54
	Ni(2)	-	0.63	0.91	8.64	10.18	0.18	0.58
α-Fe[58]		-	0.64	0.74	6.62	8.00	0.00	2.26
β-Ni[60]		-	-	-	-	-	-	0.61

As shown in Table 3.6, the total number of electrons of α-Fe is 8.00e, which has nearly the same electronic configuration as in the previous work[58]. In the Fe_9Ni_1 alloy, charge interaction between Ni and Fe atoms appears, showing a transfer of electrons from Fe atoms to Ni atoms and resulting in a decrease in electrons for Fe. For the Ni atom, in contrast, the total number of electrons increases to 10.19e, compared to the initial value of 10.00e. Furthermore, the charge transfer trend is connected with the atomic spacing between Fe atoms and the dopant Ni atoms. The total number of electrons for Fe at sites 1 and 2 around Ni is changed by –0.04e and for Fe at sites 3 and 9 with a larger spacing distance of –0.03e, and so on. For Fe atoms at site 6 with the largest distance with Ni, their electronic structure is virtually unchanged and has nearly the same electronic distribution with α-Fe. For Fe_8Ni_2, it has the same situation of the electronic configuration with Fe_9Ni_1, and the increased 0.36e of the electrons for two Ni atoms is distributed by the Fe atoms according to the distance with Ni. The magnetic moment of a ferromagnetic material mainly comes from spintronics theory, and a decrease in the number of Fe electron leads to an increase in spin polarization per Fe atom[59] and ultimately results in the variation of the magnetic properties for Fe_9Ni_1 and Fe_8Ni_2.

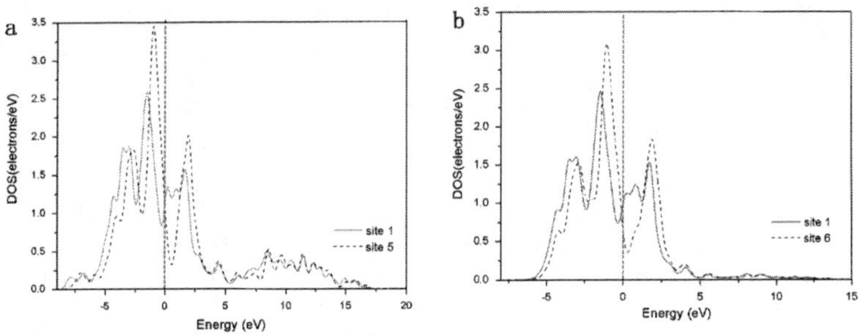

Figure 3.26 DOS of Fe atoms at different positions in (a) Fe_9Ni_1 and (b) Fe_8Ni_2, and the Fermi energy is marked by the dashed blue vertical line.

3.3.5 Magnetic Properties

The calculated magnetic moment per Fe atom of Fe_9Ni_1 and Fe_8Ni_2 is listed in Table 3.6. Compared to bcc Fe (2.32 μB per atom), the magnetic moment of Fe around Ni enhances, showing the magnetic moment of 2.83, 2.48, 2.46, 2.40, and 2.38 μB per atom for different sites in Fe_9Ni_1 and 2.86, 2.50, 2.48, and 2.54 μB per atom in Fe_8Ni_2. In conclusion, compared to 23.14 μB for the supercell (5 × 1 × 1) of pure Fe, the whole magnetic moments of Fe_9Ni_1 and Fe_8Ni_2 are 23.34 and 22.61 μB, respectively. In Fe_9Ni_1, the enhancement of the magnetic moment of Fe atoms can make up for the reduction of the magnetic moment of doped Ni atoms and eventually result in an increase of the saturation magnetization for Fe_9Ni_1. In contrast, compared to the magnetic enhancement of Fe atoms in Fe_8Ni_2, the reduction of the magnetic moment of the two doped Ni atoms plays a major role. So Fe_8Ni_2 has a relatively low magnetic moment than the α-Fe supercell (5 × 1 × 1). The experimental results have shown that the relationships between the micromagnetic moment and macrosaturation magnetization and the calculated total moments fit the experimental values[61]. Besides, in the process of experimenting, the saturation magnetization of α-Fe (Ni, Co) made by mechanical alloying (153 emu/g) is found to be higher than that of the raw sample (147 emu/g), which can be attributed to the formation of a solid solution of α-Fe (Ni, Co) and charge transfer between the Fe, Ni, and Co atoms.

Figure 3.27 shows the DOS of Fe atoms at different sites in Fe_9Ni_1 and Fe_8Ni_2, as well as α-Fe. Compared with α-Fe, you can see that the asymmetry between the up-spin and down-spin states of the Fe atoms is different with different atomic spacing to Ni around the Fermi level. Here, the spin polarization ratio P is introduced and calculated as the value of $N\downarrow - N\uparrow/N\downarrow + N\uparrow$, where $N\downarrow$ and $N\uparrow$ are the spin-down and spin-up DOS at E_f, respectively[8]. The value of P represents the spin polarizability and ultimately decides the value of the magnetic moment. The Fe atoms with the smallest atomic spacing to Ni have the largest P value, corresponding to the biggest Bohr magnetons. The Fe atoms at site 6 and site 7 away from Ni own almost the same P with α-Fe and they have less Bohr magnetons. These changes in the occupancy of both spin-up and spin-down states of Fe atoms can be considered the origin of the enhanced ferromagnetic moment.

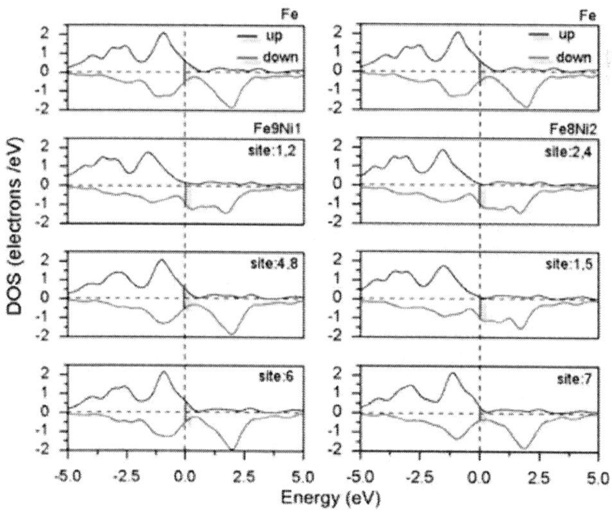

Figure 3.27 Spin-up and spin-down DOS of different Fe atoms in α-Fe, Fe_9Ni_1, and Fe_8Ni_2. The blue line shows the energy gap between the spin-up and spin-down electrons.

3.4 Fe-Si-Al Alloy

Among multitudinous ferroalloy powders prepared by the mechanical alloying method, Fe-Si-Al alloys are attracting more and more attention due to their high saturation magnetization, low coercivity, and matched permittivity[62]. During the past decade, studies have shown that the composition, preparation methods, and subsequent treatments related to mechanical alloying could influence the magnetic properties of Fe-Si-Al alloys.

Figure 3.28 shows the morphological evolution of Fe-Si-Al alloy powders at increasing milling times. It could be seen that powder shapes change significantly with the milling time increasing from 0 to 90 h. The raw powders (0 h) are in nearly fusiform shape with different sizes (Fig. 3.28a). After milling of 10 h (Fig. 3.28b), the powders are crushed mildly and change into flake or platelet shape partly, which leads to an aspect ratio increase and possibly changes the electromagnetism parameter simultaneously[63]. Moreover, both the volume and the size are much smaller than those of the original granules. With increasing milling time, the fusiform shape powders disappear gradually and the majority of powders exhibit a round shape with very small sizes (Fig. 3.28c and Fig. 3.28d). When the milling time is further increased to 90 h, the powders begin to agglomerate together after the processes of deformation and cold welding (Fig. 3.28e).

The changes of the microstructure (average grain size, internal strain, and phase transition) of the powders in the ball-milling process can be observed in XRD patterns, as given in Fig. 3.29. It is noticeable that unmilled powders show the characteristic reflections of α-$Fe_3(Si)$ associated with (110), (220), and (221), and the super lattice phase DO3 ($Fe_3Al_{0.3}Si_{0.7}$) associated with (111), (200), (220), (400), and (422). Moreover, those diffraction peaks are difficult to distinguish due to their similar structures. With the milling time increase, all diffraction peaks of α-Fe(Si) and DO3 have a tendency of broadening and a decrease in the intensity, which suggests a reduction of grain size and the growth of internal strain during the ball-milling process, as shown in Table 3.7. Also, the phases of the powders are changed simultaneously, for example, the diffraction peaks of DO3 in low angles would disappear after 25 h milling. Moreover, the composition of the alloy would shift to that of pure iron and that lead to changes in the magnetic properties[64]. When the milling time is prolonged to 90 h, the peaks of pure α-Fe are observed near the peaks of α-Fe(Si). In the drawing of partial enlargement of XRD patterns, the (110) peak of pure Fe shifts to lower angles compared with that of α-Fe(Si) due to the desolvation of a small amount of iron out from the alloy[65] and similarly in the higher angles associated with (200) and (211). In addition, it should be noted that the oxide (Al_2O_3, SiO_2) could be observed after milling of 90 h.

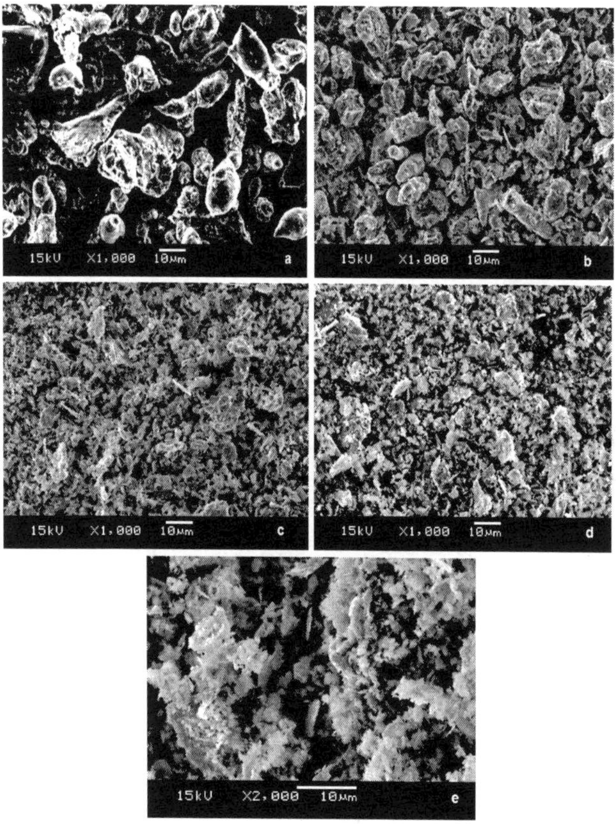

Figure 3.28 SEM micrographs of $Fe_{87.5}Si_7Al_{5.5}$ (wt%) powders following the milling time: (a) 0 h, (b) 10 h, (c) 25 h, (d) 55 h, and (e) 90 h.

The average grain size and internal strain of powders with different milling times are calculated according to the Hall–Williamson equations:[66]

$$\beta_{hkl} = [(\beta_{hkl})^2_{measured} - (\beta_{hkl})^2_{instrumental}]^{1/2} \tag{3.10}$$

$$\frac{\beta_{hkl} \cdot \cos\theta_{hkl}}{\lambda} = \frac{K}{D} + \frac{4\varepsilon \cdot \sin\theta_{hkl}}{\lambda} \tag{3.11}$$

where β_{hkl} is the full width at half-maximum (FWHM) of the peaks, $(\beta_{hkl})_{measured}$ is the measured FWHM, $(\beta_{hkl})_{instrumental}$ is the FWHM of the standard silicon sample for calibration[67], K is the Scherrer constant (K = 1), D is the grain size, λ is the wavelength of the X-ray used (λ = 0.154178 nm) in the X-ray diffract meter, ε is the internal strain introduced by milling, and θ_{hkl} the Bragg angle. Then the 4 sin θ_{hkl}/λ and (β_{hkl} cos θ_{hkl})/λ are made as the X axis and the Y axis, respectively, and the average crystallite dimension and internal strain are obtained according to the intercept of the Y axis and the slope of the line, as listed in Table 3.7.

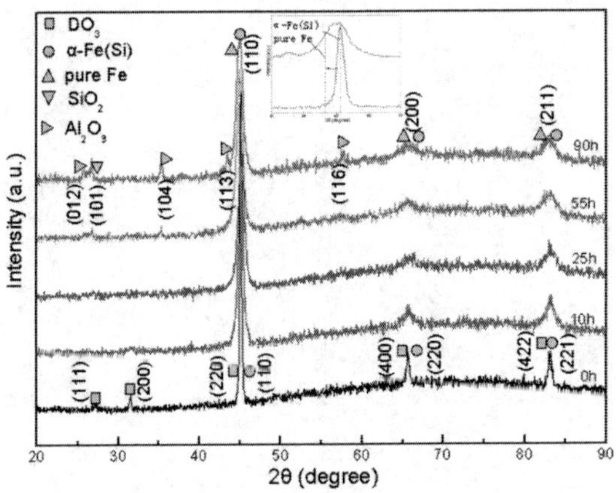

Figure 3.29 XRD patterns of $Fe_{87.5}Si_7Al_{5.5}$ (wt%) powders milled for different times (0 h, 10 h, 25 h, 55 h, and 90 h).

It is seen that the powders in the milling period from 10 to 25 h show a reduction of the average grain size from 27.5 to 6.8 nm. It could be deduced that new defects such as dislocations are formed and these dislocation pile up the grain boundaries, which leads to the formation of subgrain structures and hence the decrease in grain size. With further ball milling, the grain size is found to be slightly larger (7.2–15.5 nm) from 55 to 90 h, which may be caused by the formation of oxides and the process of dynamic recrystallization[68].

For the internal strain of powders, it increases with milling and gets the maximum value after 25 h and then decreases. At the initial stage, the internal strain increases result from a rapid increase in dislocation density caused by plastic deformation. When the internal strain reaches the critical strain level locally, it declines with the grain size increase, and the powders that have the smallest grain size (6.8 nm) show the largest lattice residual strain (0.4718%).

Table 3.7 Average grain size (D) and internal strain (ε) of $Fe_{87.5}Si_7Al_{5.5}$ (wt %) powders milled at different times

Milling time (h)	Average grain size, *D* (nm)	Internal strain (%)
10	27.5	0.2744
25	6.8	0.4718
55	7.2	0.4064
90	15.5	0.1933

3.4.1 Saturation Magnetization (M_S)

Saturation magnetization (M_S) is the property which involves the atomic origin of magnetism. It strongly depends on the chemical composition of the local environment of atoms and their electronic structures, such as charge exchange, crystal field interaction, interatomic hopping, and spin–orbit coupling[69]. Figure 3.30 shows the hysteresis loop of the powders with different milling times, and these samples exhibit excellent soft magnetic properties (coercivity of 30 Oe milled for 55 h and saturation magnetization of 131 emu/g milled for 10 h). As shown in Fig. 3.30a, the hysteresis loop distorts obviously under ball-milling stress. M_S of the powders milled for 10 h has the maximum value (131 emu/g), which then decreases from 96.14 to 63.75 emu/g corresponding to 25 and 90 h, respectively. Besides, it is interesting that hysteresis loops of powders after 0 and 25 h milling are difficult to distinguish, as seen in the inset in Fig. 3.30a.

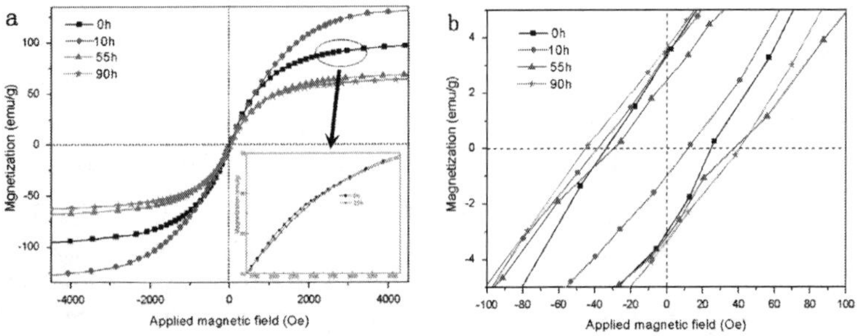

Figure 3.30 (a) Hysteresis loops of $Fe_{87.5}Si_7Al_{5.5}$ (wt%) powders milled for different times (0–90 h) and (b) the corresponding magnification of (a) near the origin. The inset in (a) shows the part of hysteresis loops of powders after 0 and 25 h milling. (a, b) Varying trend in saturation magnetization (M_S) and coercivity (H_C).

To investigate the relationship between the variations of magnetic properties, especially M_S and the phase change during the ball-milling period, the magnetic properties of pure iron (α-Fe) and solid solution α-Fe(Si) are analyzed using first principles DFT within the GGA. The spin-polarized and electron structure calculations are performed in CASTEP, version 4.3. The following points should be considered when in the modeling process, for energy-minimum requirement, one of the Fe atoms in body-centered positions is substituted by a Si atom and a supercell ($2 \times 1 \times 1$) is created to satisfy the crystal periodic, as shown in Fig. 3.31. It is well known that ferromagnetism essentially originates from electron spin. To confirm this point, the DOS of Fe is calculated. Figure 3.32 shows the TDOS and PDOS of α-Fe, and it clearly indicates that spin magnetic moments are mainly attributed to the partially filled 3d orbital. When Si ([Ar] $3s^23p^2$) atoms are dissolved into the Fe ([Ar] $4s^23d^6$) lattice, the outermost electron of Si would occupy the 3d shell of Fe and result in the reduction of Bohr magnetons of Fe. Table 3.8 shows that an average electronic number existed in each orbit (s, p, d, and f) and the total Bohr magnetons of each atom. When an Fe atom is substituted by a Si atom, the electron configuration of Fe in the unit cell would change. To be specific, the electronic number of the 3d orbit in Fe^{I} near the Si atom (as shown in Fig. 3.31) changes from 6.62 to 6.87, and consequently the magnetic moments decrease to

1.54 μB from 2.26 μB. On the contrary, the other Fe $^{\Pi}$ in the body-centered position shows bigger magnetic moments (2.72 μB) as a result of the decrease of electronic number (6.57) in the 3d shell. Overall, the supercell (2 × 1 × 1) α-Fe_3(Si) shows relatively smaller magnetic moments (5.61 μB) than those of pure α-Fe (9.01 μB).

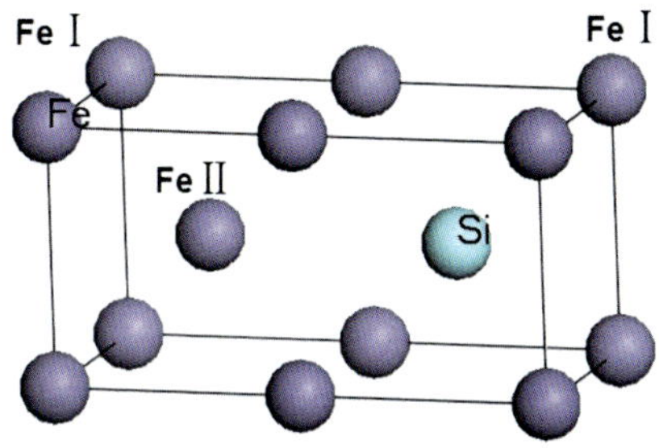

Figure 3.31 Supercell (2 × 1 × 1) of α-Fe_3 (Si), where the Si atom substitutes the Fe atom in the body-centered position allowing for the energy-minimum requirement. Fe^{I}represents the two Fe atoms near the Si atom, and Fe^{II}represents the Fe atom in another body-centered position.

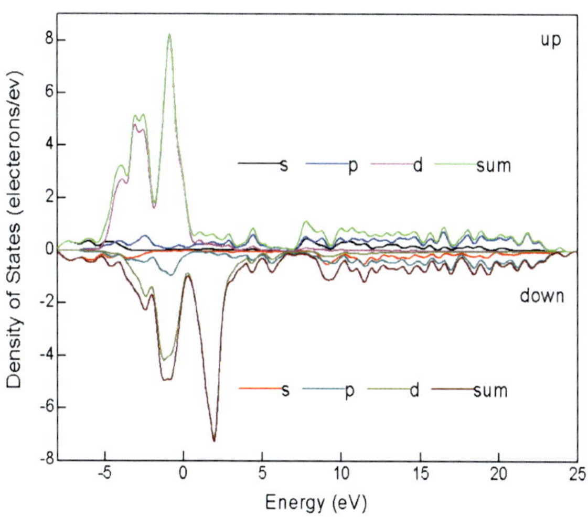

Figure 3.32 Total density of states (TDOS) and partial density of states (PDOS) of pure α-Fe calculated by generalized gradient approximation (GGA). Fermi energy is marked by the dashed vertical line, which indicates the magnetic moments of iron.

The Fe-Si-Al powders exhibit the maximum M_S of 131 emu/g after milling of 10 h, which is caused by the shift in composition to excess Fe. On the basis of the above analysis and Table 3.8, pure α-Fe has more Bohr magnetons than α-Fe_3(Si) and the internal strain is not so much at this time, so consequently it makes the value of M_S reach its maximum value when milled for 10 h. From 10 to 25 h, M_S decreases with the milling time. This falling can be attributed to the reinforcement of the internal strain (as seen in Table 3.7). At this milling stage, the average grain size of powders reduces dramatically from 27.5 to 6.8 nm. In contrast, the internal strain increases from 0.2744% to 0.4718%. Namely, large quantities of disordered regions (the grain boundaries) and extended imperfections (especially point defects and dislocations) are induced at this period and result in the decline of M_S. With the extension of milling time, M_S decreases and reaches the minimum value 63.75 emu/g at 90 h, which is ascribed to the appearance of a large amount of metal oxides.

Table 3.8 Average electronic number occupancies over each orbit (s, p, d, and f) and the total magnetons of each atom

Species	Average electronic number /e					Magnetic moment
	s	p	d	f	Total	(μ_B/atom)
α-Fe	0.64	0.74	6.62	0.00	8.00	1.13
Fe^{I}	0.47	0.72	6.87	0.00	8.06	0.77
Fe^{I}	0.47	0.71	6.87	0.00	8.06	0.77
Fe^{II}	0.61	0.80	6.57	0.00	7.98	1.36
Si	1.24	2.66	0.00	0.00	3.90	-0.09

3.4.2 Coercivity (H_C)

Magnetic coercivity (H_C) is highly structure sensitive and originates from the displacement of the domain wall. Because of the interaction between the domain wall and the elastic stress field induced by the dislocation, the movement of the domain wall is blocked by the dislocation, resulting in the change of H_C. The relationship between H_C and the related structural parameters can be interpreted by the following formula:

$$H_c \propto \frac{\lambda_s \varepsilon \delta}{M_s} \tag{3.12}$$

where λ_S is the magnetostriction constant, ε is the internal strain, and δ represents the width of the wall. As previously described, the dislocation and grain boundaries resisted the displacing of the domain wall, which blocked the transformation of magnetic moments in the magnetic domain to the external magnetic field. This needed a larger reversing field to make the magnetic flux density reach zero. Figure 3.30b illustrates the variation of the H_C with the changed milling time. At an early stage up to 10 h, H_C increases from 34 to 39 Oe, and this behavior could be explained by the increasing of internal stress and grain refinement (Table 3.7). When the milling time extends from 55 to 90 h, H_C increases due to a decrease in M_S, as shown in Fig. 3.30a. It should be noted that H_C decreases from 39 to 30 Oe at the stage from 10 to 55 h, as seen in Fig. 3.30b. One theory for interpretation of how this might have happened says that, when the grain size is small enough, H_C theoretically relates to the grain size ($H_C \propto d^6$)[70], meaning H_C is proportional to the sextic of diameter. For this reason H_C would decrease, and that we need to study further.

3.4.3 Complex Permeability

The permeability characteristics (μ) of Fe-Si-Al powders milled at different times following the frequency are shown in Fig. 3.33. Figure 3.33a represents the tendency of the real part of complex permeability (μ') and Fig. 3.33b is the imaginary part of complex permeability (μ''). As shown in Fig. 3.33a, the μ' values of powders with flake structures are all obviously superior to those of the precursor powders. The results can be explained by the following facts: (i) The electrical charge of flaky powders can be more easily polarized, (ii) the space-charge polarization and the exchange coupling reaction of the magnetic moment between particles enhance with the increase of surface area, and (iii) the eddy current loss reduces with the particle shape changing from spherical to flaky[7]. Furthermore, the Fe-Si-Al alloy milled for 10 h shows a maximum μ' of about 2.2 at a frequency of 2 GHz, and the μ' for all milled samples decreases following the frequency. The difference is that the value of μ' decreases with increasing milling time from 10 to 90 h at relatively lower frequency (2–8 GHz) but has an opposite variation tendency at higher frequency (8–18 GHz).

The frequency dependence of the imaginary part of complex

permeability (μ'') with different milling times is shown in Fig. 3.33b. At the test frequency range (2–18 GHz), the line of μ'' presents an open downward parabolic shape, and the peaks of the μ'' move toward a high-frequency region from 4.5 GHz to about 6.5 GHz after milling compared to that of the raw powders. In addition, the value of μ'' decreases at the same frequency with the extension of milling time, the maximum μ'' of the powders is only about 0.4 after 90 h milling compared to 0.85 of the raw powders. This demonstrates that the long milling time (>10 h) is adverse for the μ'' of the atomized $Fe_{87.5}Si_7Al_{5.5}$ (wt%). In later research, we shorten the milling time and find that the μ'' of the powders has increased to some degrees after milling (2–8 h), and when the time is extended to 15 h, the value of μ'' becomes lower than that of the precursor sample, as shown in Fig. 3.33c, which gives more evidence that a long milling time (>10 h) is not beneficial to improve the μ'' of the $Fe_{87.5}Si_7Al_{5.5}$ (wt%) alloy.

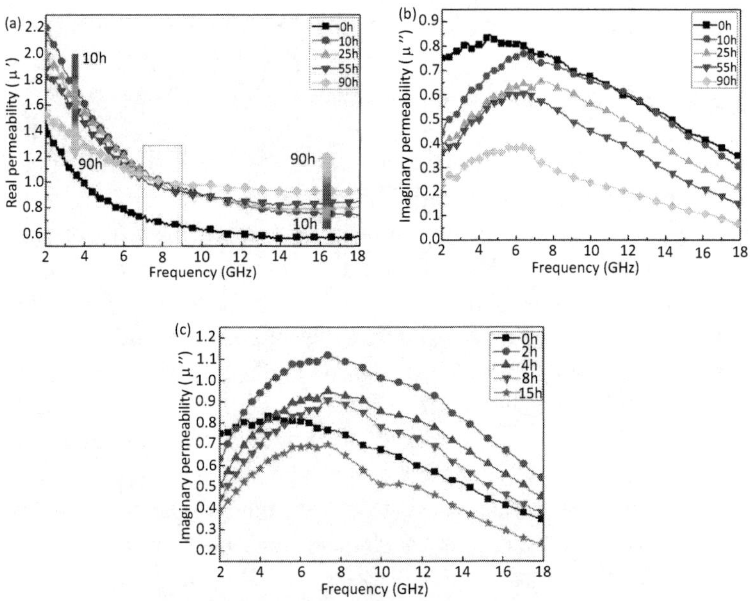

Figure 3.33 (a) Real part (a) and (b) imaginary part of complex permeability in the frequency range of 2 GHz–18 GHz for $Fe_{87.5}Si_7Al_{5.5}$ (wt%) powders milled for 0, 10, 25, 55, and 90 h. (c) Imaginary part of complex permeability of powders after short-time ball milling (0 h, 2 h, 4 h, 8 h, and 15 h).

3.4.4 Microwave Absorption Property

Figure 3.34 shows the frequency dependency of calculated reflection loss for powders with different milling times. It is clearly shown that the powders milled for 10 h, 25 h, and 55 h exhibited a better absorption property than raw powders. Especially, the powders after 25 h milling show adequate microwave absorption (RL < –10 dB) at a frequency from 9.5 to 15.5 GHz and an RL_{MIN} of –22.2 dB is obtained at 12.6 GHz. Furthermore, the frequency bandwidth of powders below –10 dB is even wider after 10 h ball milling. These results could be attributed to the improvement of impedance matching (μ, ε) after ball milling. In addition, the powders with flake shape as a layer utilize a combination of loss and resonant cancellation for microwave absorption[71]. On the contrary, the absorption properties of samples deteriorate with milling up to 90 h due to agglomeration and the oxidation phenomenon.

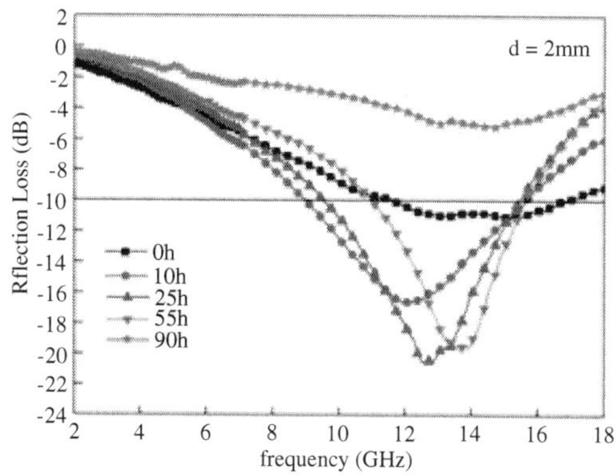

Figure 3.34 Frequency dependence of the reflection loss for $Fe_{87.5}Si_7Al_{5.5}$ (wt%) powders after different ball-milling times (0 h, 10 h, 25 h, 55 h, and 90 h). The powders with flaky shape after 10 h, 25 h, and 55 h have better microwave absorbing ability.

References

1. Han M, Qin J, Deng L (2012). Microwave permeability of nanocrystalline alloys $(Fe_{0.67}Co_{0.33})_{78}Nb_6B_{15}Cu_1$ and $(Fe_{0.67}Co_{0.33})_{78}Nb_6Al_6B_9Cu_1$, *J Alloys Compd*, **543**, 79–83.

2. Gong C, Zhang J, Zhang X (2010). Strategy for ultrafine Ni fibers and investigation of the electromagnetic characteristics, *J Phys Chem C*, **114**(22), 10101–10107.
3. Wei J, Zhang Z, Wang B (2010). Microwave reflection characteristics of surface-modified $Fe_{50}Ni_{50}$ fine particle composites, *J Appl Phys*, **108**(12), 123908.
4. Do Kim Y, Chung JY, Kim J (2000). Formation of nanocrystalline Fe–Co powders produced by mechanical alloying, *Mat Sci Eng A*, **291**(1), 17–21.
5. Murakami Y, Shindo D, Oikawa K (2002). Magnetic domain structures in Co–Ni–Al shape memory alloys studied by Lorentz microscopy and electron holography, *Acta Mater*, **50**(8), 2173–2184.
6. Koohkan R, Sharafi S, Shokrollahi H (2008). Preparation of nanocrystalline Fe–Ni powders by mechanical alloying used in soft magnetic composites, *J Magn Magn Mater*, **320**(6), 1089–1094.
7. Wang X, Gong R, Li P (2007). Effects of aspect ratio and particle size on the microwave properties of Fe–Cr–Si–Al alloy flakes, *Mat Sci Eng A*, **466**(1), 178–182.
8. Luo H, Meng F, Cai Y (2011). Effect of low-valent atom substitution on electronic structure and magnetic properties of $Fe_{1.5}M_{0.5}CoSi$ (M = V, Cr, Mn, Fe) Heusler alloys, *J Magn Magn Mater*, **323**(17), 2323–2327.
9. Bowler N (2006). Designing dielectric loss at microwave frequencies using multi-layered filler particles in a composite, *IEEE Trans Dielectr Insul*, **13**(4), 703–711.
10. Kim SS, Kim ST, Yoon YC (2005). Magnetic, dielectric, and microwave absorbing properties of iron particles dispersed in rubber matrix in gigahertz frequencies, *J Appl Phys*, **97**(10), 10F905.
11. Rousselle D, Berthault A, Acher O (1993). Effective medium at finite frequency: theory and experiment, *J Appl Phys*, **74**(1), 475–479.
12. Stoyanov AJ, Fischer EC, Überall H (2001). Effective medium theory for large particulate size composites, *J Appl Phys*, **89**(8), 4486–4490.
13. Lu B, Dong XL, Huang H (2008). Microwave absorption properties of the core/shell-type iron and nickel nanoparticles, *J Magn Magn Mater*, **320**(6), 1106–1111.
14. Liu L, Duan Y, Liu S (2010). Microwave absorption properties of one thin sheet employing carbonyl–iron powder and chlorinated polyethylene, *J Magn Magn Mater*, **322**(13), 1736–1740.

15. Tsutaoka T (2003). Frequency dispersion of complex permeability in Mn–Zn and Ni–Zn spinel ferrites and their composite materials, *J Appl Phys*, **93**(5), 2789–2796.

16. Liu SH, Liu JM, Dong XL (2006). *Electromagnetic Wave Shielding and Absorbing Materials*, Chemical Industry Press, Beijing.

17. Cho HS, Kim AS, Kim SM (2004). Study of electromagnetic wave-absorbing materials made by a melt-dragging process, *Phys Status Solidi A*, **201**(8), 1942–1945.

18. Zhang L, Zhu H (2009). Dielectric, magnetic, and microwave absorbing properties of multi-walled carbon nanotubes filled with Sm_2O_3 nanoparticles, *Mater Lett*, **63**(2), 272–274.

19. Sunny V, Kurian P, Mohanan P (2010). A flexible microwave absorber based on nickel ferrite nanocomposite, *J Alloy Compd*, **489**(1), 297–303.

20. Liu L, Duan Y, Guo J (2011). Influence of particle size on the electromagnetic and microwave absorption properties of FeSi/paraffin composites, *Physica B*, **406**(11), 2261–2265.

21. Chen N, Mu G, Pan X (2007). Microwave absorption properties of $SrFe_{12}O_{19}/ZnFe_2O_4$ composite powders, *Mat Sci Eng B*, **139**(2), 256–260.

22. Duan Y, Liu S, Wen B (2006). A discrete slab absorber: absorption efficiency and theory analysis, *J Compos Mater*, **40**(20), 1841–1851.

23. Schlesinger Z, Fisk Z, Zhang HT (1997). Is FeSi a Kondo insulator, *Physica B*, **237**, 460–462.

24. Qing YC, Zhou WC, Jia S (2010). Electromagnetic and microwave absorption properties of carbonyl iron and carbon fiber filled epoxy/silicone resin coatings, *Appl Phys A*, **100**(4), 1177–1181.

25. Wadhawan A, Garrett D, Pérez J M (2003). Nanoparticle-assisted microwave absorption by single-wall carbon nanotubes, *Appl Phys Lett*, **83**(13), 2683–2685.

26. Kondo T, Aoki T, Asano K (1999). Fabrication of the composite electromagnetic wave absorber, in *International Symposium on Electromagnetic Compatibility*, Vol. 19, Tokyo, 397–400.

27. Guo J, Duan Y, Liu L (2011). Electromagnetic and microwave absorption properties of carbonyl-iron/$Fe_{91}Si_9$ composites in gigahertz range, *J Electromagn Anal Appl*, **3**(5), 140–146.

28. Zhang XF, Dong XL, Huang H (2006). Microwave absorption properties of the carbon-coated nickel nanocapsules, *Appl Phys Lett*, **89**(5), 3115.

29. Cheng YL, Dai JM, Wu DJ (2010). Electromagnetic and microwave absorption properties of carbonyl iron/$La_{0.6}Sr_{0.4}MnO_3$ composites, *J MagnMagn Mater*, **322**(1), 97–101.

30. Li BW, Shen Y, Yue ZX (2006). Enhanced microwave absorption in nickel/hexagonal-ferrite/polymer composites, *Appl Phys Lett*, **89**(13), 2504.

31. Osaka T, Takai M, Hayashi K (1998). A soft magnetic CoNiFe film with high saturation magnetic flux density and low coercivity, *Nature*, **392**(6678), 796–798.

32. Hosseini HRM, Bahrami A (2005). Preparation of nanocrystalline Fe–Si–Ni soft magnetic powders by mechanical alloying, *Mat Sci Eng B*, **123**(1), 74–79.

33. Karimi L, Shokrollahi H (2011). Structural, microstructural and magnetic properties of amorphous/nanocrystalline $Ni_{63}Fe_{13}Mo_4Nb_{20}$ powders prepared by mechanical alloying, *J Alloy Compd*, **509**(23), 6571–6577.

34. Wang CX, Lv ZQ, Fu WT (2011). Electronic properties, magnetic properties and phase stability of alloyed cementite $(Fe,M)_3C(M=Co,Ni)$ from density-functional theory calculations, *Solid State Sci*, **13**(8), 1658–1663.

35. Duan Y, Gu S, Wen M (2013). The evolution of electronic configuration and magnetic characterization of Fe_9Ni_1, Fe_8Ni_2 alloy in theoretical calculation, *Eur Phys J B*, **86**(10), 1–7.

36. Dan NH, Thanh PT, Yen NH (2012). Inducing anisotropy in bulk Nd–Fe–Co–Al–B nanocrystalline alloys by quenching in magnetic field, *J Magn Magn Mater*, **324**(7), 1435–1439.

37. Nie Y, He HH, Gong RZ (2007). The electromagnetic characteristics and design of mechanically alloyed Fe–Co particles for electromagnetic-wave absorber, *J Magn Magn Mater*, **310**(1), 13–16.

38. Wen F, Zhang F, Liu Z (2011). Investigation on microwave absorption properties for multiwalled carbon nanotubes/Fe/Co/Ni nanopowders as lightweight absorbers, *J Phys Chem C*, **115**(29), 14025–14030.

39. Yanagimoto K, Majima K, Sunada S (2004). Effect of powder compositions on GHz microwave absorption of EM absorbing sheets, *J Jpn Soc Powder Metall*, **51**(4), 293–296.

40. Zhonglun Z, Zhijiang J, Yuping D (2013). The superior electromagnetic properties of carbonyl-iron/$Fe_{91.2}Si_{3.1}P_{2.9}Sb_{2.8}$ composites powder and impedance match mechanism, *J Mater Sci: Mater El*, **24**(3), 968–973.

41. Yu KC, Ma CC M, Teng CC (2011). Preparation and microwave absorbency of Fe/epoxy and $FeNi_3$/epoxy composites, *J Alloy Compd*, **509**(33), 8427–8432.
42. Shirakata Y, Hidaka N, Ishitsuka M (2011). High permeability and low loss Ni–Fe composite material for high-frequency applications, *IEEE Trans Magn*, **44**(9), 2100–2106.
43. Neamţu BV, Geoffroy O, Chicinaş I (2012). AC magnetic properties of the soft magnetic composites based on Supermalloy nanocrystalline powder prepared by mechanical alloying, *Mat Sci Eng B*, **177**(9), 661–665.
44. Aas CJ, Szunyogh L, Chen JS (2011). Magnetic anisotropy of FePt: effect of lattice distortion and chemical disorder, *Appl Phys Lett*, **99**(13), 132501.
45. Zeng HT, Read D, Petit D (2009). Combined electrical and magneto-optical measurements of the magnetization reversal process at a domain wall trap, *Appl Phys Lett*, **94**(10), 103113.
46. Ohnuma M, Herzer G, Kozikowski P (2012). Structural anisotropy of amorphous alloys with creep-induced magnetic anisotropy, *Acta Mater*, **60**(3), 1278–1286.
47. Chun BS, Kim SD, Kim YS (2010). Effects of Co addition on microstructure and magnetic properties of ferromagnetic CoFeSiB alloy films, *Acta Mater*, **58**(8), 2836–2842.
48. Mishin Y, Mehl MJ, Papaconstantopoulos DA (2005). Phase stability in the Fe–Ni system: Investigation by first-principles calculations and atomistic simulations, *Acta Mater*, **53**(15), 4029–4041.
49. Segall MD, Lindan PJD, Probert MJ (2002). First-principles simulation: ideas, illustrations and the CASTEP code, *J Phys: Condens Mat*, **14**(11), 2717.
50. Chen Y, Song Q, Yan H (2011). Electronic structure and magnetic properties of the N monodoping and (Li, N)-codoped ZnO, *Solid State Commun*, **151**(8), 619–623.
51. Cai Y, Huang ZF, Meng X (2011). Pressure-induced phase transformation and magnetism transition in $BaRuO_3$: a first-principles study, *Solid State Sci*, **13**(2), 350–355.
52. Rahman G, Kim IG (2008). A first-principles study on magnetic and electronic properties of Ni impurity in bcc Fe, *J Magn*, **13**(4), 124–127.
53. Basha SM, Ramasubramanian S, Thangavel R (2010). Magnetic properties of Ni doped gallium nitride with vacancy induced defect, *J Magn Magn Mater*, **322**(2), 238–241.

54. Jia X, Qin M, Yang W (2011). Magnetism in Cr doped ZnO: density-functional theory studies, *J Magn Magn Mater*, **323**(11), 1423–1427.

55. Dou Y, Jin H, Cao M (2011). Structural stability, electronic and optical properties of Ni-doped 3C–SiC by first principles calculation, *J Alloy Compd*, **509**(20), 6117–6122.

56. Deng SH, Duan MY, Xu M (2011). Effect of La doping on the electronic structure and optical properties of ZnO, *Physica B*, **406**(11), 2314–2318.

57. Koohkan R, Sharafi S, Shokrollahi H (2008). Preparation of nanocrystalline Fe–Ni powders by mechanical alloying used in soft magnetic composites, *J Magn Magn Mater*, **320**(6), 1089–1094.

58. Duan Y, Gu S, Zhang Z (2012). Characterization of structures and novel magnetic response of $Fe_{87.5}Si_7Al_{5.5}$ alloy processed by ball milling, *J Alloy Compd*, **542**, 90–96.

59. Chen YF, Song QG, Yan HY (2011). Effect of codoping with C or N on electronic structures and magnetic properties of ZnO: Cu, *Physica B*, **406**(20), 3901–3905.

60. Monkhorst HJ, Pack JD (1976). Special points for Brillouin-zone integrations, *Phys Rev B*, **13**(12), 5188.

61. Lee JH, Shishidou T, Freeman AJ (2002). Improved triangle method for two-dimensional Brillouin zone integrations to determine physical properties, *Phys Rev B*, **66**(23), 233102.

62. Hirota K, Takano Y, Yoshinaka M (2000). A new composite material with high saturation magnetization density and high electrical resistivity, *Mater Res Bull*, **35**(7), 1137–1141.

63. Zhou TD, Zhou PH, Liang DF (2009). Structure and electromagnetic characteristics of flaky FeSiAl powders made by melt-quenching, *J Alloy Compd*, **484**(1), 545–549.

64. Miraghaei S, Abachi P, Madaah-Hosseini HR (2008). Characterization of mechanically alloyed $Fe_{100-x}Si_x$ and $Fe_{83.5}Si_{13.5}Nb_3$ nanocrystalline powders, *J Mater Process*, **203**(1), 554–560.

65. Duan Y, Gu S, Zhang Z (2012). Characterization of structures and novel magnetic response of $Fe_{87.5}Si_7Al_{5.5}$ alloy processed by ball milling, *J Alloy Compd*, **542**,90–96.

66. Burton AW, Ong K, Rea T (2009). On the estimation of average crystallite size of zeolites from the Scherrer equation: a critical evaluation of its application to zeolites with one-dimensional pore systems, *Micropor Mesopor Mat*, **117**(1), 75–90.

67. Biju V, Sugathan N, Vrinda V (2008). Estimation of lattice strain in nanocrystalline silver from X-ray diffraction line broadening, *J Mater Sci*, **43**(4), 1175–1179.

68. Varin RA, Bystrzycki J, Calka A (1999). Characterization of nanocrystalline Fe–45 at% Al intermetallic powders obtained by controlled ball milling and the influence of annealing, *Intermetallics*, **7**(8), 917–930.

69. Karimi L, Shokrollahi H (2011). Structural, microstructural and magnetic properties of amorphous/nanocrystalline $Ni_{63}Fe_{13}Mo_4Nb_{20}$ powders prepared by mechanical alloying, *J Alloy Compd*, **509**(23), 6571–6577.

70. Ma T, Yan M, Wang W (2008). The evolution of microstructure and magnetic properties of Fe–Si–Al powders prepared through melt-spinning, *Scripta Mater*, **58**(4), 243–246.

71. Zhang C, Jiang J, Bie S (2012). Electromagnetic and microwave absorption properties of surface modified Fe–Si–Al flakes with nylon, *J Alloy Compd*, **527**, 71–75.

Chapter 4

Conductive Polyaniline

Conjugated polymers and their composites have attracted much attention due to their unique electrical properties. They have wide use in a number of applications, such as in solar cells, lightweight batteries, light-emitting diodes, polymer actuators, corrosion protection agents, sensors, molecular electronic devices, and stealth and electromagnetic interference (EMI) shields[1–3]. Among these polymers, the members of the polyaniline (PANI) family show much expectation due to their simple routes of synthesis, environmental stability, and ability to be doped for altering electrical conductivity, either by protonic acids or by oxidants/reductants. In the self-doped state, they show high electrical conductivity over a wide range of pH[3,4]. Typically, metals have good mechanical and shielding properties, but their heavy weight, easy corrosion, and poor processibility are disadvantages when used as a shielding material. Conducting polymers are attractive because of their light weight, corrosion resistance, good processibility, low processing cost, and simple control of conductivity. PANI is an environmentally stable conducting polymer with excellent electrical, magnetic, and optical properties, so it remains one of the most intensely studied polymer[5–9]. It has been generally regarded as one of the conducting polymer with very high potential in commercial applications, such as batteries, electromagnetic shielding, and gas sensors[10–13]. Duan and coworkers[14] reported a PANI composite with an EMI shielding

Microwave Absorbing Materials
Yuping Duan and Hongtao Guan

ISBN 978-981-4745-10-9 (Hardcover), 978-981-4745-11-6 (eBook)
www.panstanford.com

performance as high as –19.3 dB in the frequency range of 3–1500 MHz. Wang[15] and Wu[16] discovered the excellent microwave absorption capability of PANI nanocomposites. Doped PANI and its composites have been widely investigated and applied. The emeraldine base (EB) form of PANI is electrically insulating and doping can made it change from insulator into conductor and the doping process can be described as Fig. 4.1. The stable forms of PANI are leucoemeraldine base (LEB, $y = 1$), EB ($y = 0.5$), and pernigraniline base (PNB, $y = 0$)[14,17].

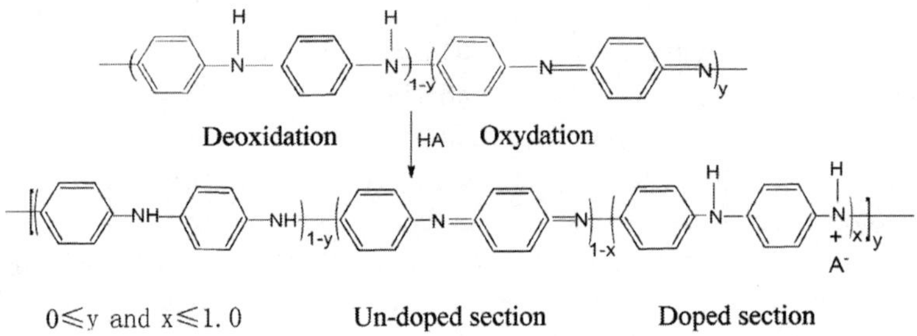

Figure 4.1 Doping process of PANI.

Here y is the extent of oxidization, depending upon synthesis of the monomer; x is the extent of doping, depending upon the doping of the polymer; and A^- is the anion of protonic acid, depending upon the dopant. In the doping process, the H^+ of protonic acid protonate N atoms connected with the quinonoid ring of the PANI molecular chain and there are electron holes generated in the valence band of the doped section of the PANI chain, that is, P-doping. With an external electric field, the electron hole moves in the whole chain by the resonance of conjugated π-electron and the electrical conductivity property of PANI-HCl is shown. The electrical conductivity level has a close relationship with the value of y and x. The larger the value of x is, and the more current carrier there will be, the better the electrical conductivity level is.

4.1 Electromagnetic Properties of Doped PANI

The special feature of polyaniline (PANI) is that the highly conducting emeraldine salt (ES) can be obtained by doping with protonic acids (H^+A^-). The protonation of the imine nitrogen atoms of the emeraldine base (EB) is a miraculous way leading to a dramatic increase in electrical conductivity, up to 12 orders of magnitude[18]. The doping level is one of the most important factors affecting the electrical conductivity and other properties of PANI. The mechanism of protonation of the EB on the grounds of theoretical and optical studies was proposed by Stafström and coworkers originally[19]. Soon afterward Heeger[20] further analyzed the doping mechanism by protonation and showed that the mechanism of protonation can be described in three steps: reaction with protonic acid adds $2H^+$ to the nitrogen atoms beside the quinoid ring, obtaining the bipolaron (ES1) form; internal redox reaction achieving the charged polaron (ES2) form; and polaron separation, that is, halving of the unit cell, yielding the final polaron (ES3) form.

4.1.1 Structure Characterization

PANI is typically prepared by chemical redox polymerization of aniline using ammonium perdisulfate ($(NH_4)_2S_2O_8$). The $(NH_4)_2S_2O_8$ HCl solution was added dropwise to the aniline-HCl solution with constant electromagnetic stirring[21]. The dark green precipitate resulting from this reaction was HCl-doped PANI and marked as doped PANI. The doped PANI was washed repeatedly in a filtering funnel with ammonia and distilled water. This dark blue powder was marked as EB-PANI. The dark blue precipitate was washed with ammonia water and distilled water and then added to HCl. The dark green precipitate was marked as redoped PANI.

Figure 4.2 displays the Fourier transform infrared (FT-IR) spectra of doped PANI, EB-PANI, and redoped PANI powders. The EB-PANI spectrum showed characteristics bands, including C=N and C=C stretching modes for the quinonoid and benzenoid ring units occurring at 1589 cm^{-1} and 1500 cm^{-1}, a band at 1305 cm^{-1} attributed to the C–N stretching mode for the benzenoid unit, a stretching of N=quinoid=N at 1162 cm^{-1}, and absorption bands near 830 cm^{-1} associated with C–H deformation in the two-substituted benzenoid ring. The observations have a good agreement with the previous experimental values[22,23]. Compared to EB-PANI, the absorption peaks of doped PANI and redoped PANI shifted simultaneously toward a lower frequency and were broadened.

This was due to delocalization of the partial positive charge on the nitrogen atom on the imido group to the aromatic ring. A resonant structure and the conjugated effect were formed under the action of electron charge delocalization. The shifted wavenumber of the N=quinoid=N absorption peak was most obvious, at about 55 cm^{-1}, and the absorption peak also became stronger and wider, which is indicative of conductivity of doped PANI[24].

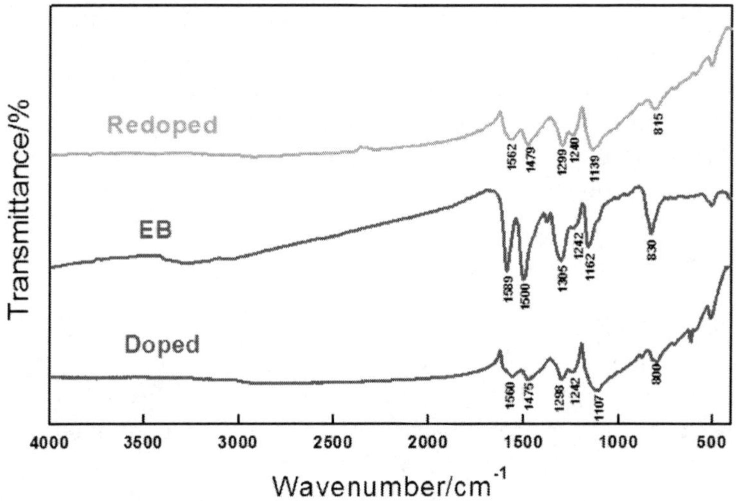

Figure 4.2 FT-IR spectra of doped PANI, EB-PANI, and redoped PANI powder.

Figure 4.3 shows X-ray diffraction patterns of doped PANI, EB-PANI, and redoped PANI powder. The EB-PANI exhibits a broadened pattern, indicating that this material is amorphous. The presence of peaks between 10° and 40° 2θ values in both the doped PANI and redoped PANI powders reveals the partial crystalline nature of PANI doped by protonic acid. The peaks are observed at 8.62°, 16.34°, 20.41°, 25.32°, and 26.92° 2θ values, implying the presence of a rigid chain and an ordered structure, resulting in partial crystallinity[23]. The increase in polarity and rearrangement of the molecular chain are likely to improve its crystallinity and order.

Compared to other crystals, the crystallinity of doped and redoped PANI is low and its crystalline region is also small. This increase in crystallinity allows for improved electrical conduction.

The crystalline domain size, or the extent of order, D_{hkl}, can be estimated from the ring angular width D (2θ) (full width at half-maximum [FWHM]) using the Scherrer formula:[25]

$$D_{hkl} = \frac{\kappa\lambda}{\Delta(2\theta)\cos\theta} \tag{4.1}$$

where λ is the X-ray wavelength used (its value is 0.145 nm), θ is the Bragg angle, and k ($k \approx 0.9$) is the outline exponential. The D_{hkl} values associated with the strongest X-ray diffraction peak are given in Table 4.1, and the calculated average grain size $\bar{D}_{hkl}$ of the doped PANI is about 4.10 nm.

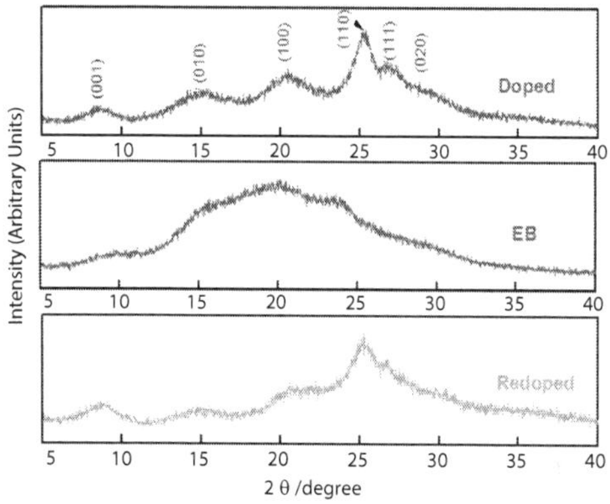

Figure 4.3 X-ray diffraction patterns of doped PANI, EB-PANI, and redoped PANI powder. The pattern of EB-PANI indicates that this material is amorphous.

Table 4.1 Indexation and domain length D_{hkl} of doped polyaniline

2θ(degree)	(hkl)	D_{hkl} (nm)	$\bar{D}_{hkl}$(nm)
8.6200	(001)	4.21	
14.7100	(010)	3.08	
20.4133	(001)	6.44	
25.3173	(110)	3.31	4.10
26.9200	(111)	4.68	
28.8800	(020)	4.00	
16.34	-	2.98	

(hkl) refers to indices of the crystallographic plane.

4.1.2 Morphologies

The transmission electron microscopy (TEM) image and selected area electron diffraction (SAED) of the redoped PANI powder are shown in Fig. 4.4. The redoped PANI particles are spherical in pattern and about 4–5 nm in diameter (Fig. 4.4a), which is in agreement with the particle size calculated by the Scherrer formula. However, the grain structure of the redoped PANI is an irregular accumulation composed of small grains. This is because polarity and the acting force between molecular chains had been enhanced after PANI was doped by HCl. Figure 4.4b shows that the crystalline region is enclosed by the amorphous portion in the overlap region of the redoped PANI grain. Diffraction rings with obvious peripheral diffraction spots can be seen in Fig. 4.4c, which verified that redoped PANI was fractionally crystalline, and these diffraction rings correspond to the X-ray diffraction peaks.

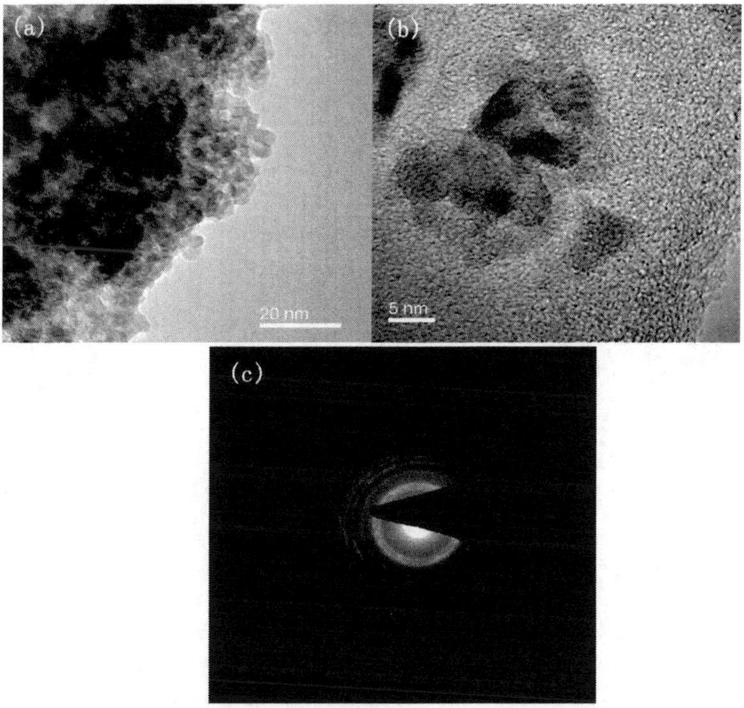

Figure 4.4 TEM image (a, b) and selected area electron diffraction (SAED) (c) of redoped PANI particles.

4.1.3 Electrical Properties of Redoped PANI

The volume resistivities of rubber composites containing different redoped PANI loadings, which are responses for the electrical properties of the redoped PANI, are shown in Table 4.2, and the curve of the volume resistivities versus redoped PANI loading is presented in Fig. 4.5. The volume resistivity decreased from 10^{13} Ω·cm to 5.18×10^{-1} Ω·cm with an increasing mass percent of the PANI powder from 0 to 50 wt%. The volume resistivity of the composites containing 50 wt% PANI decreased by about 14 orders of magnitude compared with that of the rubber matrix without PANI, which indicates that the redoped PANI has a high conductivity and that the conductivity of PAN-EB increases greatly after doping. The half-filled polar band and large conjugate π-band that formed in the redoped PANI molecular chain increased the degree of electron charge delocalization on the molecular chain of the redoped PANI, which reduced the energy gap of the electron transition. This then facilitated the electron charge transition under the electric field, thus enhancing the electrical conduction.

Table 4.2 Conductivity (σ) of redoped PANI/silicon rubber composites

No.	Mass percent of PANI (%)	Current (A)	Voltage (V)	Volume resistivities (ρ_v/Ω·cm)	Conductivity (σ/S·cm^{-1})
1	0	–	–	~10^{13} (supplied by manufacturers)	10^{-13}
2	10	3.68×10^{-7}	2.52×10^{-4}	1.67×10^{5}	5.99×10^{-6}
3	20	5.8×10^{-5}	6.74	3.03×10^{4}	3.30×10^{-5}
4	30	4.95×10^{-3}	4.50	2.56×10^{2}	3.91×10^{-3}
5	40	6.88×10^{-3}	3.79	1.49×10^{2}	6.69×10^{-3}
6	50	2.11×10^{-1}	4.12	5.18×10^{-1}	1.93×10^{0}

PANI: HCl-redoped polyaniline

From Fig. 4.5, it can also be seen that volume resistivity does not decrease linearly with increase in the redoped PANI fraction. Volume resistivity of rubber composites containing 10 wt% decreases sharply by about 8 orders of magnitude compared with that of the rubber matrix without PANI. For PANI loading changes from 10 to 40 wt%, the volume resistivity decreases gradually by about 2 orders of magnitude. For a further increase in the PANI fraction from 40 to 50 wt%, volume resistivity decreases sharply again by about 3 orders of magnitude.

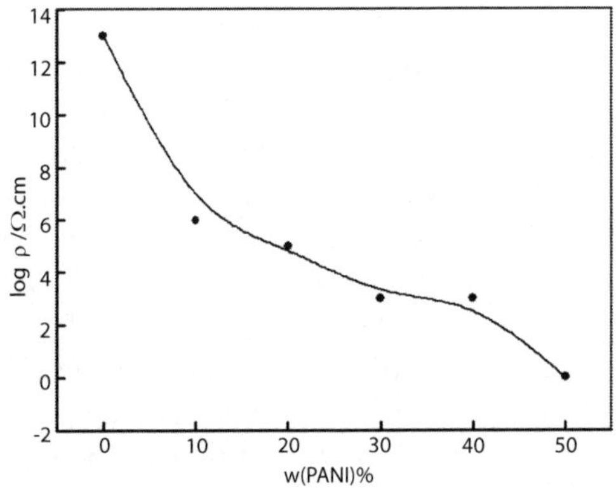

Figure 4.5 Effect of different mass percent of redoped PANI on conductivity of the composite.

The possible explanation for this variation in volume resistivity is the following. From typical scanning electron microscopy (SEM) micrographs of the sample fracture surface (as shown in Fig. 4.6), the PANI powder had good dispersibility in the matrix and there was better disaggregation as the content increased. When the PANI loading was about 10 wt%, there were many enrichment zones in the conductive PANI, in the form of "isolated conductive islands," and there were a few conductive particles in other areas. However, Fig. 4.5 shows the resistivity of approximately 10^5 Ω·cm, which contributes to self-conduction of redoped PANI and conduction via hopping or tunneling[26]. When PANI loading was between 20 and 40 wt%, isolated conductive islands became more frequent and gradually linked with each other, creating more conductive chains or networks in the matrix. Consequently, the volume resistivity decreased gradually by 2–3 orders of magnitude. However, when the PANI filler approached 50 wt%, more extensive conductive chains or networks were able to form, so the volume resistivity decreased sharply by about 3 orders of magnitude.

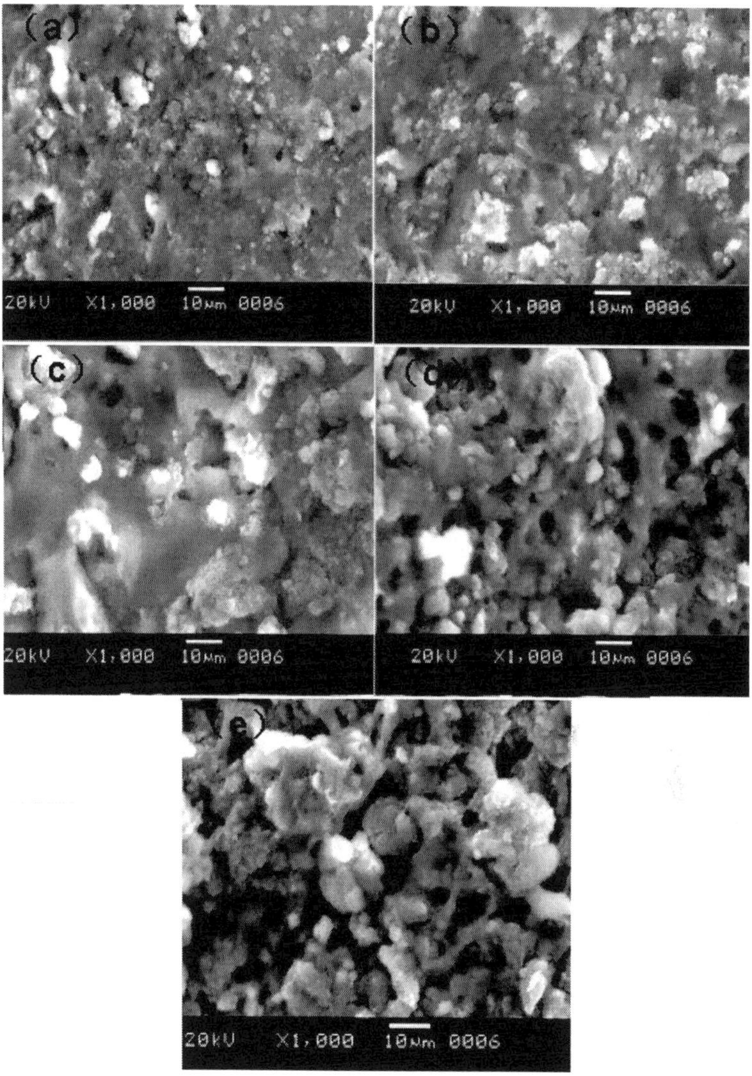

Figure 4.6 SEM micrograph of the sample fracture face: (a) 10 % PANI, (b) 20 % PANI, (c) 30% PANI, (d) 40% PANI, and (e) 50 % PANI.

The redoped PANI products were again heated for 24 h in a vacuum drying oven at 60°C, 80°C, 100°C, 120°C, 140°C, and 160°C. The prepared PANI powder was then added to a rubber matrix and the electrical conductivity of 50 wt% PANI/rubber composites was researched. The curve of the electrical conductivity versus the temperature of heat treatment is shown in Fig. 4.7. The electrical conductivity of redoped PANI was nearly invariant when the heat treatment was below 100°C. However, when the heat treatment was above 120°C, the electrical conductivity decreased sharply, reaching 2.66×10^{-7} S·cm^{-1} at a temperature of 160°C, on the basis of which it is deduced that deprotonation of the redoped PANI is responsible for the conductivity loss. Besides the deprotonation, additional changes in the polymer structure, like oxidative degradation, aromatic ring substitution, and crosslinking, are likely to induce the irreversible decrease in the conductivity[27]. On the basis of this analysis, the applied range of temperatures of the HCl-redoped PANI should not exceed 100°C.

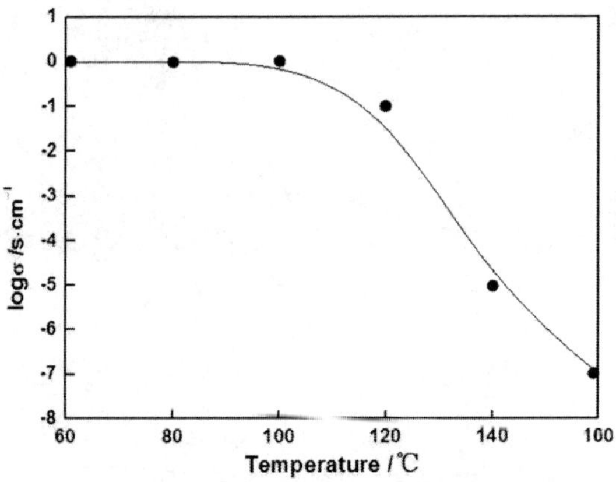

Figure 4.7 Curve of the electric conductivity of a PANI composite versus heat treatment temperature.

4.1.4 Dielectric Properties

The permittivity ε of a dielectric material is a quantified response of molecules polarizing themselves along an electric field due to attraction or repulsion between a set of charges. Normally, $\varepsilon = \varepsilon_0\varepsilon_r$, where $\varepsilon_0 = (1/36\pi) \times 10^{-9}$F/m is the absolute permittivity of free space (vacuum) and ε_r is a dimensionless number that specifies relative permittivity of the material. It can be written as $\varepsilon_r = \varepsilon_r' - j\varepsilon_r''$, where ε_r' and ε_r'' are the real and imaginary parts of ε_r, respectively.

ε_r'', associated with electric field loss, results primarily from electric dipole polarization at microwave frequencies[28].

The electric loss tangent (tan δ_ε) and the magnetic loss tangent (tan δ_μ) of a material can be written as tan $\delta_\varepsilon = \varepsilon_r''/\varepsilon_r'$ and tan $\delta_\mu = \mu_r''/\mu_r'$, respectively, which means that the ratio of the energy loss/unit radian in the dielectric to the energy storage in a dielectric can be determined[28].

Figure 4.8 illustrates the real and imaginary components of the complex relative dielectric permittivity versus frequency for the prepared EB-PANI and redoped PANI powder–paraffin wax composites. Figure 4.8 shows that values of ε_r' and ε_r'' for the redoped PANI powder declined from 9.31 to 5.95 and 3.44 to 2.10, respectively, with increasing frequency in the 2–18 GHz range. However, for the EB-PANI powder, ε_r' and ε_r'' had variable values within a narrow range are about 3.5 and 0.3, respectively. Figure 4.8a,b also shows that the values for relative dielectric permittivity versus frequency are higher for the redoped PANI powder than for the EB-PANI powder. According to the theory of complex permittivity[29], when electromagnetic radiation is incident on metallic surfaces, the electric field induces two types of electrical currents within the materials; that is, the conduction and displacement currents. The former arises from free electrons, and it will give rise to electric loss (imaginary permittivity). The latter comes from the bound charges, that is, polarization effect (real permittivity). Compared to EB-PANI, the increased complex permittivity can be ascribed to dielectric relaxation and space charge polarization effects of redoped PANI powder under electromagnetic radiation. The increase in the imaginary component of complex permittivity can be attributed to enhanced electrical conductivity of the redoped PANI powder. The high value of complex permittivity of the redoped powder also indicates high displacement and conduction currents in the material and suggests that the redoped powders are suitable for use as microwave absorption materials in the measured frequency range.

As already mentioned, dielectric relaxation is one of important reasons for increased complex permittivity of the redoped PANI powder. Consequently, there must be relaxation polarization loss under alternating electromagnetic fields for the redoped PANI powder. Permittivity can be represented by the Debye dipolar relaxation expression:[30]

$$\varepsilon_r = \varepsilon_\infty + \frac{\varepsilon_s - \varepsilon_\infty}{1 + j2\pi f\tau} = \varepsilon'(f) - j\varepsilon''(f) \tag{4.2}$$

where f, τ, ε_s, and ε_∞ are the frequency, relaxation time, stationary dielectric constant, and optical dielectric constant, respectively. From Eq. 4.2, it can be deduced that

$$\varepsilon'(f) = \varepsilon_\infty + \frac{\varepsilon_s - \varepsilon_\infty}{1 + (2\pi f)^2 \tau^2} \tag{4.3}$$

$$\varepsilon''(f) = \frac{2\pi f \pi(\varepsilon_s - \varepsilon_\infty)}{1 + (2\pi f)^2 \tau^2} \tag{4.4}$$

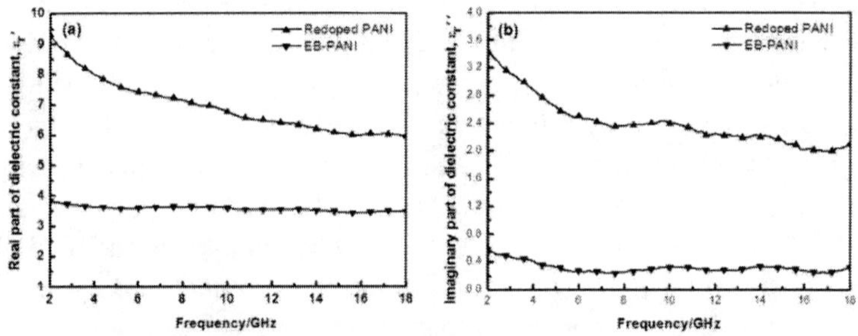

Figure 4.8 Dependence of microwave frequency on the real part (a) and imaginary part (b) of dielectric permittivity of PANI.

According to Eqs. 4.3 and 4.4, a standard semicircle for single relaxation time can be obtained, namely

$$\left(\varepsilon' - \frac{\varepsilon_s - \varepsilon_\infty}{2}\right)^2 + \varepsilon''^2 = \left(\frac{\varepsilon_s - \varepsilon_\infty}{2}\right)^2 \tag{4.5}$$

Equation 4.5 is usually defined as the Cole–Cole semicircle[31]. Figure 4.9 shows the curve characteristics of ε' versus ε'' for the as-prepared redoped PANI. Note that the as-obtained redoped PANI presents a clear segment of three overlapped semicircles. The presence of three semicircles suggests that there are multiple dielectric relaxation processes with dissymmetrical distributing of relaxation time, with each semicircle corresponding to a Debye dipolar relaxation. There are ordered microareas at the nanometer scale in the unordered relaxation body of redoped PANI, and the scale of the ordered microarea plays a key role in the relaxation time[32]. When discontinuous relaxation times are distributed around several special relaxation times, τ_0, the Cole–Cole plots will be composed of several overlapped semicircles. In other words, the dissymmetrical distribution of the relaxation time will result in deflection of the Cole–Cole plots. Therefore, Fig. 4.9 shows that multiple dielectric relaxations include grain inner relaxation, grain boundary relaxation, and crystal defect relaxation. Lastly, it can be seen that the as-prepared redoped PANI is indubitably a good dielectric loss material.

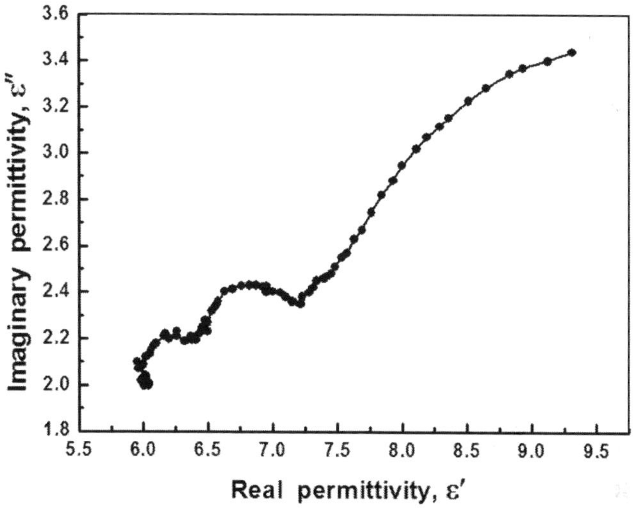

Figure 4.9 Cole–Cole plots of the imaginary versus real permittivity parts for doped PANI powder.

4.1.5 Microwave Absorbing Properties

To further confirm dependence of microwave absorption properties on permittivity and permeability, reflection losses (RLs) (dB) of redoped PANI/paraffin wax samples were calculated according to transmission line theory[33]. The normalized input impedance Z_{in} of a microwave absorber is given by

$$Z_{in} = Z_0\sqrt{\frac{\mu_r}{\varepsilon_r}}\tanh\left(j\frac{2\pi f}{c}\sqrt{\mu_r\varepsilon_r}d\right) \tag{4.6}$$

where μ_r and ε_r are the relative permeability and permittivity of the composite medium, respectively; c is the velocity of electromagnetic waves in free space; f is the frequency of the microwave; and d is the thickness of the absorber. The RL is related to Z_{in} as

$$\text{RL (dB)} = 20\log\left|\frac{Z_{in} - Z_0}{Z_{in} + Z_0}\right| \tag{4.7}$$

where Z_0 is the characteristic impedance of free space. The calculated RLs as a function of frequency for samples with different thickness are shown in Fig. 4.10. The as-prepared redoped PANI had excellent microwave absorption capability. The maximum RL of the samples with 2.0 mm thickness reaches −14.3 dB at 16.4 GHz, and the effective absorption band (RL under −10 dB) ranges from 10.0 to 13 GHz. The 1.5 mm thickness sample shows a maximum RL of −13.2 dB at 16 GHz and an effective absorption band from 14 to 18 GHz. However, at a sample thickness of 5 mm, the maximum RL (−9.2 dB at 4 GHz) and effective absorption band is greatly weakened. The increasing thickness shifted the absorbing peak value to the lower-frequency band, as shown in Fig. 4.10. This shift can be attributed mainly to wave superposition of incident and backward waves with variation in sample thickness. On the basis of analysis of the electrical conduction and dielectric properties mentioned above, the excellent microwave absorption capability may be attributable primarily to resistance loss and multiple dielectric relaxation loss.

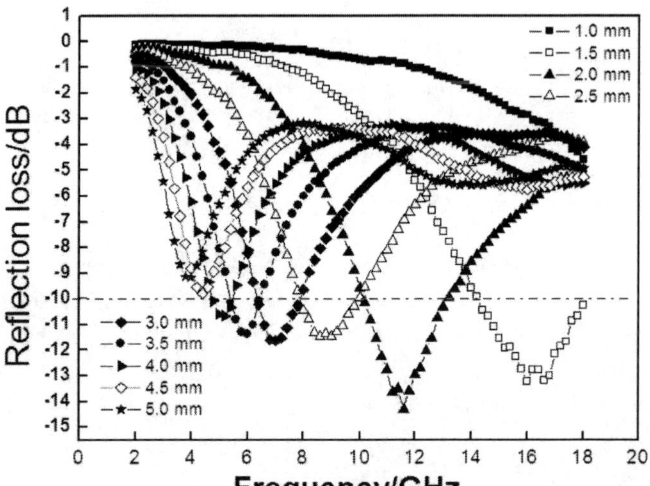

Figure 4.10 Reflection loss of the 60 wt% redoped PANI/paraffin wax composites with different thickness versus frequency (2–18 GHz).

4.2 Electromagnetic Properties of γ-MnO_2/Polyaniline Composites

Manganese dioxide is one of the most attractive inorganic materials because of its wide range of applications in catalysts, molecular sieves, and electromagnetic materials[34-36]. The microwave absorption property of commercial MnO_2 powder was first discovered by Guan et al.[37] Inorganic nanocomposites containing conducting polymers have attracted a great deal of interest. In this area, PANI is one of the most promising conducting polymers. It is easy to prepare and often exhibits unusual electromagnetic properties combined with moderate electrical conductivity. In the present work, γ-MnO_2/PANI nanocomposites are prepared by an in situ polymerization method[38].

A calculated amount of γ-MnO_2 was first dissolved in HCl. The ammonium persulfate HCl solution is added dropwise to the aniline-HCl solution with constant electromagnetic stirring. The γ-MnO_2 dispersion solution was then poured rapidly into the aniline and ammonium persulfate solution. The obtained precipitate was washed and dried in a vacuum drying oven. Black powdered MnO_2/PANI composites were obtained and labeled as samples 1#, 2#, 3#, and 4# corresponding to the γ-MnO_2 and aniline mass ratios of 1:1, 2:1, 2:1, and 3:1, respectively. Sample 2# was reimmersed in the same aniline-HCl solution for 12 h.

4.2.1 FT-IR Spectral Analysis

FT-IR spectra of samples 1#, 2#, 3#, 4#, MnO_2, and PANI are shown in Fig. 4.11. The typical absorption peaks of PANI are noticeable in MnO_2/PANI composites with different components, indicating that the PANI structure was obtained in a different polymerization system. As increasing MnO_2 in the MnO_2/PANI composite samples 1#, 3#, and 4#, the spectral characteristic peaks shifted to a higher frequency. The higher the content of MnO_2, the more evident the blue shift to a high frequency, and the larger the intensity ratio of the C=C stretching vibration characteristic bands on the quinone-imine and phenylene-diamine ring. Moreover, due to the high activity and oxidizability of MnO_2, polymerization of aniline was induced, which would result in more EB form in the structure of the PANI of the samples. Meanwhile, the typical absorption peaks of samples 3# and 4# in Fig. 4.11 at 610 cm^{-1} were attributable to the Mn–O stretching vibration for MnO_2. For samples 2# and 3# which had the same mass ratio of reactants of γ-MnO_2 and aniline, after sample 2# was again dealt with by dipping in the same aniline-HCl solution for 12 h, the characteristic peaks significantly generated a red shift to a lower frequency compared with that of sample 3#. The intensity ratio of the C=C stretching vibration characteristics bands on the quinone-imine and phenylene-diamine ring was 1:1, which indicated that the obtained PANI structure was the ideal EB structure. Moreover, comparing with sample 3#, the broader characteristic peaks at 1135 cm^{-1}, corresponding to N=Q=N (Q was the quinoid ring) stretching modes, showed that the doped degree of sample 2# was higher. So it was concluded that the high-oxidation-state compounds of PANI were obtained in the reduction reaction through the posttreatment of the aniline-HCl solution to a MnO_2/PANI composite, and it was useful to improve the emeraldine structure of the obtained MnO_2/PANI. In addition, the above results indicated that γ-MnO_2 was successfully decorated with doped PANI.

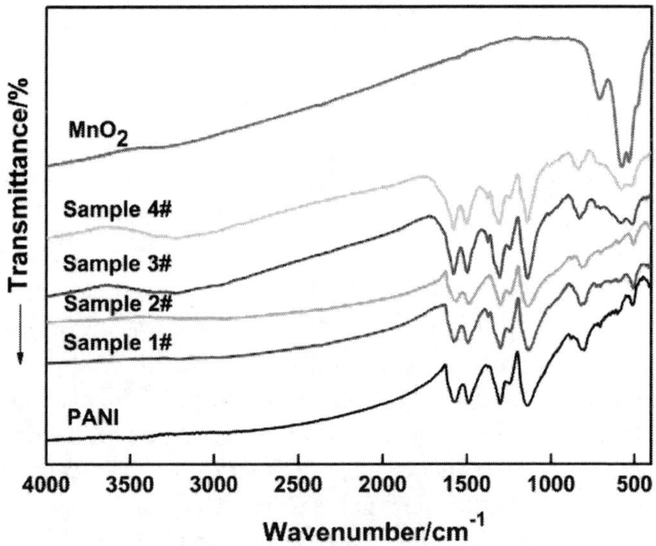

Figure 4.11 FT-IR spectra analysis of MnO_2/PANI composites.

4.2.2 XRD Analysis

Figure 4.12 depicts the XRD patterns of the MnO_2/PANI composite and MnO_2. As shown in Fig. 4.12, the peaks of sample 1# occurred at 10–30° 2θ values, which is similar with the characteristic peaks of PANI. For samples 3# and 4#, with the amount of MnO_2 increasing, the characteristic peaks of PANI became gradually weak, but the characteristic peaks of MnO_2 became visible by degrees, also indicating that the PANI in the samples was mostly amorphous. Moreover, comparing to sample 3#, sample 2# revealed again the characteristic peaks of PANI, which included the characteristic broad peaks of MnO_2 at 22.31°, and the diffraction intensity became stronger. It was obvious that the MnO_2 had a great influence on the XRD patterns of the MnO_2/PANI. The reason should be that MnO_2 made PANI form the high-oxidation-state structure for great activity of γ-MnO_2 when PANI deposited on the MnO_2 surface. Furthermore, MnO_2 decreased the doped degree of PANI by protonic acid, which impeded the crystallization of PANI and destroyed the intrinsic crystallization of MnO_2. But through the posttreatment of the aniline-HCl solution (sample 2#), the high-oxidation-state structure on the molecular chain of PANI was reduced and the doping of protonic acid was replenished, making the peaks of PANI stronger.

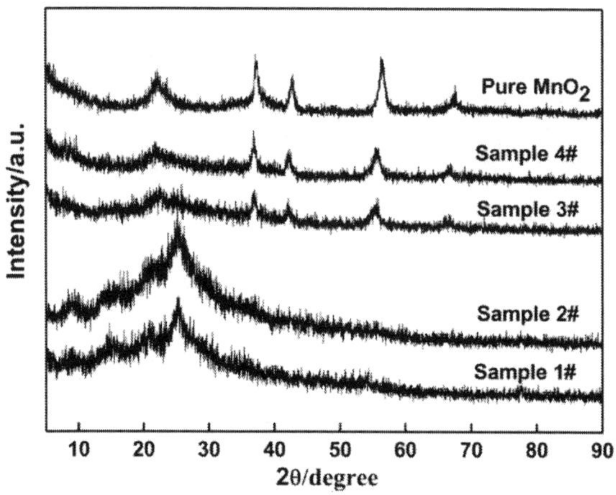

Figure 4.12 X-ray diffraction patterns for MnO_2/PANI composites and MnO_2 powders.

4.2.3 Morphologies

The TEM image of MnO_2/PANI-2# is shown in Fig. 4.13. It can be seen from Fig. 4.13a that the MnO_2/PANI composite particles are spherical in pattern with the grain size 50–70 nm, and MnO_2 particles were successfully decorated with doped PANI. As can be seen from Fig. 4.13b, the distribution of the composite particles was flocculent. It is the reason that the polarity of PANI on the surface of the MnO_2 particles leads to the intermolecular attraction and particle agglomeration.

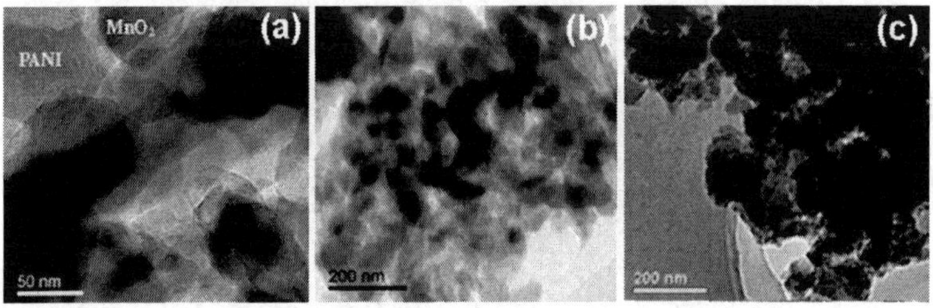

Figure 4.13 TEM images of MnO_2/PANI (a, b) and PANI (c).

4.2.4 Electrical Properties

The volume resistivities of MnO_2/PANI composites can be tested using the test method according to Ref. [3]. The tested and calculated values are shown in Table 4.3. MnO_2 has poor electrical conductivity. However, when MnO_2 was decorated with doped PANI, the conductivity of the MnO_2/PANI composites was greatly improved. With the increase of MnO_2 content in the reaction, the conductivity of the MnO_2/PANI decreased from 4.46×10^{-4} to 3.18×10^{-6} S·cm^{-1}. The conductivity of sample 2# increased about 2 orders of magnitude to 9.78×10^{-4} S·cm^{-1} compared with sample 3# and MnO_2. Variations in conductivity may be explained by analyzing the structures of the composites. The ideal emeraldine structure formed in the composite system and PANI doped with HCl possessed higher conductivities, as seen in samples 1# and 2#. A decline in the conductivities of samples 3# and 4# may be attributed to the large amount of MnO_2 available, which may have resulted in the formation of high-oxidation and amorphous structures of PANI on the surface of MnO_2. Furthermore, there is an increase in the number of molecular chain faults of PANI.

Posttreatment of sample 2# with the aniline-HCl solution, the chain faults decreased, improving the molecular structure and increasing the conductivity of the MnO_2/PANI composite.

Table 4.3 Electric conductivity of MnO_2/PANI composites

Sample	1#	2#	3#	4#	MnO_2
m(PANI): m(MnO_2)	1:1	1:2*	1:2	1:3	–
Volume resistivity $\rho_v/\Omega\cdot$cm	2.24×10^3	1.02×10^3	2.81×10^5	3.14×10^5	4.04×10^5
Conductivity $\sigma/S\cdot cm^{-1}$	4.46×10^{-4}	9.78×10^{-4}	3.56×10^{-6}	3.18×10^{-6}	2.47×10^{-6}

*Sample 2# is again dipped in the same aniline-HCl solution for 12 h, washed, and dried in the same condition.

4.2.5 Electromagnetic Parameters

Figure 4.14 is the complex permittivity and permeability versus frequency for the prepared MnO_2/PANI-2# powder–paraffin wax composites with loading levels of 40%. An increase in the real, ε', and imaginary, ε'', parts of the complex permittivity was found after the MnO_2 was decorated with doped PANI, as shown in Fig. 4.14a,b. The increase of the real part, ε', can be attributed to the improved electrical conductivity of the doped PANI and the enhanced polarizability of the composite particles in the electromagnetic field. According to the formula $\varepsilon'' = \sigma/\omega\varepsilon_0$, it is evident that the imaginary part, ε'', of the composite particles increases with the increasing electric conductivity. Moreover, the ε' and ε'' values of the MnO_2/PANI–paraffin wax composites decreased when the frequency was increased, showing good frequency dispersion effects. This feature is only observed in wave absorbing materials with wide absorption bands. Figure 4.14c,d shows that ε' and ε'' of the MnO_2/PANI samples were approximately 1 and 0, respectively, indicating that the magnetism of the composite particles is very poor. The dielectric loss tangents $\tan\delta_\varepsilon$ of the MnO_2/PANI composites were improved compared to pure MnO_2, as seen in the curves in Fig. 4.15. These results reveal that MnO_2/PANI composites are good dielectric loss materials.

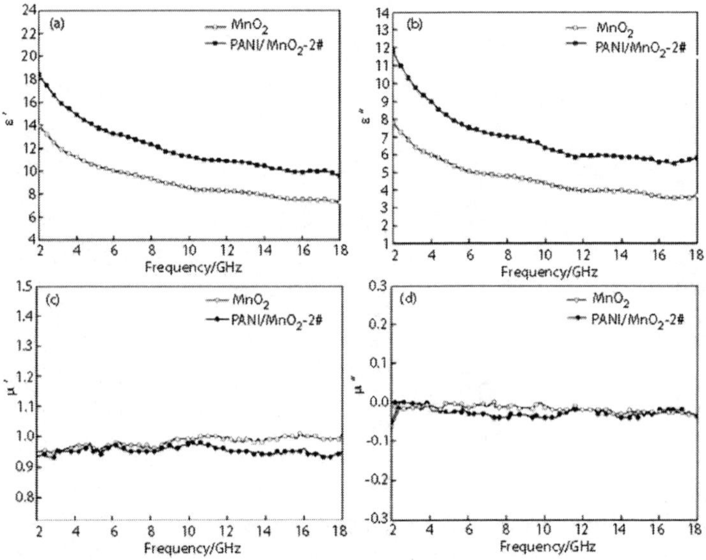

Figure 4.14 Dependence of the microwave frequency on the (a, b) dielectric permittivity and (c, d) permeability of PANI/MnO_2 and MnO_2.

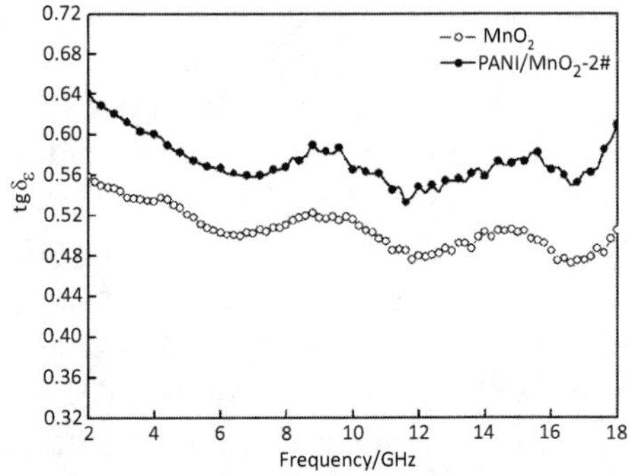

Figure 4.15 Comparison of the loss tangents tan δ_ε of MnO_2/PANI and MnO_2.

4.2.6 Microwave Absorbing Properties

To further prove the dependence of their microwave absorption properties on the permittivity, the RLs (dB) of the MnO_2/PANI–

paraffin wax composites were calculated according to transmission line theory[32] at a loading level of 40% MnO_2/PANI and a thickness of 2 mm. As is shown in Fig. 4.16, the as-prepared MnO_2/PANI composites had excellent microwave absorption capabilities. An effective absorption band under −10 dB was obtained at 5 GHz, and the maximum RL was −21 dB at 13.56 GHz. In comparison, the effective absorption band under −10 dB was 3 GHz and the maximum RL was only −14.20 dB at 11.5 GHz for pure MnO_2. As an inorganic semiconducting material, the MnO_2 in the MnO_2/PANI composite destroys the conductive network of PANI to a certain extent, thus lowering the conductivity. As such, the input impedance Z_{in} of the composite increased and became close to the impedance of the free-space Z_0. The electromagnetic wave could then be transmitted to the composite interiors adequately, a necessary condition for the loss of incident electromagnetic waves. After MnO_2 was decorated with doped PANI, the electrons and electron charges on the molecular chain aggregated on the surface of MnO_2 particles. Under the alternating electromagnetic field, interfacial and space charge polarization were easily formed on the interface of the MnO_2 and PANI. The RL may be attributed to the resistance loss of local conductive particles, the dielectric loss of MnO_2, and polarization relaxation losses of the interface. Under these various loss mechanisms, the microwave absorbing properties of the MnO_2/PANI composites were better than those of pure MnO_2, thus showing the good potential of MnO_2/PANI composites in absorber applications.

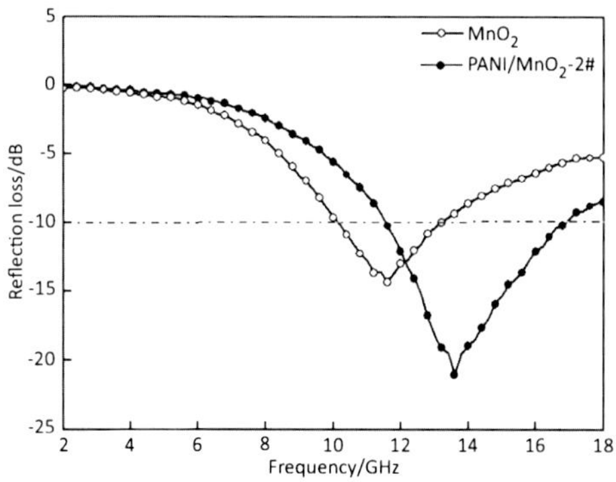

Figure 4.16 Comparison of reflection losses between MnO_2/PANI and MnO_2.

4.3 Theoretical Investigation of Polyaniline

The excited-state hydrogen-bonding dynamics of camphorsulfonic acid–doped PANI and the protonation mechanism are introduced, and the polarizable continuum model (PCM) with time-dependent density functional theory (TDDFT)/density functional theory (DFT) is used. The geometry optimizations were carried out using B3LYP and employing the 6-31G+(d, p) basis set and all the simulations were performed with the development version of the Gaussian09 software.

4.3.1 Excited-State Hydrogen-Bonding Dynamics of Camphorsulfonic Acid–Doped Polyaniline

The Ph_2NH (DPA) and (4-iminocyclohexa-2, 5-dienylidene) phenylamine cation (ICPA) can be used to model the distinct amine and iminium nitrogen cation environments in ES respectively, while CSA is modeled by dimethylsulfoxide (DMSO), and the H-bond can be formed between the S=O group of DMSO and the H–N groups of DPA as well as ICPA monomers, which are denoted as DPA-DMSO and ICPA-DMSO hydrogen-bonded complexes respectively[39]. The chemical structures of the monomers as well as hydrogen-bonded complexes are shown in Fig. 4.17a.

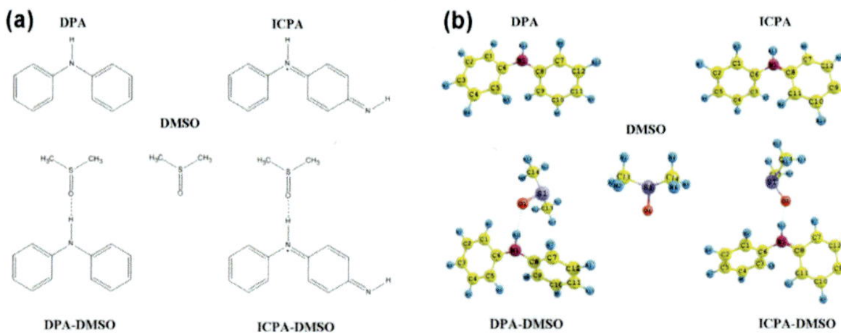

Figure 4.17 Chemical structures (a) and geometric structures (b) of monomers as well as hydrogen-bonded complexes. The dotted lines represent intermolecular hydrogen bonds.

The equilibrium geometries of the hydrogen-bonded DPA-DMSO and ICPA-DMSO complexes as well as the involved monomers in ground state are shown in Fig. 4.17b, and the important structure parameters are collected in Table 4.4. The lengths of the H-bond

S1=O1···H11-N1 in DPA-DMSO and ICPA-DMSO complexes are 1.9680 Å and 1.6482 Å, respectively, and the shorter H-bond in ICPA-DMSO complex will show relative stronger hydrogen bond interaction. There is an obvious phenomenon that bond lengths of H11-N1 and S1-O1 in the complexes are all elongated simultaneously, and the stronger hydrogen bond interaction causes much more obvious stretch. In other words, when the atoms are involved in the formation of the hydrogen bond the bands connected by these atoms will be elongated.

Table 4.4 Bond lengths (Å) and bond angles (°) for the isolated monomers and the hydrogen-bonded complexes in their ground states

	DPA	ICPA	DMSO	DPA-DMSO	ICPA-DMSO
C6-N1	1.4004	1.4305	-	1.3964	1.4264
C8-N1	1.4004	1.3293	-	1.3959	1.3220
H11-N1	1.0092	1.0175	-	1.0189	1.0532
O1-H11	-	-	-	1.9680	1.6482
C13-S1	-	-	1.8363	1.8317	1.8214
C14-S1	-	-	1.8363	1.8304	1.8222
S1-O1	-	-	1.5190	1.5297	1.5499
C13-S1-C14	-	-	96.4773	97.2521	98.5812
C6-N1-C8	129.6500	129.7435	-	129.1043	128.1074

The electronic excitation energies and oscillation strengths of the DPA-DMSO, ICPA-DMSO hydrogen-bonded complexes as well as the isolated DPA, ICPA monomers are predicted using the TDDFT method and collated in Table 4.5.

The S2 state of the hydrogen-bonded DPA-DMSO complex possesses the largest oscillator strength. As a result, the DPA-DMSO complex will be photoexcited to the S2 state directly. It is obvious that the S1–S8 states electronic excitation energies of the DPA-DMSO complex are lower compared with the corresponding states of the DPA monomer. We can draw the conclusion that the intermolecular hydrogen bonding S1=O1···H11–N1 is strengthened for the S1–S8 states of the DPA-DMSO complex and can induce a red shift. In the ICPA-DMSO complex, all the electronic excitation energies of the calculated excited states except for the S1–S3 states are under those of the ICPA monomer; therefore, the hydrogen bond S1=O1···H11–N1 in the ICPA-DMSO complex is strengthened for the S4–S8 states and will bring a red shift. On the contrary, the electronic spectra of the S1–S3 states are shifted toward blue due to the weakened hydrogen bond. Furthermore, the ICPA-DMSO hydrogen complex maybe primarily excited to the S3 state on account of its largest oscillator strength.

Table 4.5 Electronic transition energies (eV) and corresponding oscillation strengths (in parentheses) of the isolated monomers and hydrogen-bonded complexes for the low-lying electronically excited states (R: red shift; B: blue shift)

	DPA	ICPA	DPA-DMSO	ICPA-DMSO
S1	4.1873(0.1055) H->L+1(52%)	2.5555(0.0013) H-1->L(70%)	4.0217(0.0171)R H->L+2(46%)	2.6974(0.0348)B H-1->L(57%)
S2	4.2395(0.2917) H->L (51%)	2.6985(0.3109) H->L (52%)	4.1455(0.2684)R H->L (56%)	2.8581(0.0053)B H-2->L (68%)
S3	4.3224(0.0210) H->L+2(52%)	2.8636(0.0321) H-2->L(52%)	4.2196(0.1258)R H->L+1(41%)	2.9004(0.2292)B H->L (50%)
S4	4.5982(0.0964) H->L+3(68%)	3.6241(0.0099) H-4->L (52%)	4.2838(0.0211)R H->L+3(57%)	3.0435(0.0713)R H-3->L (61%)
S5	5.0717(0.0399) H->L+5(70%)	4.3498(0.4518) H-3->L (54%)	4.3015(0.0270)R H->L+3(59%)	3.6249(0.0084)R H-4->L (58%)
S6	5.0778(0.0003) H->L+4(66%)	5.0512(0.0125) H->L+1(52%)	4.5006(0.0507)R H->L+4(69%)	3.6881(0.0055)R H-6->L (45%)
S7	5.2354(0.0064) H->L+6(63%)	5.0821(0.0032) H-5->L (52%)	4.7581(0.0292)R H->L+6(64%)	4.3890(0.4694)R H-6->L (46%)
S8	5.5357(0.0097) H->L+7(69%)	5.2226(0.0043) H-6->L+1(52%)	4.7828(0.0004)R H->L+5(62%)	5.1311(0.0013)R H-8->L (48%)

The infrared spectra and vibration model of the isolated DPA, ICPA monomers as well as the hydrogen-bonded DPA-DMSO, ICPA-DMSO complexes are shown in Fig. 4.18. From Fig. 4.18a, it's easy to see that the presence of the N–H absorption peak at 3635 cm^{-1} in the DPA monomer shifts to 3475 cm^{-1} in the DPA-DMSO complex, and the intensity of the vibration is enhanced in the complex. From Fig. 4.18b, the N–H absorption peak of the ICPA-DMSO complex is also enhanced and shifts to 2895 cm^{-1} from 3542 cm^{-1} compared to the

ICPA monomer. The vibration model of the isolated monomers and hydrogen-bonded complexes are simulated, as shown in Fig. 4.18c,d.

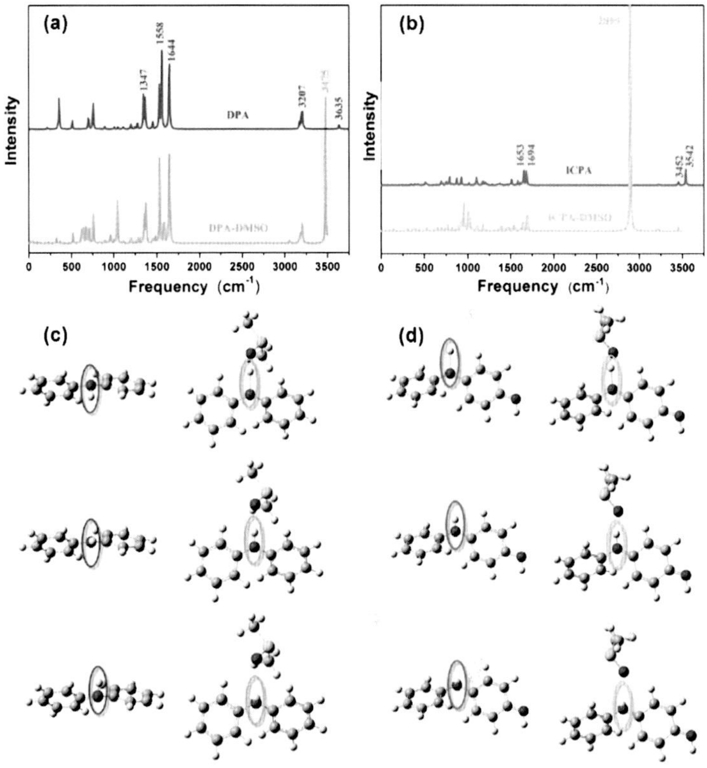

Figure 4.18 Infrared spectra and vibration model of the isolated DPA and ICPA monomers and the hydrogen-bonded DPA-DMSO and ICPA-DMSO complexes. (a) Infrared spectra of DPA and DPA-DMSO. (b) infrared spectra of ICPA and ICPA-DMSO, (c) vibration model of DPA and DPA-DMSO, and (d) vibration model of ICPA and ICPA-DMSO.

The analysis of molecular orbitals (MOs) will contribute to a further understanding of the nature of excited states. As mentioned before, the discussion has been focused on the S2 state of the hydrogen-bonded DPA-DMSO complex and the S3 state of the ICPA-DMSO hydrogen complex. The two states of the DPA-DMSO and ICPA-DMSO complexes are associated with the highest occupied molecular orbital (HOMO) and lowest unoccupied molecular orbital (LUMO) orbitals. The frontier MOs of the DPA-DMSO and ICPA-DMSO complexes as well as the DPA and ICPA monomers are shown in Fig. 4.19. It is obvious that in the MOs of ICPA-DMSO, the electron density is focused on the benzenoid ring in the HOMO orbital, whereas the quinoid ring in the LUMO orbital possesses higher electron density, and the electron density is transferred from the benzenoid ring to the quinoid ring during the HOMO to LUMO orbital transition.

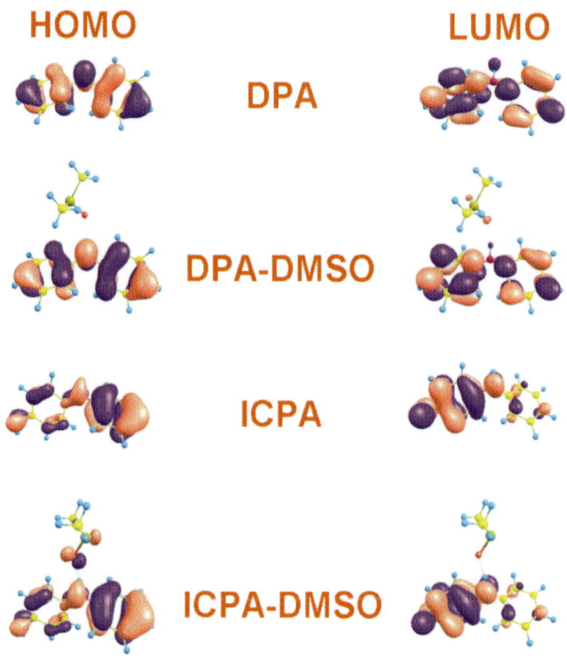

Figure 4.19 Frontier molecular orbitals (MOs) of the hydrogen-bonded DPA-DMSO and ICPA-DMSO complexes. The MOs of the isolated DPA and ICPA are also shown for comparison.

4.3.2 Theoretical Investigation of the Protonation Mechanism

The models used for the description of the doping progress are referenced from the mature model of Stafström et al.[19] Meanwhile, the different spin states are also taken into account, and each configuration is described in its singlet and triplet state when the stability is investigated. The inorganic HCl is utilized as the doping agent. The systems are guaranteed neutral equilibrium: the positive charge due to protonation is balanced by chloride counterions. Such models are appropriate for estimating the reason for the notable increase in electrical conductivity through doping[40]. The models are shown in Fig. 4.20.

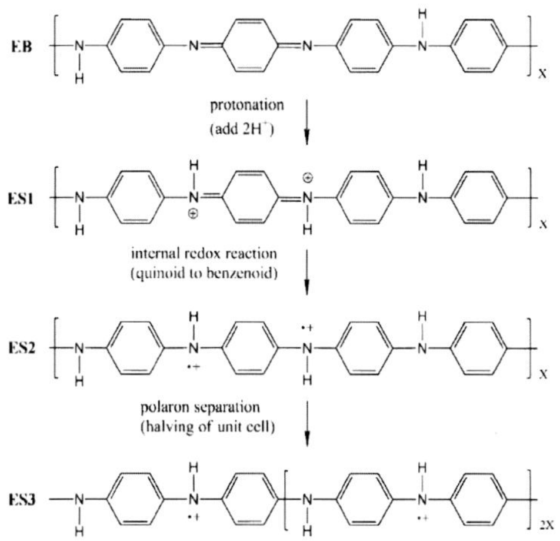

Figure 4.20 Protonation mechanism of PANI. EB, before protonation; ES1, formation of bipolarons; ES2 and ES3, separation of polarons, which results in a polaron lattice.

The equilibrium geometries of the structures involved in the mechanism are predicted using B3LYP. Figure 4.21 shows the equilibrium geometries. The bond length alternation (BLA), valence angles, and torsion angles are analyzed as general descriptions of equilibrium geometries. Figure 4.22 presents BLA data for the tetramer of the geometries. The results visualize ring4 (phenyl terminus), display the lowest BLA in all cases, partly as an expression of the boundary effect, which shows that ring4 remains practically unaffected by protonation and the polarons do not delocalize efficiently on the terminal segments. In the case of the EB, ring2 exhibits a quinoid-like with the outstandingly stronger bond alternation while the other three aromatic rings are benzenoid-like with lower BLA values. The proton doping of the EB reveals the tendency to reduce the quinoid character of the protonated ring (ring2) with decreased BLA. The BLA of the polaron lattice ES3 shows that proton interaction becomes much stronger with a more formidable loss of quinonoid character in the protonated ring2. From what has been discussed earlier, it would be reasonable to draw the conclusion that doping of the EB shows a tendency toward bipolaron delocalization, which is in accordance with the experimental findings[41]. During the doping progress, the H^+ and Cl^- anions emerge from protonic acid decomposition and blend in with the backbone, and the charged ions interact with nitrogen atoms of amine and imine, forming a polaron and bipolaron, which infuse to the delocalized π-bond of the entire chain. Consequently, PANI displays the higher electrical conductivity.

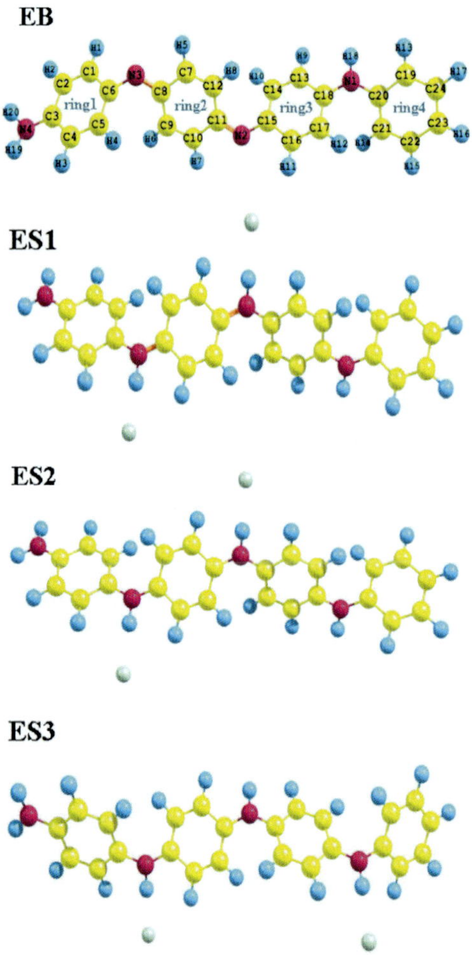

Figure 4.21 Equilibrium geometries of emeraldine base (EB) and emeraldine salts: ES1 (bipolaron), ES2 (polarons), and ES3 (polaron).

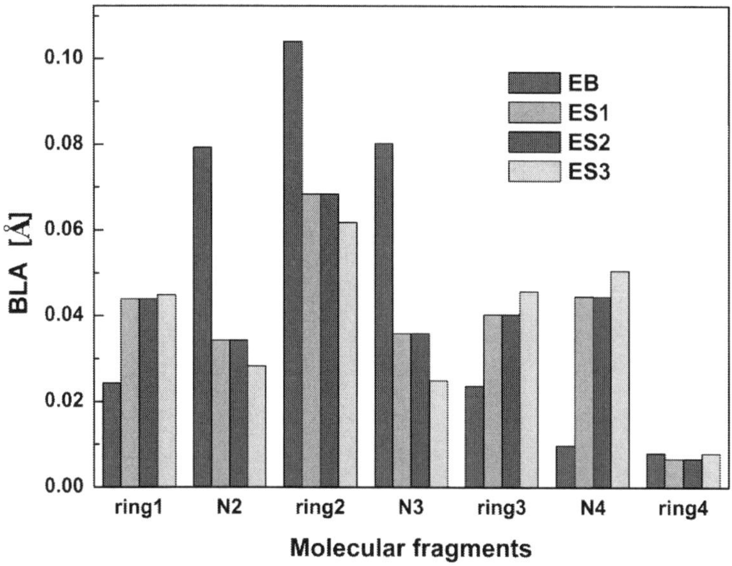

Figure 4.22 Bond length alternation (BLA) values of emeraldine base (EB) and emeraldine salts (ES1, ES2, and ES3) fragments in the tetramer.

The charge distribution is described by means of natural bond orbital (NBO) group charge, which is obtained by summation of the atomic charges of all atoms in a fragment. The charge distribution in the tetramer salt also allows an estimate of the delocalization of the bipolaron. The NBO group charges for the EB and ESs are collected in Fig. 4.23. The charges of the chloride ions reach reasonable values (~0.900) as the absolute value. The repeating units in the bipolaronic structures extend four rings and the polaronic structure extends two rings, leading to an alternative higher and lower NBO charge distribution. So the transformation between bipolaronic and polaronic would bring a redistribution of the quinonoid ring charge to other rings in the bipolaronic repeating unit. In the ES3 polaron form, the positive charges of rings and the negative charges of nitrogens are well distributed along the molecular skeleton, and the equably distributed charges should be conducive to electrical conductivity.

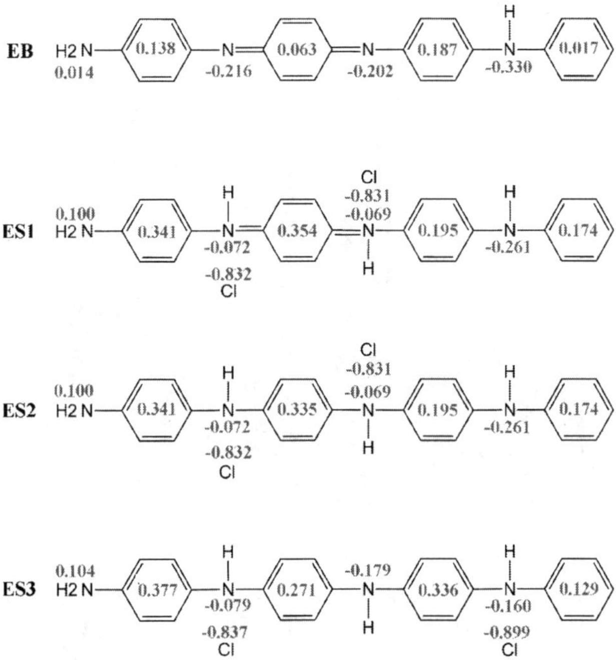

Figure 4.23 Natural bond orbital (NBO) group charge distribution of emeraldine and emeraldine hydrochloride.

The investigation of polarons' and bipolarons' relative stability in conjugated polymers has not reached an agreement. Different theoretical methods lead to different conclusions[42–44]. One of the most recent studies has been reported by de Oliveira and dos Santos[45], whose results were consistent with the preference of polarons in long chains and the coexistence of polarons and bipolarons in short oligomers. For further discussion on the existence form of ES, the energy relations among ESs (ES1, ES2, and ES3) both in singlets and triplets are described in Fig. 4.24. The ES1 tetramer of ESs in its bipolaron spin-less form is the most stable configuration. However, the energy difference of the same multiplicity is so small that spin configurations can coexist and easily convert into each other. Additionally, the triplet state shows much higher energy than the singlet state in all configurations. The bipolaronic lattice ES1 is the most stable structure and should be responsible for the notable increase in the electrical conductivity of doped PANI.

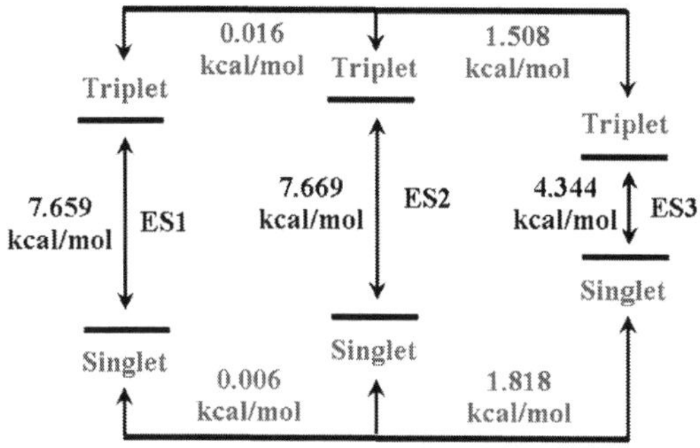

Figure 4.24 Energy of different emeraldine hydrochloride tetramer configurations.

References

1. Ameen S, Ali V, Zulfequar M, Haq MM, Husain M (2007). Electrical conductivity and dielectric properties of sulfamic acid doped polyaniline, *Curr Appl Phys*, **7**(2), 215–219.
2. Sudha JD, Sivakala S, Prasanth R, Reena VL, Nair PR (2009). Development of electromagnetic shielding materials from the conductive blends of polyaniline and polyaniline-clay nanocomposite-EVA: preparation and properties, *Compos Sci Technol*, **69**(3), 358–364.
3. Malinauskas A (2004). Self-doped polyanilines, *J Power Sources*, **126**(1), 214–220.
4. Kalasad MN, Gadyal MA, Hiremath RK, Mohamed Ikram I, Mulimani BG, Khazi IM, Anantha Krishnan SK, Rabinal MK (2008). Synthesis and characterization of polyaniline rubber composites, *Compos Sci Technol*, **68**(7), 1787–1793.
5. Yang G, Hou W, Feng X, Xu L, Liu Y, Wang G, Ding W (2007). Nanocomposites of polyaniline and a layered inorganic acid host: polymerization of aniline in the layers, conformation, and electrochemical studies, *Adv Func Mater*, **17**(3), 401–412.
6. Kumar SA, Singh AP, Saini P, Khatoon F, Dhawan SK (2012). Synthesis, charge transport studies, and microwave shielding behavior of nanocomposites of polyaniline with Ti-doped γ-Fe_2O_3, *J Mater Sci*, **47**(5), 2461–2471.
7. Harish C, Sreeharsha VS, Santhosh C, Ramachandran R, Saranya M, Vanchinathan TM, Govardhan K, Grace AN (2013). Synthesis of polyaniline/graphene nanocomposites and its optical, electrical and electrochemical properties, *Adv Sci Eng Med*, **5**(2), 140–148.

8. Yoo JE, Bucholz TL, Jung S, Loo YL (2008). Narrowing the size distribution of the polymer acid improves PANI conductivity, *J Mater Chem*, **18**(26), 3129–3135.

9. Gong C, Zhang J, Zhang X, Yu L, Zhang P, Wu Z, Zhang Z (2010). Strategy for ultrafine Ni fibers and investigation of the electromagnetic characteristics, *J Phys Chem C*, **114**(22), 10101–10107.

10. Manuel J, Kim JK, Matic A, Jacobsson P, Chauhan GS, Ha JK, Cho KK, Ahn JH (2012). Electrochemical properties of lithium polymer batteries with doped polyaniline as cathode material, *Mater Res Bull*, **47**(10), 2815–2818.

11. Duan YP, Liu SH, Guan HT (2006). Investigation of electromagnetic characteristics of polyaniline composite, *J Compos Mater*, **40**(12), 1093–1104.

12. Chen X, Wong CKY, Yuan CA , Zhang G (2012). Impact of the functional group on the working range of polyaniline as carbon dioxide sensors, *Sensor Actuat B*, **175**, 15–21.

13. Cho S, Hwang SH, Kim C, Jang J (2012). Polyaniline porous counter-electrodes for high performance dye-sensitized solar cells, *J Mater Chem*, **22**(24), 12164–12171.

14. Duan Y, Liu S, Guan H (2005). Orientation effects of high magnetic field on grain shape of doped polyaniline, *Sci Technol Adv Mater*, **6**(5), 513–518.

15. Wang ZZ, Bi H, Liu J, Sun T, Wu XL (2008). Magnetic and microwave absorbing properties of polyaniline/γ-Fe_2O_3 nanocomposite, *J Magn Magn Mater*, **320**(16), 2132–2139.

16. Wu KH, Ting TH, Wang GP, Yang CC, Tsai CW (2008). Synthesis and microwave electromagnetic characteristics of bamboo charcoal/polyaniline composites in 2–40 GHz, *Synth Met*, **158**(17), 688–694.

17. Luo X, Chung DDL (1999). Electromagnetic interference shielding using continuous carbon-fiber carbon-matrix and polymer-matrix composites, *Compos Part B*, **30**(3), 227–231.

18. Sariciftci NS, Heeger AJ, Cao Y (1994). Paramagnetic susceptibility of highly conducting polyaniline: Disordered metal with weak electron-electron interactions (Fermi glass), *Phys Rev B*, **49**(9), 5988–5992.

19. Stafström S, Bredas JL, Epstein AJ, Woo HS, Tanner DB, Huang WS, MacDiarmid AG (1987). Polaron lattice in highly conducting polyaniline: theoretical and optical studies, *Phys Rev Lett*, **59**(13), 1464–1467

20. Heeger AJ (2001). Semiconducting and metallic polymers: the fourth generation of polymeric materials, *J Phys Chem B*, **105**(36), 8475-8491.

21. Yuping D, Guangli W, Xiaogang L, Zhijiang J, Shunhua L, Weiping L (2010). On the correlation between structural characterization and electromagnetic properties of doped polyaniline, *Solid State Sci*, **12**(8), 1374–1381.

22. Liu P (2009). Synthesis and characterization of organo-soluble conducting polyaniline doped with oleic acid, *Synth Met*, **159**(1), 148-152.

23. Athawale AA, Kulkarni MV, Chabukswar VV (2002). Studies on chemically synthesized soluble acrylic acid doped polyaniline, *Mater Chem Phys*, **73**(1), 106–110.

24. Motheo AJ, Santos JR, Venancio EC, Mattoso LHC (1998). Influence of different types of acidic dopant on the electrodeposition and properties of polyaniline films, *Polymer*, **39**(26), 6977–6982.

25. Pouget JP, Jozefowicz ME, Epstein AJ, Tang X, Macdiarid AG (1991). X-ray structure of polyaniline, *Macromolecules*, **24**(3), 779–789.

26. Ghosh P, Chakrabarti A (2000). Conducting carbon black filled EPDM vulcanizates: assessment of dependence of physical and mechanical properties and conducting character on variation of filler loading, *Eur Polym J*, **36**(5), 1043–1054.

27. Prokeš J, Stejskal J (2004). Polyaniline prepared in the presence of various acids: 2. Thermal stability of conductivity, *Poly Degrad Stab*, **86**(1), 187–195.

28. Du JH, Sun C, Bai S, Su G, Ying Z, Cheng HM (2002). Microwave electromagnetic characteristics of a microcoiled carbon fibers/paraffin wax composite in Ku band, *J Mater Res*, **17**(5), 1232–1236.

29. Shi S, Zhang L, Li J (2009). Complex permittivity and electromagnetic interference shielding properties of Ti_3SiC_2/polyaniline composites, *J Mater Sci*, **44**(4), 945–948.

30. Tian S (ed) (2006). *Materials Physical Performance*, Beijin University of Aeronautics and Astronautics Press, Beijing.

31. Lisheng Z, Chuanxiang X (1996). Studying on temperature Cole-Cole diagrams by dielectric temperature spectrum, in *Electrical Insulation and Dielectric Phenomena*, paper presented at IEEE Annual Report of the Conference on Electrical Insulation, Millbrae, CA.

32. Lines ME, Glass AM (eds) (2001). *Principles and Applications of Ferroelectrics and Related Materials*, Oxford University Press, Oxford.

33. Guan H, Zhao Y, Liu S, Lv S (2006). Application of manganese dioxide to electromagnetic wave absorber: effective permittivity and absorbing property, *Eur Phys J Appl Phys*, **36**(03), 235–239.

34. Reddy RN, Reddy RG (2004). Synthesis and electrochemical characterization of amorphous MnO_2 electrochemical capacitor electrode material, *J Power Sources*, **132**(1), 315–320.

35. Chen Y, Liu C, Li F, Cheng H (2005). Preparation of single-crystal α-MnO_2 nanorods and nanoneedles from aqueous solution, *J Alloy Compd*, **397**(1), 282–285.

36. Yuping D, He M, Xiaogang L, Shunhua L, Zhijiang J (2010). The microwave electromagnetic characteristics of manganese dioxide with different crystallographic structures, *Physica B*, **405**(7), 1826–1831.

37. Guan H, Liu S, Zhao Y, Duan Y (2006). Electromagnetic characteristics of nanometer manganese dioxide composite materials, *J Electron Mater*, **35**(5), 892–896.

38. Jianjun H, Yuping D, Jia Z, Hui J, Shunhua L, Weiping L (2011). γ-MnO_2/polyaniline composites: preparation, characterization, and applications in microwave absorption, *Physica B*, **406**(10), 1950–1955.

39. Zhang Y, Duan Y, Wang T (2014). Excited-state hydrogen-bonding dynamics of camphorsulfonic acid doped polyaniline: a theoretical study, *Phys Chem Chem Phys*, **16**(47), 26261–26265.

40. Zhang Y, Xi Q, Chen J, Duan Y (2014). Theoretical investigation of the protonation mechanism of doped polyaniline, *J Clust Sci*, **25**(6), 1501–1510.

41. Shacklette LW, Wolf JF, Gould S, Baughman RH (1988). Structure and properties of polyaniline as modeled by single-crystal oligomers, *J Chem Phys*, **88**(6), 3955–3961.

42. Romanova J, Petrova J, Tadjer A, Gospodinova N (2010). Polyaniline-water interactions: a theoretical investigation with the polarisable continuum model, *Synth Met*, **160**(9), 1050–1054.

43. Petrova JN, Romanova JR, Madjarova GK, Ivanova AN, Tadjer, AV (2011). Fully doped oligomers of emeraldine salt: polaronic versus bipolaronic configuration, *J Phys Chem B*, **115**(14), 3765–3776.

44. Varela-Álvarez A, Sordo JA, Scuseria GE (2005). Doping of polyaniline by acid-base chemistry: density functional calculations with periodic boundary conditions, *J Am Chem Soc*, **127**(32), 11318–11327.

45. De Oliveira ZT, Dos Santos MC (2000). Relative stability of polarons and bipolarons in emeraldine oligomers: a quantum chemical study, *Solid State Commun*, **114**(1), 49–53.

Chapter 5

Other Absorbents

5.1 Electromagnetic Properties: Barium Titanate Powder

For significant absorption of electromagnetic (EM) waves, the microwave absorption materials (MAMs) should have electric and/or magnetic dipoles, which interact with the EM fields in the radiation. The complex permeability and permittivity of materials determine the reflection and attenuation characteristics of the EM wave absorbers[1,2]. A significant MAM should have suitable values of complex permeability and permittivity. Barium titanate ($BaTiO_3$ or BT) as one of the perovskite-type structure is widely used in multilayer ceramic capacitors (MLCCs), positive temperature coefficient (PTC), ferroelectric nonavolatile memory (FRAM), dynamic random access memory (DRAM), transducers, thermistors, and electro-optic devices as the dielectric, due to its high dielectric constant, good ferroelectric properties, and nonlinear optical properties[3–7]. The interaction of the temperature, phase transitions, grain size, and dielectric properties of BT and its compound must be concerned. It is well known that BT is a kind of dielectric material. The relaxation presence of BT occurred in the gigahertz (GHz) frequency, which can be characterized by a decrease in the dielectric constant (K) and a peak in the dielectric loss ($\tan\delta$) with increasing frequency. One may speculate whether BT can serve as a lossy filler[8–10].

Microwave Absorbing Materials
Yuping Duan and Hongtao Guan

ISBN 978-981-4745-10-9 (Hardcover), 978-981-4745-11-6 (eBook)
www.panstanford.com

The shape of the microwave absorbent plays an important role on the microwave absorption properties. Moreover, numerical calculations on EM constants of carboxyl iron powder and other materials have been carried out, and it indicates that the absorbing ability of microwave absorbent with a needle or flake shape is higher than that of microwave absorbent with sphere shape. Barium titanate is prepared by a modified sol-gel method.

The acetate derived barium and titanium-containing sol was prepared in the following route: Solution A was prepared by dissolving barium carbonate in the mixed solution of acetic. Solution B was formed by mixing $(C_4H_9O)_4Ti$ and ethanol, and then was added to Solution A. The light yellowish transparent sol was obtained. The sol was smeared on the glass slabs to make xerogel. Annealing of the flake-shaped barium (FBT) powders was done at different temperatures, respectively, and then FBT with luster was obtained[11].

5.1.1 Thermal Analysis (DTA and TG)

Figure 5.1 shows the DTA-TG curves of FBT xerogel. The DTA revealed clear exotherms at about 360°C and 650°C. The maximal weight loss (–21.06%) at approximately 210°C–390°C is attributed to oxidizing combustion of the ethyoxyl and acetate in the flake-shaped xerogel, which is confirmed by the corresponding clear exotherms at about 360°C in the DTA curve. A further weight loss (–11.85%) at approximately 620°C–670°C is due to the formation of $BaTiO_3$, corresponding to another clear exothermic peak (650°C) in the DTA curve.

5.1.2 Microstructure and Morphology Analysis

Figure 5.2 shows the XRD patterns of FBT annealing at different temperatures from 600°C to 1200°C for 3 h. It can be seen from Fig. 5.2 that no new phase is created when the temperature is 600°C. It is indicated that the products annealing at 600°C are amorphous. When the temperature increases to 700°C, the $BaTiO_3$ phase appears. This is consistent with the weight loss in the TG curve. During the annealing temperature from 700°C to 900°C, the intensity of $BaTiO_3$ increases with increasing temperature and there is almost no other phase in the powder except a little $BaCO_3$, which is nearly not detected by XRD. When the temperature increases to 1000°C, a new phase Ba_2TiO_4 is detected and as the heat treatment temperature increases, the intensity of Ba_2TiO_4 increases, too. Beauger et al.[12,13] proposed that Ba_2TiO_4 were formed as products of the following reactions:

$$BaTiO_3 + BaCO_3 \rightarrow Ba_2TiO_4 + CO_2\uparrow \quad (5.1)$$

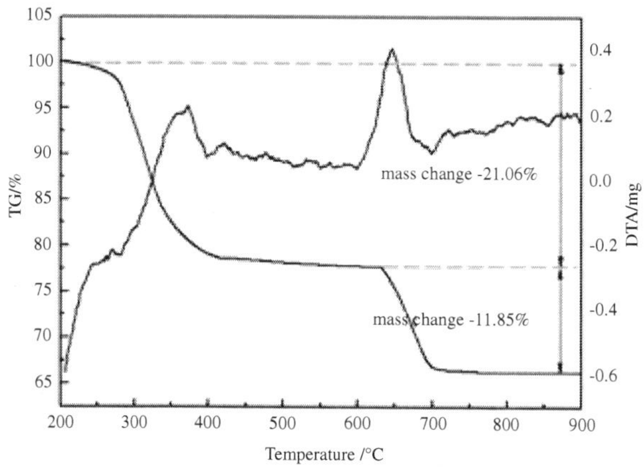

Figure 5.1 DTA and TG curve of a flake-shaped precursor.

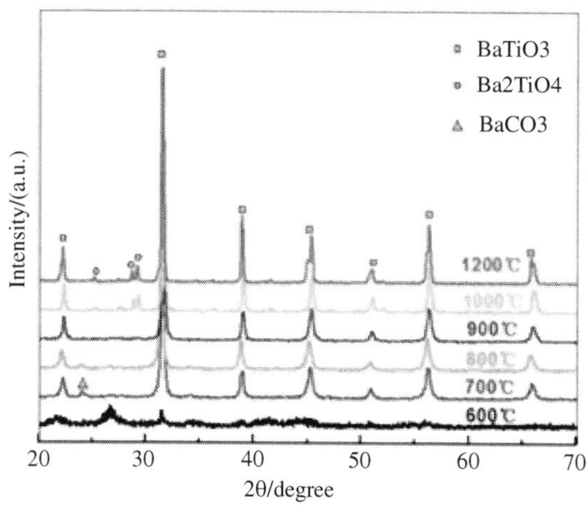

Figure 5.2 XRD patterns of FBT annealing at different temperatures from 600°C to 1200°C.

They found that after the formation of a barium titanate layer, Ba_2TiO_4 is formed at the interface of $BaCO_3$–$BaTiO_3$. The formation of Ba_2TiO_4 continues until $BaCO_3$ is consumed completely.

To help understand the evolution of the powder in terms of the $BaTiO_3$ phase, precise XRD (Fig. 5.3) observations were performed. When the temperature is 700°C, cubic $BaTiO_3$ (PDF No.75-0211) is detected. As the annealing temperature increases to 900°C, diffraction peak at $2\theta \approx 45°$ slightly widens, as is shown in Fig. 5.3, which can be indicative of the split of the peak. This implies that a small quantity of tetragonal $BaTiO_3$ is formed in the sample annealing at 900°C. However, it is still difficult to distinguish between (002) and (200) due to their overlap[14]. A similar XRD pattern is observed at the annealing temperature 1000°C. When the temperature increases to 1200°C, the split of diffraction peaks is much more easily discriminated, which means the tetragonal structure $BaTiO_3$ becomes the main component at this annealing temperature.

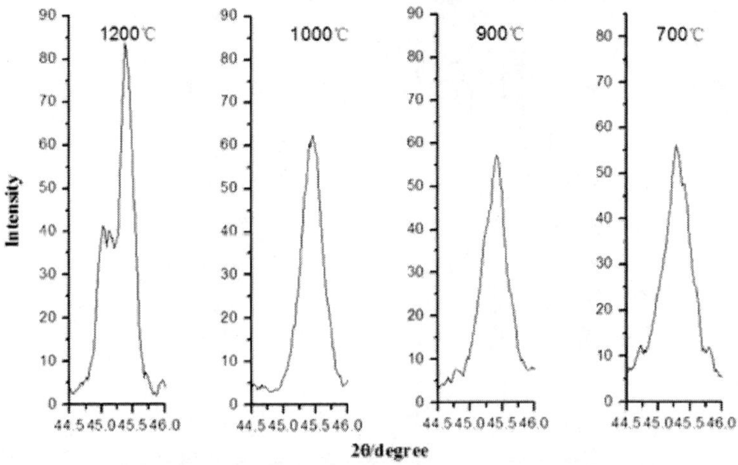

Figure 5.3 Precise XRD patterns annealing at temperatures from 700°C to 1200°C ($2\theta \approx 45°$).

Figure 5.4 shows micrographs of FBT particles obtained by the modified sol-gel method. These 80°C xerogel particles (Fig. 5.4a) and particles annealing at different temperatures from 700°C to 1200°C (Fig. 5.4b–f) all had the flake shape, indicating that the flake shape was preserved during the reaction in the annealing process. It can be seen in Fig. 5.4a that these particles before annealing are fleshly, which is attributed to organic substances existing in the xerogel. These particles in Fig. 5.4b–f after annealing look dry compared with 80°C xerogel particles, which is due to the oxidizing combustion of ethyoxyl, acetate, and other compounds in the flake-shaped xerogel. According to the figure, the average particle size of the FBT powder is about 40–50 um and the thickness is about 3–5 um.

Figure 5.4 SEM images of FBT powder annealing at different temperatures: (a) 80°C xerogel, (b) 700°C, (c) 800°C, (d) 900°C, (e) 1000°C, and (f) 1200°C.

5.1.3 Microwave Electromagnetic Properties of FBT

Figure 5.5a shows the permittivity for FBT annealing at 700°C. The real part $\varepsilon'$of the permittivity remains almost invariable with frequency until 16.1 GHz, while the imaginary part ε'' is close to zero. It is demonstrated that cubic $BaTiO_3$ hardly has any dielectric loss. It can be seen in Fig. 5.5b–d that the values of $\varepsilon'$and ε'' for FBT annealing at different temperatures (800, 900, and 1000°C) all exhibit a clear wave between 8 and 18 GHz, and the width of wave becomes broader with the increasing temperature, which is attributed to the increase in tetragonal $BaTiO_3$. It is indicated that a small quantity of tetragonal $BaTiO_3$ in the cubic $BaTiO_3$ only can be polarized on the high frequency, and the more the content of tetragonal $BaTiO_3$, the higher the permittivity. These results can be well explained by the theory. Cubic $BaTiO_3$ shows paraelectric properties, but tetragonal $BaTiO_3$ shows ferroelectric properties; and the high dielectric constant and ferroelectric properties of $BaTiO_3$ come from the tetragonal structure[15,16]. In addition, it can

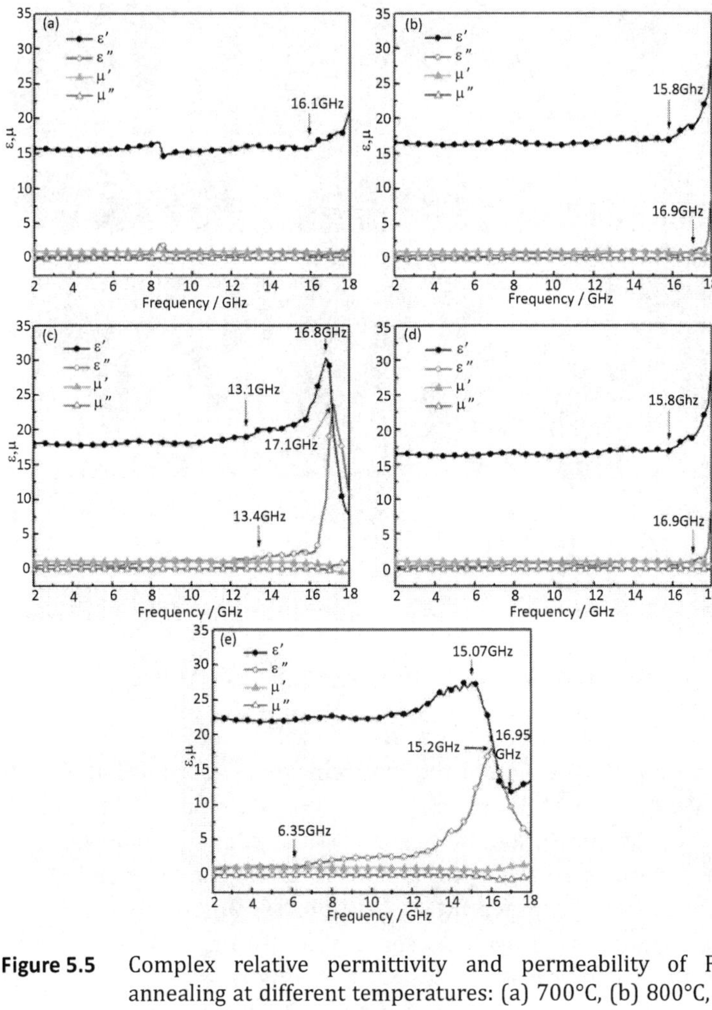

Figure 5.5 Complex relative permittivity and permeability of FBT annealing at different temperatures: (a) 700°C, (b) 800°C, (c) 900°C, (d) 1000°C, and (e) 1200°C.

be found that the values of ε' always increase before the values of ε'', and the maximum of ε'' corresponds to the degressive region of ε'. It is because of the high frequency, first, the polarizability and electric displacement can keep up with the variation in the EM field, then the value of ε'comes to the maximum, and subsequently on one frequency, orientation polarization cannot keep up with the variational EM field, so the value of ε' decreases with increasing frequency and the value of ε''comes to the maximum. Figure 5.5e shows the permittivity for FBT annealing at 1200°C. The maximums of $\varepsilon'$and ε''all decrease compared to that of FBT annealing at 1000°C, which is due to the increasing Ba_2TiO_4. Figure 5.5a–e also shows the permeability for FBT annealing at different temperatures. It can be seen that the real part $\mu'$and the imaginary part μ'' remain almost invariant with frequency. These results indicate that BT hardly has any magnetic loss.

Figure 5.6 shows the dielectric loss tangent tan δ of FBT on the frequency 2–18GHz. According the values of dielectric loss factor (tan $\delta = \varepsilon''/\varepsilon'$), it can be known that FBT powders can cause high dielectric loss at high frequency. To further reveal the microwave absorption properties of the FBT, the reflection loss (RL) of the FBT/paraffin composite sample with 70wt.% FBT was calculated at a given frequency and sample thickness using the relative complex permeability and permittivity according to the transmission line theory.

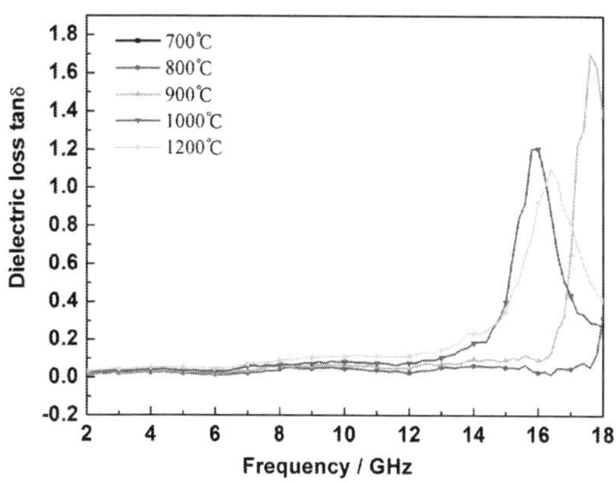

Figure 5.6 Dielectric loss factor of FBT–paraffin wax sample verse frequency.

Figure 5.7 shows a typical relationship between the RL and frequency for FBT (annealing at different temperatures from 700°C to 1200°C)/paraffin composite samples in the 2–18 GHz range with a thickness of 1 mm. It is observed that the width of the absorption band (RL < 10 dB) becomes broad and the maximal reflection loss becomes large with the increasing temperature until 1000°C. When the temperature increases to 1200°C, microwave absorption property of the FBT goes down due to increasing Ba_2TiO_4. It is worth noticing that FBT annealing at 1000°C exhibits the best microwave absorption property, and the maximum of RL reaches –13.6 dB at 17.6 GHz; while the absorption band (RL < –10 dB) is in the 16.5–18 GHz range. To investigate the relationship between the RL and sample thickness, FBT annealing at 1000°C, which exhibits the best microwave absorption property when the thickness of all samples is 1 mm, was picked as an absorbent to prepare the samples. The relationship between RL and frequency for samples with different thicknesses from 1 mm to 5 mm in the 2–18 GHz range was studied, as is shown in Fig. 5.8. The sample thickness has some effects on the EM RL of a wave absorbing material, and thickness can influence the absorbing peak values and bandwidth[17,18] of the material. When the sample thickness is beyond the matching thickness at a certain frequency band, the microwave reflection loss will not increase but drop gradually. So it can be seen from Fig. 5.8 that the maximal reflection loss reaches –29.6 dB at 12 GHz with a matching thickness of 4 mm, –16.45 dB at 9.58 GHz with a matching thickness of 5 mm, and –13.56 dB at 17.62 GHz with a matching thickness of 1 mm. In addition, it can be observed that when the thickness of the sample increases, the location of the absorbing peak is shifted toward a lower frequency.

The propagating wavelength in a material is expressed by Ref. [19]:

$$\lambda_m = \frac{\lambda_0}{(|\mu\varepsilon|)^{1/2}} \tag{5.2}$$

Where λ_0 is the free space wavelength. When the thickness of the sample is equal to an odd number multiple of the quarter wavelength, the incident and reflected waves in the material are out of phase by 180°, resulting in total cancellation of the reflected waves at the air–material interface. As the frequency increases, wavelength becomes short, so the location of the absorbing peak is shifted to a higher frequency for a smaller value of sample thickness (Fig. 5.8). The dielectric relaxation occurring in the gigahertz frequency region of BT will lead to dielectric loss[20–22]. The relaxation of BT came from the inertia of the boundary leaded and piezoelectric resonance and presence of ferroelectric or 90° domain wall in BT. FBT powders with large surfaces have large absorption and dispersion sections. Moreover, dipoles of big strength are formed in FBT powders. These dipoles have a long relaxation time, giving rise to dielectric loss and the microwave absorption property. In addition, the composite can be polarized repeatedly in a high-frequency electric field when the absorber filler content is high correspondingly, which causes the electric energy to be changed into other forms of energy, such as heat energy.

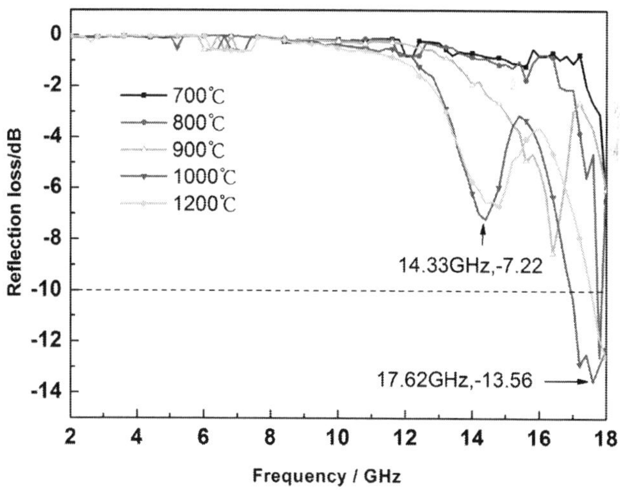

Figure 5.7 Frequency dependence of the RL of FBT–paraffin wax with a thickness of 1 mm.

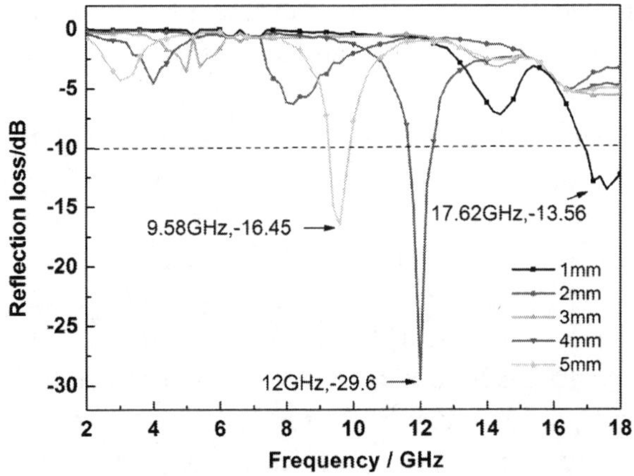

Figure 5.8 Frequency dependence of the RL of FBT (1000°C)/paraffin with different thickness.

5.2 Mn_4N Absorber

Magnetic nitride absorption materials made by dispersing magnetic fillers in an insulating matrix continue to play an important role in the investigation and application of MAMs[23–27]. Magnetic nitride as new types of EM wave absorption materials has attracted intensive interest on account of the following facts: First, the nitrides are harder, better wearing, have higher resistance, and are less sensitive toward corrosion than the parent metals, because N atoms can interstitially enter the lattice of magnetic 3D metal in a covalent or metallic state. Furthermore, the interstitial insertion can reduce its density efficiently. Moreover, due to the magnetovolume effect, some low nitrogen contents, such as $Fe_{16}N_2$, Fe_4N, and Fe_2Ni_2N, have higher saturation magnetization than parent metals[24,25,28,29].

Manganese nitrides, as one of the magnetic metal nitrides group, have attracted much attention because of their attractive physical properties, especially their magnetic properties. Considering the outstanding EM properties, it is, therefore, conceivable that excellent EM wave absorption property applications could be achieved in manganese nitrides.

The pure Mn powder and pure ammonia gas are used as the starting materials to synthesize the manganese nitride powder by solid–gas reaction in a tubular furnace.

The crystal structure of Mn_4N is sketched, as shown in Fig. 5.9. Mn_4N belongs to the perovskite structure, and the chemical formula can be written as $(Mn_{II})_3Mn_IN$, where Mn_I is the corner atom, Mn_{II} is the face-centered atoms, and N atom are located at body center. Figure 5.9 shows the X-ray diffraction patterns of the raw manganese and the as-milled powder. It can be noted that raw manganese exhibits a single α-Mn phase (JCPDS No.89-4085). The crystal structure of the as-milled powder is identified as Mn_4N with the corresponding (111), (200), (211), (220), (311), and (222) peaks. Trace peaks of MnO are detected, which cannot be eliminated though in the atmosphere of Ar[30]. There are no peaks of Mn, which demonstrate that Mn turns to Mn_4N completely. The lattice parameters for the Mn_4N phase are computed using the d-spacings' value and the respective (*hkl*) parameters. The lattice constant is a = 3.844 Å, which is in good agreement with those reported in JCPDS card number 89-4804.

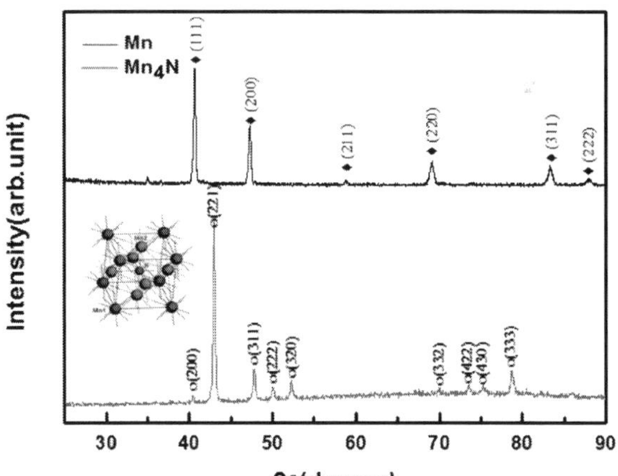

Figure 5.9 XRD patterns of Mn_4N (a) and raw manganese (b); the inset image is the crystal structure of Mn_4N.

Figure 5.10 shows SEM morphologies of the raw manganese and as-milled Mn_4N. The pure α-Mn powders are flaky with the corner angle, while the Mn_4N powders are flaky with a smooth surface. The particles are almost the same, both in microsize ranging from 5 to 50 μm, which can neglect the effect of size on EM properties[31,32].

Figure 5.10 SEM image of samples: (a) Mn_4N and (b) Mn.

5.2.1 Magnetic Properties

Figure 5.11 exhibits magnetization (*M*) versus the applied magnetic field (*H*) for Mn_4N. The saturation magnetization (*Ms*) and coercivity (*Hc*) are 11.7 emu/g and 85 Oe. So Mn_4N is a well-known soft ferrimagnetic material with low coercivity and moderate magnetization.

The magnetic hysteresis (*M–H*) loops of Mn are hardly measured because of its antiferromagnetism with poor magnetic properties.

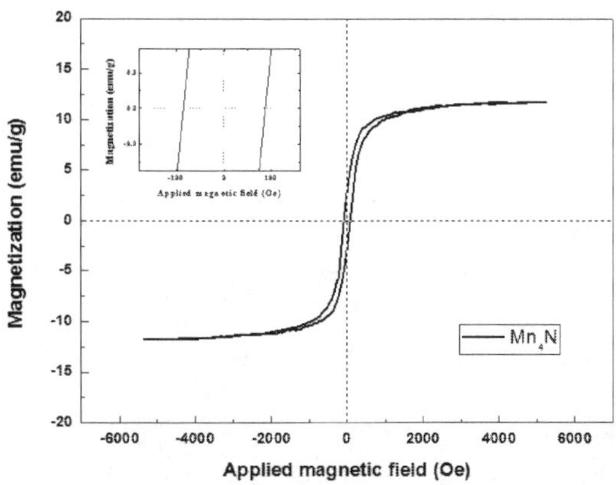

Figure 5.11 *M–H* hysteresis loop for Mn_4N at room temperature.

5.2.2 Microwave Electromagnetic Properties

5.2.2.1 Effect of temperature on microwave electromagnetic properties of Mn_4N

The prepared manganese nitride powders of series heating temperature (400°C–900°C) in steps of 100°C are prepared by the solid-gas reaction method giving various products denoted as A-400, A-500, A-600, A-700, A-800, and A-900, respectively[33].

Figure 5.12 shows microwave EM properties of samples A-400 to A-900 as a function of frequencies in the range of 2–18 GHz. From Fig. 5.12a, the real permittivity (ε') and the imaginary permittivity (ε'') of all samples are almost constant, but with notable fluctuations at 10–12 GHz.

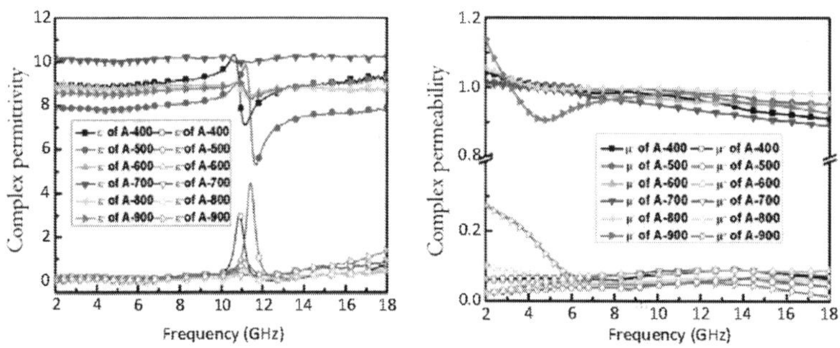

Figure 5.12 Complex relative permittivity (a) and complex relative permeability (b) of the samples A-400, A-500, A-600, A-700, A-800, and A-900 from 2 to 18 GHz.

The Debye dipolar relaxation expression is a suitable way to resolve the mechanism of dielectric loss and can be expressed by the Cole–Cole equation[34]. From the Cole–Cole equation, the diagram of ε'–ε'' should be a semicircle instead of an integral circle. Semicircles of the Cole–Cole graphical plots for A-400 to A-900 are recorded in Fig. 5.13, which demonstrate that the relaxation-type dielectric loss takes place at the microwave frequency range. The relaxation loss may be attributed to interfacial polarization resonance due to the electronegativity difference between conductor manganese nitrides and insulator paraffin and to the permanent electric dipoles resulting from defects in nitrogen doping manganese. However, odd circles also emerge from the Cole–Cole circles for all the samples, which come from the dielectric nonlinear resonances, drastic oscillations for the real permittivity ε', and a sharp peak for the imaginary permittivity ε'' at 10–12 GHz, especially for A-400 and A-500. At the same frequency range, similar resonances are found in permeability frequency spectrum. The analogous phenomenon is also found in TiO_2 nanoparticle metamaterial[35] and Ni nanoparticles[36]. In the alternate electric fields, the charged ions are driven to move with the external fields. When the wavelength of the electric field is equal to the displacement of charged ions, dielectric resonance occurs. Meanwhile, alternate interaction between electric fields and magnetic fields leads to a resonant magnetic permeability. After all, the resonance is beneficial for excellent microwave absorption properties of manganese nitrides.

According to Fig. 5.12b, we can see that μ' decreases with increase in frequency. It is noted that μ'' of A-900 shows a peak near 0.3 at 2–4 GHz and then decreases, while the others perform at constant lower than 0.1. Meanwhile, it is in good agreement with the static magnetic performance described earlier. It demonstrates that Mn_4N shows a better magnetic loss than other manganese nitrides. It is well known that the magnetic loss arises from micron magnetic particles at microwave frequency regions, mainly from the eddy current loss and natural resonance. Referring to free-electron theory[34], the imaginary permittivity ε'' is in direct proportion to conductivity σ. The low ε'' also indicates high electrical resistivity. Paraffin, an insulator, intensively decreases the conductivity of composites, meaning that the composites have high eddy current–limiting frequency. So the eddy current loss is related to thickness d and electric conductivity σ of the composite and can be described by the following:

$$\mu''(\mu')^{-2} f^{-1} = 2\pi\mu_0 d^2\sigma/3 \quad (5.3)$$

If magnetic loss merely results from eddy current loss, the $\mu''(\mu')^{-2} f^{-1}$ value should be constant, as the frequency changes. Figure 5.14 shows the value of $\mu''(\mu')^{-2} f^{-1}$ for the samples A-400 to A-900. As you can see, the values of A-400, A-800, and A-900 decrease sharply at a low frequency range and then remain nearly constant at a higher frequency, while the other samples are nearly constant at 2–18 GHz. Hence, you can deduce that the eddy current loss could contribute to magnetic loss at high frequencies. However, the decrease in 2–6 GHz for A-400, A-800, and A-900 may be caused by a natural resonance loss of Mn_4N.

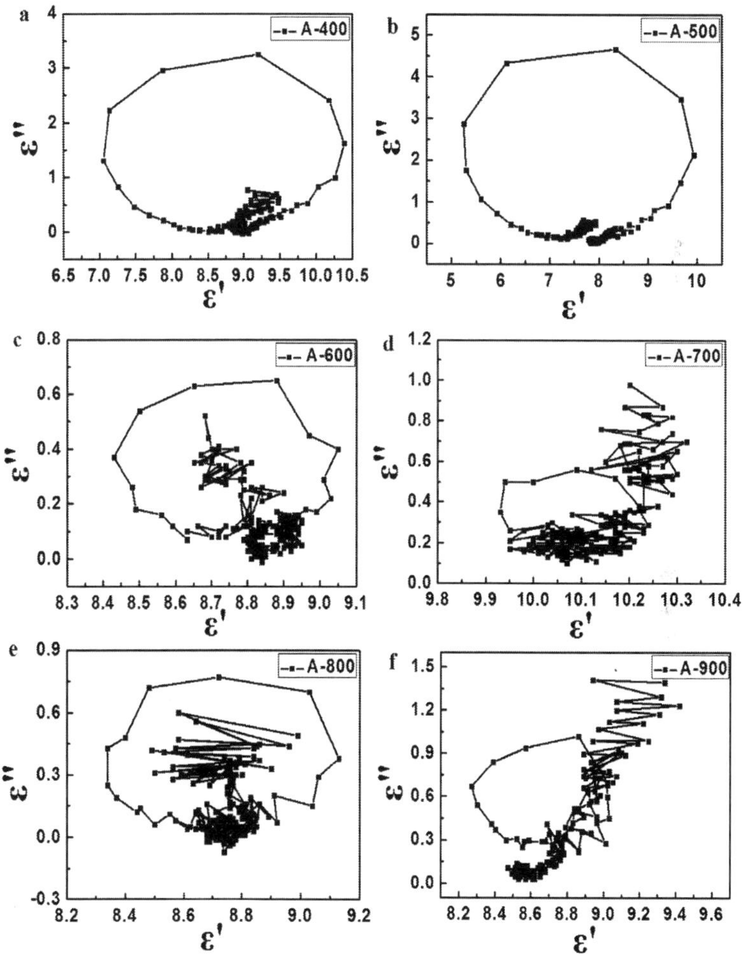

Figure 5.13 Typical ε'–ε'' curves of the composites A-400, A-500, A-600, A-700, A-800, and A-900.

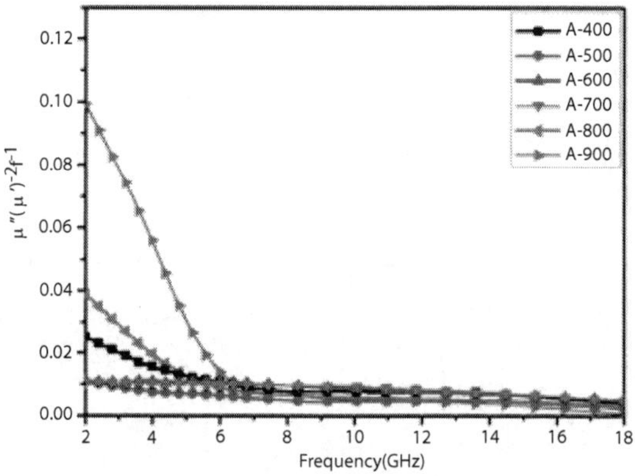

Figure 5.14 Value of $\mu''(\mu')^{-2} f^{-1}$ as a function of frequency for the samples A-400, A-500, A-600, A-700, A-800, and A-900.

Further, from the perspective of natural resonance loss, the resonance frequency depends on the crystalline anisotropy, magnetic particle geometry, particle size, demagnetizing field, and interaction of magnetic particles. For sphere magnetic particles, the complex permeability spectra could be fitting, using the Landau–Lifshitz–Gilbert (LLG) equation[37]:

$$\mu' = 1 + \frac{A\left[1-(1-\alpha^2)\left(\frac{f}{f_r}\right)^2\right]}{\left[1-(1+\alpha^2)\left(\frac{f}{f_r}\right)^2\right]^2 + 4\alpha^2\left(\frac{f}{f_r}\right)^2} \tag{5.4}$$

$$\mu'' = \frac{\alpha A\left(\frac{f}{f_r}\right)\left[1+(1+\alpha^2)\left(\frac{f}{f_r}\right)^2\right]}{\left[1-(1+\alpha^2)\left(\frac{f}{f_r}\right)^2\right]^2 + 4\alpha^2\left(\frac{f}{f_r}\right)^2} \tag{5.5}$$

where $A = \frac{M_s}{H_a}$ M_s is saturation magnetization, and H_a is the anisotropy field. f is frequency, f_r is the resonance frequency (intrinsic natural resonance), and α is the damping coefficient. Figure 5.15 shows the typical curve-fitting results for A-900. It is clear that the theoretical calculation curve of μ'' is in good agreement with the experimental results. To verify the correctness of the results, the μ' curve is calculated using the obtained parameters. According to the calculation results, the value of A, α, and f_r are 0.2742, 0.5829, and 3.0695, respectively. It implies that the fitting curve of μ' is almost in good agreement with the experimental results. There is slight change between the calculation and experiment curves, which may come from the geometry particle discrepancy. So, the magnetic loss at 2–6 GHz mainly comes from nature resonance of Mn_4N.

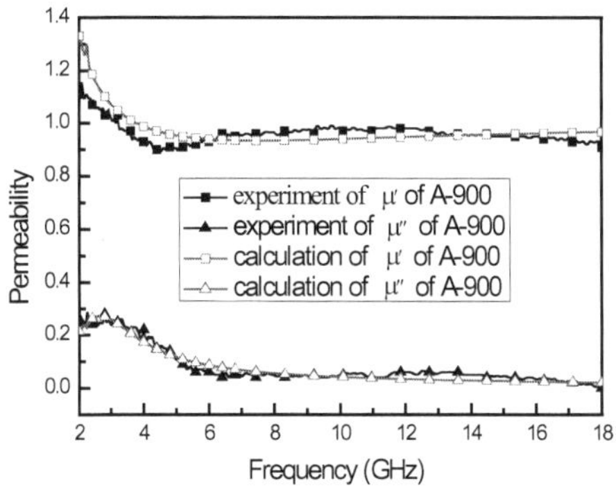

Figure 5.15 Permeability of A-900 in comparison between experiment and calculation.

The RL (dB) is an effective evaluation standard of the microwave absorption capacity of materials. It has been proven that the directly measured and calculated RL are in remarkable resemblance due to the same underlying physical origin. To further obtain the microwave absorption properties of samples A-400 to A-900 composite, the RL is calculated according to the transmission line theory[38]:

$$RL(dB) = 20\log\left|\frac{Z_{in} - Z_0}{Z_{in} + Z_0}\right| \quad (5.6)$$

$$Z_{\text{in}} = Z_0\sqrt{\frac{\mu}{\varepsilon}}\tanh\left[j\left(\frac{2\pi fd}{c}\right)\sqrt{\mu\varepsilon}\right] \quad (5.7)$$

where Z_{in} is the input impedance of the absorbing material, Z_0 is the impedance of air, μ is the relative complex permeability, ε is the relative complex permittivity, f is the frequency of EM wave, d is the thickness, and c is the velocity of light. The RL can be used to indicate the microwave absorption properties of metal-backed slabs of materials and low reflection corresponding to high absorption.

Figure 5.16 shows the color map of RL values calculated from the measured EM parameters of samples. It can be seen that the minimum RL value moves toward the lower-frequency region with increasing thickness. The large RL values, exceeding−10 dB, are obtained from 4.5–18, 5.7–18, 4.7–18, 4.2–18, 4.3–18, and 2–18 GHz for A-400 to A-900 by tuning a matching thickness of 2–20 mm, respectively. The strong microwave absorption in the 2–4 GHz range can be attributed to high magnetic loss, which is caused by natural resonance, as explained above. The value reaches −11.83 dB at 3.6 GHz for the coating thickness of 7 mm. The wider bandwidth of A-900 comes from the magnetic natural resonance. The results indicate that Mn_4N may be a potential microwave absorbing filler and have a wider frequency bandwidth for RL <−5 dB by adjusting the matching thickness ranging from 2 GHz to 18 GHz.

5.2.2.2 Effect of grain size on microwave electromagnetic properties of Mn_4N

The synthesized manganese nitride powder is grinded with a different time, and then sieved with different meshes of 70, 100, 140, 270, and 400. The sample of 400 meshes divided into three parts, and two parts are further grinded with 15 and 30 minutes, respectively.

Figure 5.17 shows the EM parameters of Mn_4N with different grain sizes in the range of 2 to 18 GHz. From Fig. 5.17a, the complex permittivity shows an obvious decreasing trend with the decrease in grain size. The results may be interpreted as follows: first, it is conducive to the formation of larger eddy current zone within the particles with the increase of grain size. The conductivity and inner eddy current increase consequently. Second, as the specific surface area of particle decreases, the contact area of the nonconduction interface with paraffin matrix diminishes accordingly, which also leads to the increase in conductivity. So higher permittivity is attained owing to improved electrical conductivity[39]. It can also be seen from Fig. 5.17a,b that the real and imaginary permittivity curves display great fluctuation versus frequency at 11–13 GHz. Moreover, ε'' curves present the accordant resonance peak. The fluctuation and peaks may be related to particle morphology and properties of samples. For Mn_4N, the dominant dipolar polarization, spontaneous polarization, and the associated relaxation phenomena constitute the microwave loss mechanisms. The irregular shape of Mn_4N particles is closely related to the intensity of spontaneous polarization. The larger the particle, the stronger is the spontaneous polarization. Therefore, stronger spontaneous polarization occurs in Mn_4N with larger grain sizes[40].

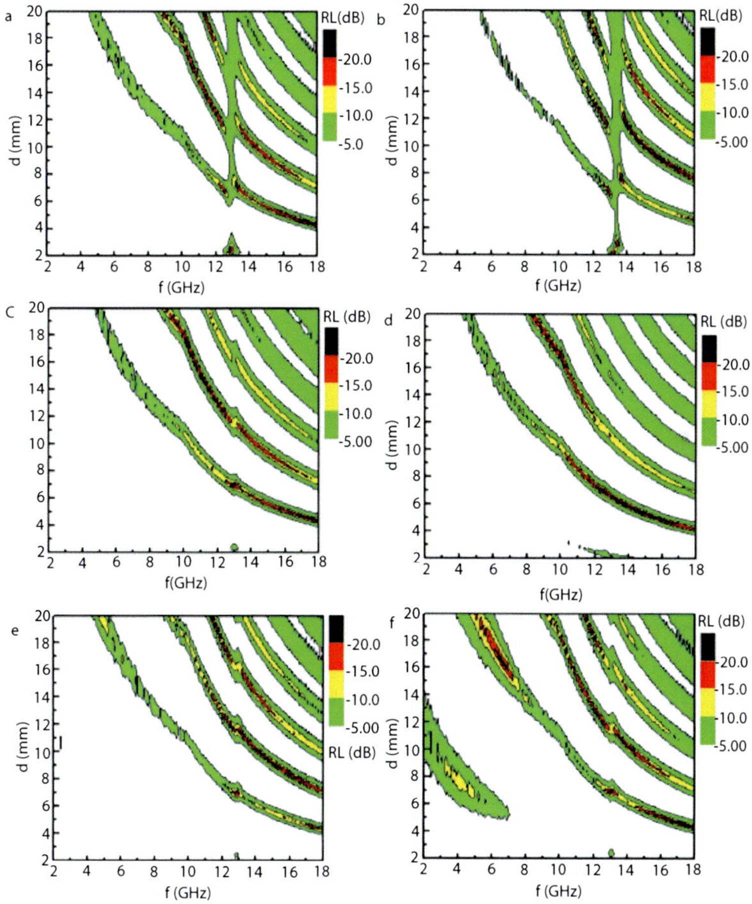

Figure 5.16 Color map of RL < −5 dB calculated from the measured EM parameters of the samples A-400 (a), A-500 (b), A-600 (c), A-700 (d), A-800 (e), and A-900 (f).

Figure 5.17c,d shows the real part (μ') and imaginary part (μ'') values of relative complex permeability for Mn_4N as a function of frequency at 2–18 GHz. As shown in the figure, the real part (μ') increases with the decrease in grain size, which is consistent with the variation of saturation magnetization. Contrarily, the imaginary part decreases with the decreasing grain size. So it is beneficial for enhancing the microwave absorbing property for Mn_4N with larger grain sizes. The reasons are the following: on the one hand, smaller size particles have higher saturation magnetization; on the other hand, compared to larger particles, smaller particles have a fewer magnetic poles, so the demagnetizing field effects in the process of magnetization is weaker, which makes it more easy to be magnetized and have a larger real part of relative complex permeability. The main reason of the higher imaginary part of the sample with larger particles is that larger particles are easier to bring the eddy current effect than smaller particles. To induce the change of the external electric field, there will be a larger eddy current on the surface of the larger particles, which makes the imaginary part increase.

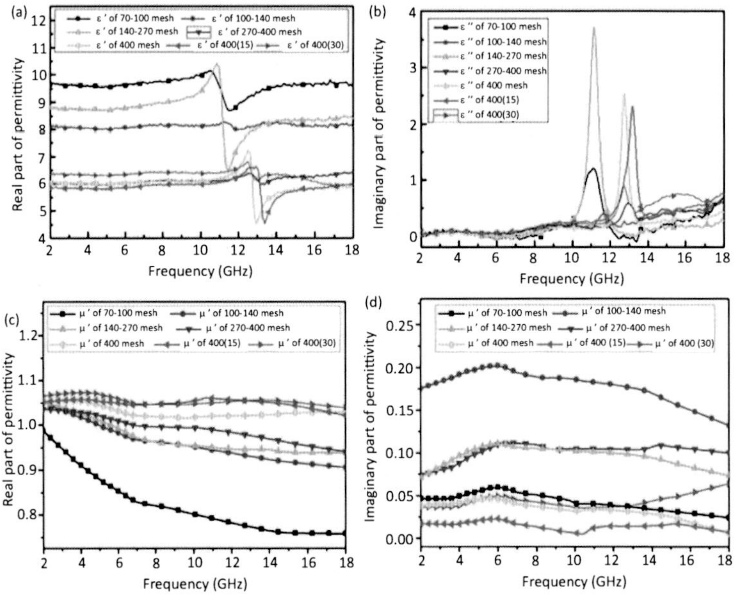

Figure 5.17 Complex permittivity (a, b) and complex permeability (c, d) of Mn_4N with different grain sizes.

Figure 5.18 shows the RL of the 75 wt% Mn_4N/ paraffin with different grain sizes in the frequency range 2–18 GHz with a thickness of 2 mm. Obviously, the 70 mesh sample, the 100 mesh sample, and the 140 mesh sample have a similar absorption peak at 11 GHz. With the decrease in grain size, EM wave loss peak shifts toward higher frequency. Besides, the 70 mesh sample has the highest RL. The bandwidth under –10 dB can reach 13–15 GHz. The largest loss peak can reach –14 dB at 13.5 GHz. In conclusion, the optimal absorbing size of Mn4N is from 70 to 100 mesh, which is attributed to stronger spontaneous polarization and better magnetic loss for larger Mn_4N particles.

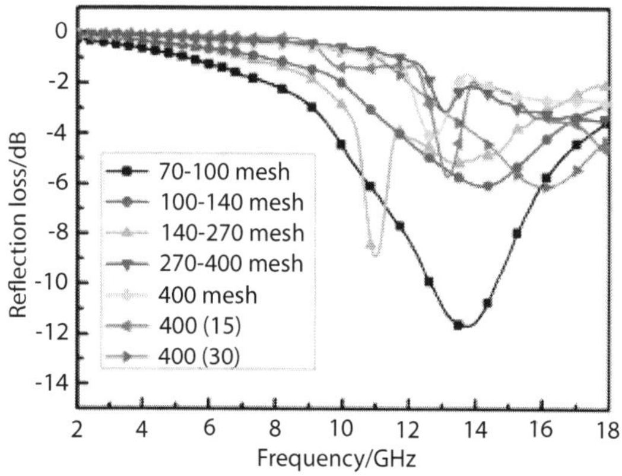

Figure 5.18 Reflection loss of 75 wt% Mn_4N/paraffin with different grain sizes at an absorber thickness of 2 mm in the frequency range of 2–18 GHz.

5.2.2.3 First-principles calculations

First-principles calculations on the basis of the density functional theory (DFT) are performed using the Cambridge Serial Total Energy Package (CASTEP) code[40] to investigate the properties of Mn_4N and α-Mn. The exchange correlation functional is treated using the generalized gradient approximation (GGA) and employing the Perdew–Burake–Enzerh (PBE) functional form[41]. The numerical integration of the Brillouin zone is performed using an 8×8×8 Monkhorst–Pack k-point sampling, and the cutoff energy is assumed to be 380 eV for the plane-wave basis. The valence electron configurations for the Mn and N atoms are chosen as 3. In the optimization process, the energy change and the maximum tolerances for the force, stress, and displacement are set at 0.29 × 10^{-4} eV/atom and 0.59 × 10^{-1} eV/Å, 0.1 GPa, and 0.2 × 10^{-2} Å, respectively[42].

On the microscopic scale, the spin electron states are calculated on the basis of t *ab* initio density function theory (DFT) method. Figure 5.19 exhibits the total density of states (DOS) of the above-mentioned specimens, respectively, which clearly show the influences of N doping on the 3d electron states around the Fermi energy level and magnetic properties. The value of atoms' magnetic moments can be obtained from the deviation areas between the spin-up occupied state and spin-down occupied state. For all samples, the asymmetrical DOS reflects the presence of magnetism. It is observed that the DOS between the spin-up and spin-down electrons of Mn is remarkably symmetrical, but otherwise significantly different near the Fermi level for the Mn_4N sample. The symmetrical shape is attributed to its weak magnetic properties. The asymmetrical DOS between the spin-up and spin-down channels near the Fermi level suggests that the Mn_4N system is ferromagnetic. The calculated total magnetic moment is $0.94\mu_B$ in the Mn_4N primitive cell.

Charge density maps serve as a complementary tool for achieving a proper understanding of the electronic structure of the system being studied. Figure 5.20 shows the spin charge density contour in the (110) plane for Mn_4N and Mn. It is obvious that appreciable charge is distributed in the outer space of Mn_2 and N atoms. The near spherical charge distribution around the Mn_1 sites is negligible. As a result, these atoms are fairly isolated, which indicates that the bonding Mn_1–Mn_2 is mainly ionic. The bonding Mn_2–N is characterized by a relatively covalent bonding due to the N-p–Mn_2-d hybridization. The homogeneous distribution of the charge density in the interstitial region can be considered as the results of metallic bonding. Recent researches have proved that microwave dielectric properties are related to the band characteristics[30,43]. The calculations of bond lengths show that the Mn_2–N, Mn_2–Mn_2, and Mn_1–Mn_2 bond lengths of Mn_4N are 1.92, 2.71, and 2.71 Å, respectively, which are all lower than the bond length of Mn with 3.49 Å. Generally, shorter bond length corresponds to higher bond energy, representing higher storing energy as well as real part of dielectric constant. The higher imaginary part of Mn_4N can contribute to the covalent bonding between Mn_2 and N atoms, which form a pair of electric dipole and attenuate wave energy. For metallic and ionic bonding, the relaxation time can be 10^{-15}–10^{-12} s, which is too short to make a response to the variation of microwave, resulting in low dielectric loss.

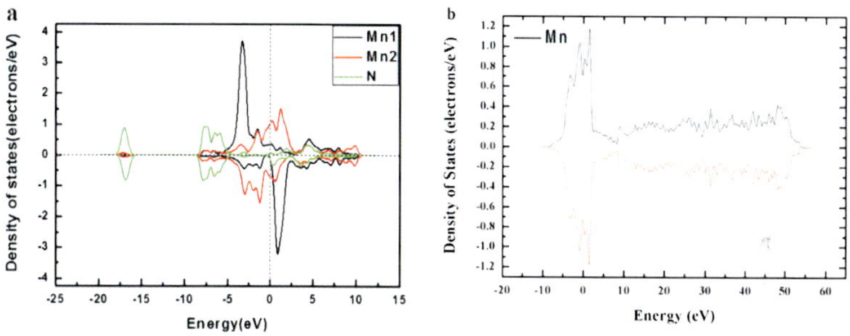

Figure 5.19 Calculated spin-up and spin-down partial density of states of (a) Mn_4N and (b) Mn.

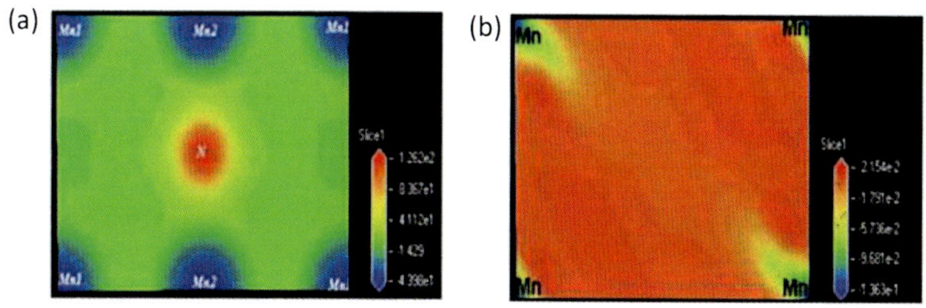

Figure 5.20 Calculated spin-polarized electronic charge density contours (111) for (a) Mn_4N and (b) Mn.

References

1. Giannakopoulou T, Kontogeorgakos A, Kordas G (2003). Single-layer microwave absorbers: influence of dielectric and magnetic losses on the layer thickness, *J Magn Magn Mater*, **263**(1), 173–181.

2. Singh P, Babbar VK, Razdan A, Puri RK, Goel TC (2000). Complex permittivity, permeability, and X-band microwave absorption of CaCoTi ferrite composites, *J Appl Phys*, **87**(9), 4362–4366.

3. Sayer M, Sreenivas K (1990). Ceramic thin films: fabricat and applications, *Science*, **247**(4946), 1056–1060

4. Janas VF, Safari A (1995). Overview of fine-scale piezoelectric ceramic/polymer composite processing, *J Am Ceram Soc*, **78**(11), 2945–2955.
5. Cohen RE (1992). Origin of ferroelectricity in perovskite oxides, *Nature*, **358**(6382), 136–138.
6. Lee MB, Kawasaki M, Yoshimoto M, Koinuma H (1995). Heteroepitaxial growth of $BaTiO_3$ films on Si by pulsed laser deposition, *Appl Phys Lett*, **66**(11), 1331–1333.
7. Roy D, Krupanidhi SB (1992). Pulsed excimer laser ablated barium titanate thin films, *Appl Phys Lett*, **61**(17), 2057–2059.
8. Powles JG (1948). Dielectric properties of titanates at ultra-high frequencies, *Nature*, **162**, 614–164.
9. Davis L Jr, Rubin LG (1953). some dielectric properties of barium-strontium titanate ceramics at 3000 megacycles, *J Appl Phys*, **24**(9), 1194–1197.
10. Lurio A, Stern E (1960). Measurements of the dielectric constant of $BaTiO_3$ single crystals in the paraelectric region at X band, *J Appl Phys*, **31**(10), 1805–1809.
11. Jing L, Wang G, Duan Y, Jiang Y (2009). Synthesis and electromagnetic characteristics of the flake-shaped barium titanate powder, *J Alloy Compd*, **475**(1), 862–868.
12. Beauger A, Mutin JC, Niepce JC (1983). Synthesis reaction of metatitanate $BaTiO_3$, *J Mater Sci*, **18**(10), 3041–3046.
13. Beauger A, Mutin JC, Niepce JC (1983). Synthesis reaction of metatitanate $BaTiO_3$, *J Mater Sci*, **18**(12) 3543–3550.
14. Kong LB, Ma J, Huang H, Zhang RF, Que WX (2002). Barium titanate derived from mechanochemically activated powders, *J Alloy Compd*, **337**(1), 226–230.
15. Terashi Y, Purwanto A, Wang WN, Iskandar F, Okuyama K (2008). Role of urea addition in the preparation of tetragonal $BaTiO_3$ nanoparticles using flame-assisted spray pyrolysis, *J Eur Ceram Soc*, **28**(13), 2573–2580.
16. Cho SD, Paik KW (2001). Relationships between suspension formulations and the properties of $BaTiO_3$/epoxy composite films for integral capacitors, paper presented at the Electronic Components and Technology Conference, Orlando.
17. Guan H, Liu S, Duan Y, Zhao Y (2007). Investigation of the electromagnetic characteristics of cement based composites filled with EPS, *Cement Concrete Compos*, **29**(1), 49–54.

18. Oh JH, Oh KS, Kim CG, Hong CS (2004). Design of radar absorbing structures using glass/epoxy composite containing carbon black in X-band frequency ranges, *Compos B*, **35**(1), 49–56.

19. Yusoff AN, Abdullah MH, Ahmad SH, Jusoh SF, Mansor AA, Hamid SAA (2002). Electromagnetic and absorption properties of some microwave absorbers, *J Appl Phys*, **92**(2), 876–882.

20. Devonshire AF (1951). CIX. Theory of barium titanate—Part II, *Philos Mag*, **42**(333), 1065–1079.

21. Arlt G, Böttger U, Witte S (1993). Emission of GHz shear waves by ferroelastic domain walls in ferroelectrics, *Appl Phys Lett*, **63**(5), 602–604.

22. Chen X, Wang G, Duan Y, Liu S (2008). Electromagnetic characteristics of barium titanate/epoxide resin composites in X and Ku bands, *J Alloy Compd*, **453**(1), 433–436.

23. Hui J, Yuping D, Zhuo L (2012). Influence of Fe-doping on the microstructure and electromagnetic performance of manganese oxides, *Physica B*, **407**(6), 971–977.

24. Zhang J, Yan C, Liu S (2012). Preparation of Fe_2Ni_2N and investigation of its magnetic and electromagnetic properties, *Appl Phys Lett*, **100**(23), 233104.

25. Pan H, Cheng X, Zhang C (2013). Preparation of Fe_2Ni_2N/SiO_2 nanocomposite via a two-step route and investigation of its electromagnetic properties, *Appl Phys Lett*, **102**(1), 012410.

26. Liu JR, Itoh M, Jiang J (2004). A GHz range electromagnetic wave absorber with wide bandwidth made of $FeCo/Y_2O_3$ nanocomposites, *J Magn Magn Mater*, **271**(2), L147–L152.

27. Zuo W, Qiao L, Chi X (2011). Complex permeability and microwave absorption properties of planar anisotropy $Ce_2Fe_{17}N_3$–δ particles, *J Alloy Compd*, **509**(22), 6359–6363.

28. Wu XL, Zhong W, Jiang HY (2004). Magnetic properties and thermal stability of γ'-Fe_4N nanoparticles prepared by a combined method of reduction and nitriding, *J Magn Magn Mater*, **281**(1), 77–81.

29. Bezdička P, Kláriková A, Paseka I (1998). Magnetic properties of α'-FeN_x and α''-$Fe_{16}N_2$ nitrides, *J Alloy Compd*, **274**(1), 10–17.

30. Duan Y, Liu Z, Zhang Y (2013). A theoretical study of the dielectric and magnetic responses of Fe-doped α-MnO_2 based on quantum mechanical calculations, *J Mater Chem C*, **1**(10), 1990–1994.

31. Huang H, Wang F, Lv B (2012). Microwave absorption of γ'-$Fe_{2.6}Ni_{1.4}N$ nanoparticles derived from nitriding counterpart precursor, *J Nanosci Nanotechnol*, **12**(4), 3040–3047.

32. Munekata H, Ohno H, Von Molnar S (1989). Diluted magnetic III-V semiconductors, *Phys Rev Lett*, **63**(17), 1849.

33. Duan Y, Wen M, Zhang Y, Chen J (2014). Effect of temperature on the structural, magnetic, and microwave electromagnetic properties of manganese nitrides, *J Supercond Nov Magn*, **27**(8), 1917–1925.

34. Moliton A (2007). *Basic Electromagnetism and Materials*, Springer Science & Business Media, Germany.

35. Yahiaoui R, Mounaix P, Vigneras V (2013). All dielectric metamaterials at millimeter wavelengths using single-size TiO_2 resonators: simulation and experiments, paper presented at Microwave Conference (EuMC), Nuremberg, Germany.

36. Zhang XF, Guan PF, Dong XL (2010). Transform between the permeability and permittivity in the close-packed Ni nanoparticles, *Appl Phys Lett*, **97**, 033107.

37. Zhang S, Cao Q, Zhang M (2013). Effects of Sr^{2+} or Sm^{3+} doping on electromagnetic and microwave absorption performance of $LaMnO_3$, *J Appl Phys*, **113**(7), 074903.

38. Guan H, Zhao Y, Liu S (2011). Application of manganese dioxide to electromagnetic wave absorber: effective permittivity and absorbing property, *Eur Phys J Appl Phys*, **36**(03), 235–239.

39. Liu L, Duan Y, Guo J (2011). Influence of particle size on the electromagnetic and microwave absorption properties of FeSi/paraffin composites, *Physica B*, **406**(11), 2261–2265.

40. Ma L, Wang G, Liu L (2010). Cobalt-coated barium titanate particles: preparation, characterization and microwave properties, *J Alloy Compd*, **505**(1), 374–378.

41. Clark SJ, Segall MD, Pickard CJ, Hasnip PJ, Probert MI (2005). First principles methods using CASTEP, *Z Kristallogr*, **220**, 567–570.

42. Duan Y, Xi Q, Zhang Y, Wen M, Wang T (2015). The thermal stability and microwave electromagnetic properties of Mn_4N, *Appl Phys A*, **120**(3), 1075–1081.

43. Chen Z, Chen J, Han P, Hao J (2014). First-principles calculation of electronic structure and microwave dielectric properties of Fe-doped o-Cr_7C_3, *Comp Mater Sci*, **83**, 298–302.

Chapter 6

Hybrid Microwave Absorbers

Microwave absorbers (MAs) are widely applied in both military and civil fields on account of their ability to eliminate electromagnetic wave pollution and to prevent information leakage. The theoretical and experimental work with MAs was initially started in the mid-1930s[1]. After that, intensive attention was drawn to studying these functional devices. The characterizations of thin thickness, wide absorbing bandwidth, light weight, and strong absorption have gradually become the focus of the study on the electromagnetic wave absorbing and shielding fields both at home and abroad[2].

At present, absorbers fabricated in the form of sheets by mixing rubber or epoxy resin (ER) with absorbent powders have been become the focus of study because of their practical and simple preparation method. The effective absorption frequency band of an MA is narrowed as just adding a wave absorbent into the ER. To improve the absorption efficiency and effective absorbing band of the slab absorber, different methods are used, such as adding an optimized absorbent, compounding different types of absorbents, using special structures, etc. In addition, coupling agents are used to improve the adhesion between fillers and matrix resins by chemical reactions, which play an important role in improving the mechanical properties of a polymer composite.

On the basis of the above analyses and summary of experiments, we can fabricate a suitable MA according to specific requirements.

Microwave Absorbing Materials
Yuping Duan and Hongtao Guan

ISBN 978-981-4745-10-9 (Hardcover), 978-981-4745-11-6 (eBook)
www.panstanford.com

6.1 Introduction

The microwave absorption property of a material is typically characterized in terms of the power reflection of a plane wave reflected from an infinite slab of the material that is backed by a metallic surface[3]. The power reflection or reflectivity of the coating, generally produced for normal incidence, is commonly expressed as reflection loss (RL)[4]

$$\mathrm{RL} = 20\log|\Gamma| = 20\log\left|\frac{Z_{\mathrm{in}} - Z_0}{Z_{\mathrm{in}} + Z_0}\right| \tag{6.1}$$

where Γ, Z_{in}, and Z_0 represent the reflection coefficient, input impedance of the coating, and intrinsic impedance of free space with a value of 377 Ω, respectively.

According to transmission theory, for a single-layer absorber backed by a perfect conductor, the input impedance of the absorber, Z_{in}, of a metal-backed microwave absorbing coating is given by

$$\mathrm{Z}_{\mathrm{in}} = \eta\tanh(\gamma \times d) \tag{6.2}$$

$$\eta = Z_0\sqrt{\mu/\varepsilon} \tag{6.3}$$

$$\gamma = j\frac{2\pi f}{c}\sqrt{\mu\varepsilon} \tag{6.4}$$

where η, γ, d, ε, and μ denote the intrinsic impedance, propagation constant, thickness, relative complex permittivity, and permeability of the composite coating, respectively. C and f are the velocity of light and the frequency of wave, respectively.

The poly(vinyl chloride) (PVC)-backed microwave absorbing coating should be considered as a quasi-double-layer structure, as shown in Fig. 6.1, the Z_{in} of which can be expressed as follows:

$$Z_{\mathrm{in}} = \eta_2\frac{\eta_1\tanh(\gamma_1 d_1) + \eta_2\tanh(\gamma_2 d_2)}{\eta_2 + \eta_1\tanh(\gamma_1 d_1)\tanh(\gamma_2 d_2)} \tag{6.5}$$

As is known, to design a microwave absorbing coating with high performance, two important conditions should be satisfied[5]. The first is impedance matching, in which the input impedance of the coating should be made approximate to the impedance of free space, ensuring the wave enters the absorption coating as much as possible. When $Z_{\mathrm{in}} = Z_0$, the impedance matching is satisfied perfectly, and the EM wave will be incident completely without any reflection, theoretically. Secondly, the incident wave should be attenuated rapidly through the material layer as far as possible.

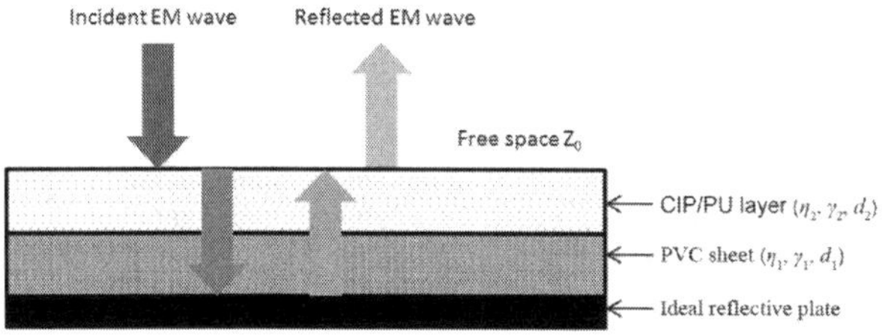

Figure 6.1 Schematic illustration of the quasi-double-layer composite coating based on the PVC sheet.

6.2 Composition and Structure of the Composite Absorbing Material

This book starts from the transmission path or channel of the electromagnetic (EM) wave inside the material or channel, that is, from the transmission path and channel of the EM wave in the absorber. Then absorbers are classified according to two major components, which are structure and distribution characteristic. Each type corresponds to a physical model, and then the absorber is designed according to its physical model. Emphasis is on the transmission path or channel constituted by the wave-transmitting material. Besides, this book tries to interpret the meanings of impedance matching from the view of the material's composition and distribution. Actually, what the impedance matching emphasizes is whether the EM wave can spread smoothly inside the material. When the EM wave is incident on an absorber with a certain thickness, only when all of the EM wave can penetrate into the material can the absorbing agent play its maximum absorption function.

To obtain an appropriate absorber in a desired frequency range, the first thing that should be considered is impedance matching, which requires the incident EM wave to enter the material by the greatest extent, because this is the prerequisite for the absorbent to efficiently function[1]. From this aspect, many designs of absorbers with special structures or optimizing processes have been carried out to improve microwave absorption properties of conventional microwave absorbers (CMAs).

When the EM wave is reflected from the incident plane of a material, we define "improperly matched" as the reflection happening inside the material and "mismatch" as the reflection happening near the surface. Thus, when the EM wave interacts with the material, impedance matching is closely related to whether to reflect or not and the magnitude and degree of reflection. The essence of this problem is whether the EM wave passage or path is unobstructed or not. The absorber in which EM wave spreads smoothly must have a transmission channel and path which is provided by the wave-transmitting material. When the transmission is disrupted, it must produce reflection or scattering. Therefore, from the perspective of the absorber, it is the precondition that the transmission channel and path be unobstructed, and the absorber can play the role of absorption. When designing the absorber, it should emphasize the design of the transmission channel and path. After analyzing the composition and distribution characteristics of multiple absorbers, the absorbers can be divided into four types. Each type corresponds to a physical model and the design should be differentiated according to different physical models.

6.2.1 Equality Distribution

Equality distribution is that the absorbing agent is dotted in an absorber and the wave-transmitting material forms a continuous matrix material. This uniform distribution is also called point distribution. The physical model which is compatible with uniform distribution is described below.

The absorbing agent particles or short fibers are uniformly distributed in the matrix material formed by the wave-transmitting material. Among them, most of the particles are isolated and there is no obvious agglomeration in the macroscale. The main characteristic of this kind of absorber is the even distribution of particles, and this characteristic should be the starting point when you design an absorber or do theoretical studies. Figure 6.2 shows the illustration diagram of uniform distribution.

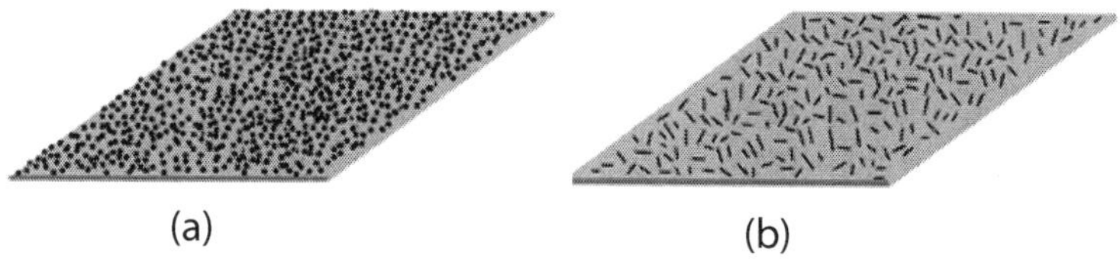

Figure 6.2 The diagram of uniform absorber distribution: (a) the cross-sectional view of equality distribution of particulate absorbents and (b) the cross-sectional view of equality distribution of fiber absorbents.

6.2.2 Layered Distribution

Layered distribution of an absorber means the distribution and composition of the adjacent plane or adjacent layer has rapid change. The plane mentioned not only includes the surface of the geometric sense but also includes the layer with a certain thickness on the engineering level. So, layered distribution is also called planar distribution. The physical model which is compatible with layered distribution is described below.

This kind of absorber is composed of multiple surfaces or multiple thin layers. There are more than two adjacent planes or adjacent layers, whose distribution and composition have rapid change. This model does not exclude cases in which the distribution and composition of the plane or layer are the same.

The absorbers which conform to this model can be multilayer membrane absorbers, multilayer board absorbers, multilayer fiber and fabric absorbers, and layered absorbers with intercalation.

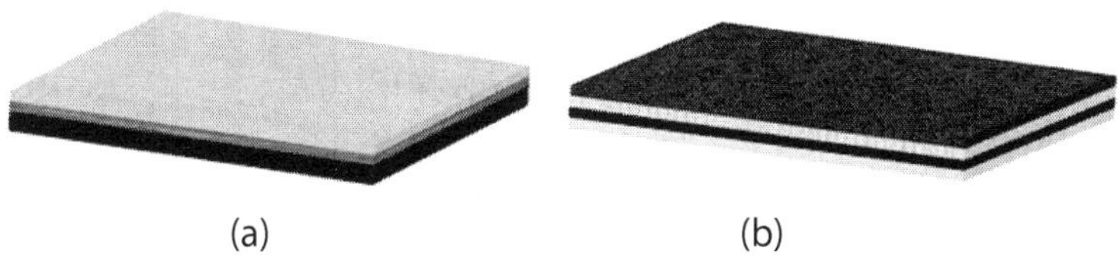

Figure 6.3 Layered distribution: (a) multilayer membrane or board absorber and (b) multilayer fiber and fabric absorber.

6.2.3 Spherical Shape Distribution

This kind of distribution is that the absorber distributes on the surface of the globular wave-transmitting material. The absorber forms a continuous conducting layer or magnetic layer on the surface and constitutes a resonant cavity. Its physical model is described as follows: The absorber consists of a number of resonant balls which accumulate together in accordance with the arbitrary shape shell. It can be binders with penetration capability or the interspace between ball and ball. These absorbers accumulated by many resonant balls form a complex resonant group. Figure 6.4 shows the illustration map of ball distribution.

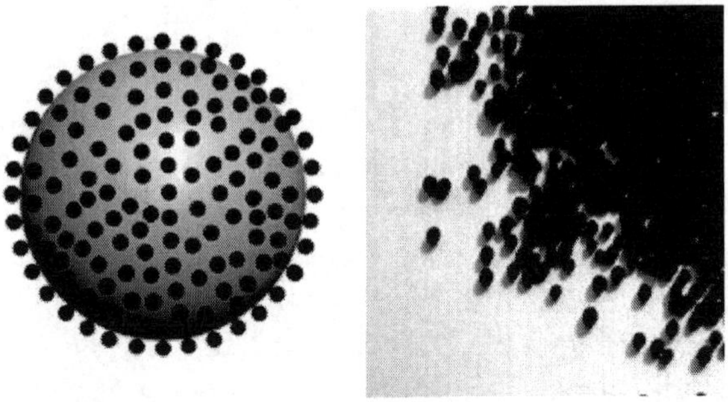

Figure 6.4 Spherical shape distribution.

Absorbers which associate with spherical shape distribution characteristics include hollow crystal ball–smearing absorbing agents on the surface, hollow polymer beads–smearing absorbing agents on the surface, and porous ceramics with closed-cell structure–smearing absorbing agents.

6.2.4 Distribution in Open Porous Foam

Numerous open tubular hollows constitute complex cavities and the absorbing agent distributes through the inner wall. As the foam absorber used in the anechoic chamber, its matrix material is the continuum consisted by the wave-transmitting material. Its physical model is shown in Fig. 6.5: There are tubular cavities or poriform cavities distributing on the continuous matrix material at random. The cavities are open continually with absorbing agents distributing through the inner wall. Its distributions can be continuous or incontinuous.

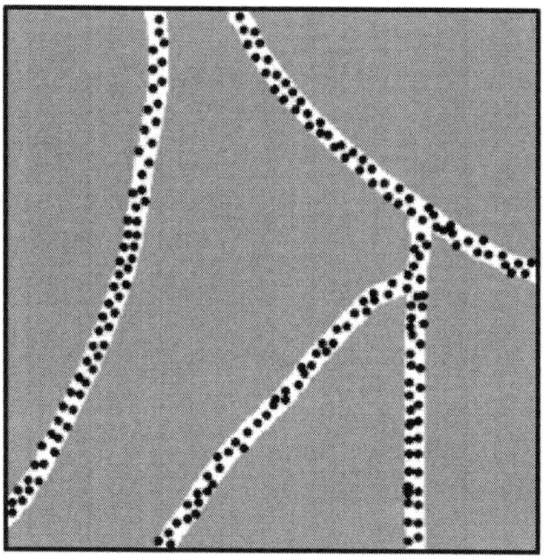

Figure 6.5 Distribution in open porous foam.

6.3 Structure Type of the Absorber

Absorbing materials can be classified into two kinds according to the forming process and carrying capacity. One of them is radar absorbing coating material, and the other is structural radar absorption material. Radar absorbing coating material is also called absorbing coating, and it can be classified as magnetic absorbing coating and dielectric absorbing coating in accordance with the kind of absorbing agent. The permeability of magnetic coating material can be improved by controlling the properties and filling ratio of the magnetic material and the thickness of the coating. By designing additives and absorbing agents, the absorbing properties can meet the demands for different application fields[6,7].

The main characteristic of an absorbing coating is the slight change to the equipment and it has a slight influence on the maneuvering and firepower performance of the weapon system. An absorbing coating is suitable for applying to existing stealth weapons, and it is easy to construct and has good adaptability to objects of any shape. An absorbing coating is usually composed of an absorbing agent and a binder. The absorbing agent has certain

EM parameters, so it is an important part and it directly decides the absorbing coating's loss ability of the incident EM wave. The binder can make the absorbing coating membraniform and thus make the absorbing coating adhere to the surface of the coated objects and form a continuous membrane[8]. The binder must be a good wave-transmitting material to make the absorption most efficient.

Absorbing coatings have made quite a progress in recent years, and new coatings are developing continually. However, absorbing coatings still have many problems, such as the narrow effective frequency band, bad cohesiveness, caducity, and heavy densities. All these factors will influence the flight performances of plane, missile, and so on. So, structural radar absorption materials still have more advantages compared to absorbing coatings. Structural radar absorption material has two functions, which are load bearing and radar cross section (RCS) reduction. So structural radar absorption material has been paid more attention and has been researched since World War II. Now it has become an important development direction of modern stealth materials and has got high attention from researchers at home and abroad[1,9].

6.3.1 Radar Absorbing Coating Material

Radar absorbing coating material can be classified as absorption type, interference type, or resonance type.

6.3.1.1 Absorbing-type radar absorbing coating: Dallenbach coating

The Dallenbach coating is the earliest and simplest absorbing coating. It is formed by a homogeneous lossy layer coating on the surface of a metal sheet, as shown in Fig. 6.6. Due to its thinner thickness, it is particularly suitable for coating on the surface of a flying target in order to reduce the RCS of the equipment.

The Dallenbach coating is a kind of single-layer structure, so the impedances of the lossy medium and the space wave have great differences, thus leading to larger reflection on the surface of the coating. It generally has a narrow absorbing frequency and can only be used at a certain frequency, while at another frequency, its absorption effect will be poorer. This disadvantage influences the absorption property of the Dallenbach coating and limits its practical application[10,11].

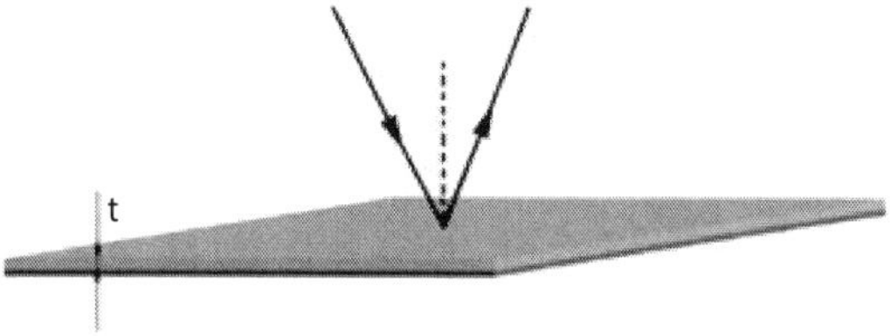

Figure 6.6 Dallenbach coating structure diagram.

6.3.1.2 Interference-type radar absorbing coating: Salisbury absorption screen

To improve the wave absorbing property of the Dallenbach coating, a lossless medium is filled between the metal sheet and the lossy medium, and the thickness of the medium is designed as a quarter of the wavelength of the EM wave in the medium. The EM wave reflected from the back of the coating and that directly reflected from the surface of the coating can be counteracted by mutual interference. This structure is called the Salisbury absorption screen[12], which is shown in Fig. 6.7. Its typical wave absorbing curve with a thickness of 7.5 mm is shown in Fig. 6.8.

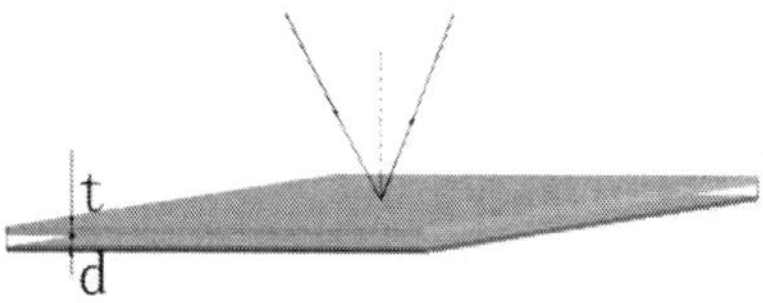

Figure 6.7 The structure of the Salisbury absorption screen.

Afterward, there appeared many studies on the absorption improvement for Salisbury screen[9–11]. An intervention layer has been added in the middle of the Salisbury absorption screen and it was found to have great influence on the EM wave absorption. But essentially, it is still a kind of resonance type. Though it has a good effect for absorbing EM waves at the center frequency, at other frequencies its effect is still not good enough. However, it broadens the absorption frequency bandwidth compared to the Dallenbach coating[12].

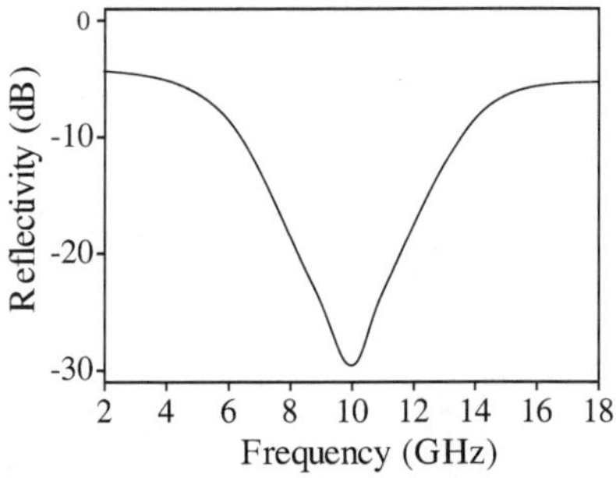

Figure 6.8 The wave absorbing curve of a typical Salisbury absorption screen.

Through adding a dielectric material with a high dielectric constant between the Salisbury absorption screen and the metal back, the absorption capacity of the hybrid structure can be optimized and improved by adjusting the dielectric layer's EM parameters, as shown in Fig. 6.9 and Fig. 6.10[13].

Figure 6.9 Hybrid Salisbury absorption screen.

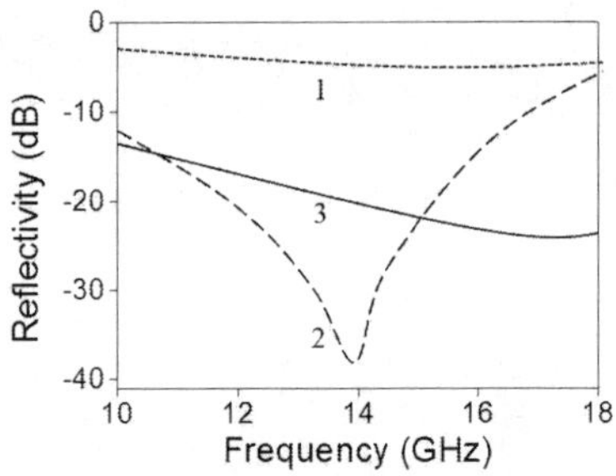

Figure 6.10 Wave absorbing curve of the hybrid Salisbury absorption screen.

In Fig. 6.10, curve 1 is the absorption curve for a single-layer Salisbury screen, curve 2 is that for a hybrid Salisbury absorption screen, and curve 3 is the absorbing curve for a hybrid Salisbury screen with optimized EM parameters. The impedance of the

Salisbury absorption screen is $R = 340\ \Omega$, $d_1 = d_2 = 4.8$ mm, and after hybridization and optimization, the Salisbury screen's parameters are designed as $R = 170\ \Omega$, $d_1 = 4.8$ mm, and $d_2 = 1.6$ mm. It can be seen that the wave absorbing capacity is greatly enhanced. Though the peak value decreases, the effective frequency band gets wider and the bandwidth in which the RL is below −20 dB increases from 3 GHz to 9 GHz. And also, the material's thickness decreases.

If the Salisbury absorption screen is designed as a symmetric structure, its wave absorbing capacity can also be improved greatly, as shown in Fig. 6.11 and Fig. 6.12[14]. The first curve in Fig. 6.12 is the Salisbury absorption screen's absorption curve. The impedance of the Salisbury absorption screen is $R = 184\ \Omega$ and the total thickness is 6 mm. The dielectric constant of the resistor layer is $\varepsilon_r = 4.0$ and its thickness is 1 mm. The dielectric constant of the dielectric layer is $\varepsilon_r = 1.1$. The second curve is the symmetric structure's absorption curve with a total thickness of 12 mm. The impedance and dielectric constant of the resistor layer are $R = 258\ \Omega$ and $\varepsilon_r = 4.0$, respectively, whereas the dielectric constant of the dielectric layer is $\varepsilon_r = 1.1$.

Figure 6.11 Symmetric Salisbury absorption screen.

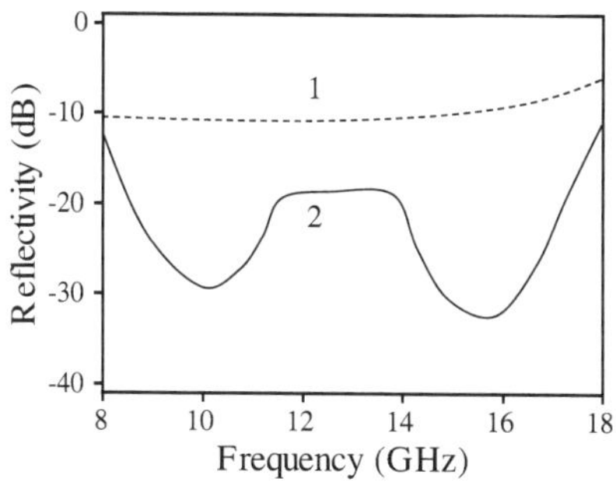

Figure 6.12 Absorbing curve of the symmetric Salisbury absorption screen.

6.3.1.3 Resonance absorbing coating

A resonance absorbing coating includes multiple absorbing units. To weaken EM waves to the extreme extent, one needs to adjust the EM parameters and the size of each unit to generate resonance to the incident EM waves.

If each absorbing unit is designed to resonate at a different frequency band, then a wave absorbing coating with a wide bandwidth can be obtained[15,16]. The structure diagram of a resonance absorbing coating is shown in Fig. 6.13.

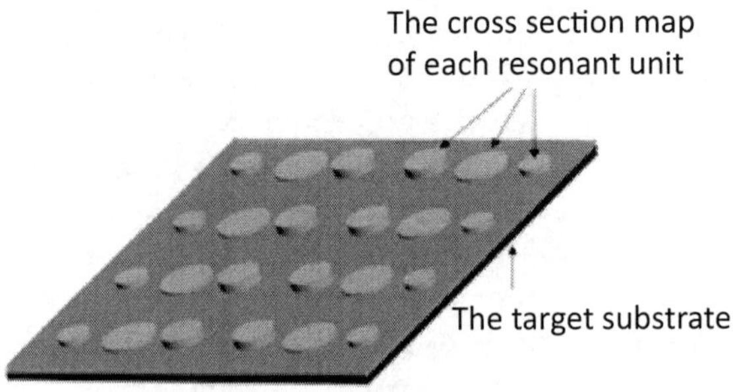

Figure 6.13 Sketch of a resonant absorbing coating.

Since the resonant structure is very rugged and it's not convenient for application, the rough part of the coating is often filled with low-dielectric-resin materials in order to make the coating stronger and easy to use.

6.3.1.4 New emerging wave absorbing coatings

With the development of stealth technology and the increasing demands for new absorbing materials with comprehensive performances, some novel EM wave absorbing coatings are emerging and their absorbing mechanisms are studied extensively and intensively.

Plasma stealth technology began in the 1960s and has seen great progress in the recent few decades. Plasma absorbing coating means coating a kind of radioactive material on the target and making the local space near the surface ionization; therefore the plasma is formed to absorb the EM waves. Under the electric field of the incident EM wave, the free electrons in the plasma start the forced oscillation with a frequency equivalent to that of the radar carrier. In the process of oscillation, the moving electrons, neutral molecules, atoms, and ions collide with each other and increase the kinetic energy of the particles, so the EM energy can be converted to the heat energy of the medium. It has been proposed that when the thickness of the coating is 0.025 mm (use of ^{210}Po as a source), the RCS can be reduced by 10%–20% for the 1.0 GHz vertical incident EM wave. If it is used in special parts where it can cause multiple reflections, the absorbing effects could be multiplied[17–20].

Chiral material is a new kind of absorbing material developed in recent years. Chirality refers to the phenomena where there is no symmetry existing between an object and its mirror image and the object cannot overlap its mirror image by any operation[21,22]. There are two advantages of chiral materials compared to ordinary absorbing materials: First, it is easier to adjust the chiral parameters of materials than the EM parameters; the second is that the frequency sensitivity of the chiral materials is smaller than the EM parameters, so it is easy to expand the frequency band. A kind of chiral absorbing coating was introduced by the United States patent US4948922 which is composed of conductive polymers or other low-loss-dielectric materials[23–27]. The chiral material was doped in the matrix with a spiral shape and the coating is efficient to reflect and absorb microwaves in a wider range of frequency[28–31].

Conductive polymer absorbing coating is a new field of polymer material science. Through the charge transfer between certain polymers with conjugated π-electrons and doping agents, impedance matching and EM losses can be achieved. Thus, the conductive polymer absorbing coating is an effective absorbent. There exist free radicals in the polymer chain structures after doping, and the dipole transition enables the polymer to be electrical conductive. Up till now, the doped polyphenylacetylene (PPA), polyacetylene, and poly(phenyl-benzene thiophene), as well as p-phenylene, PPY, polythiophene, PANI, etc., have been reported as conductive polymer microwave absorbing materials[32–34].

6.3.2 Structural Absorbing Materials

6.3.2.1 Absorbing layer with board structure

This absorbing structure is composed of multiple layers with different parameters and different thicknesses of the plate in order to improve the absorbing performance of the absorbing materials as far as possible. The absorbing structure was usually composed of three layers with different structures: transparent layer, absorbing layer, and reflecting layer. The absorbing property of the board structure layers is determined by the total admittance of the lossy layer, the parameter of the dielectric layer, and the thickness of each layer.

The Jaumann absorber is the earlier absorbing layer structure and among them the Salisbury absorbing screen is its simplest form. The Jaumann absorber originated from Germany during World War II, and it is composed of multilayer resistance cards alternately stacked with dielectric materials. The resistance of the resistors reduces from surface to bottom and the bandwidth of the whole absorbent is related to the number of resistors[1], as shown in Fig. 6.14.

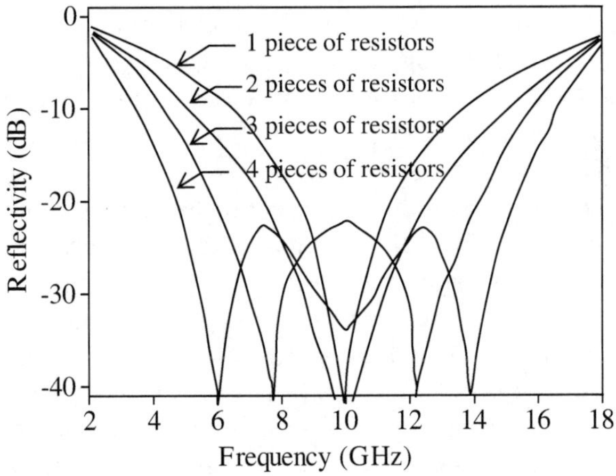

Figure 6.14 Absorption curve of the Jaumann absorber.

6.3.2.2 Sandwich absorber

To further increase the absorption efficiency of the absorber, the absorbing layer can also be designed into more complex structures, for example, the transparent layer and absorbing layer can be designed as sandwich structures filled with absorbing material, as shown in Fig. 6.15. The absorbing materials can be flocculent, foamy, globular, or fibrous. In addition, hollow microspheres are much better for the absorbing material.

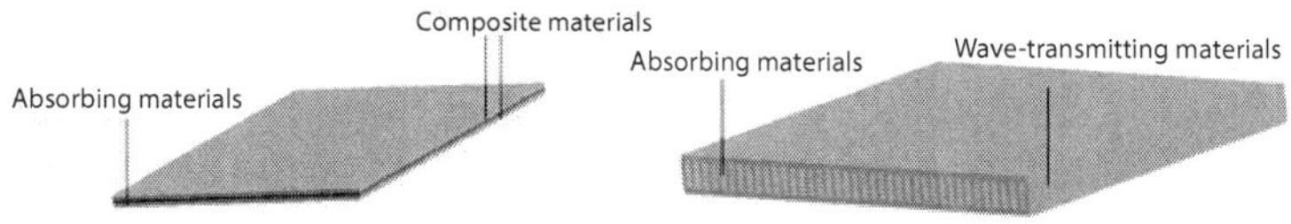

Figure 6.15 Absorbing material filling structure.

Sandwich absorbing material is generally made of composite materials. The radar absorbing composites not only have a light weight, high strength, and high mechanical modulus but also can effectively weaken the radar wave. Thus, this kind of composites can efficiently absorb radar waves; meanwhile, the material itself is a kind of structure material, playing the role of bearing and weight loss.

The absorbing layer of a board-type structure can be designed as a corrugated sandwich structure or a pyramidal structure, as shown in Figs. 6.16 and 6.17. The structure of a corrugated sandwich plate has two intersected slopes and four pyramids with inclined plane intersections. Pyramids with different heights have different effective absorption bandwidths. To have better absorption, the tip angle can be designed as 40°[16].

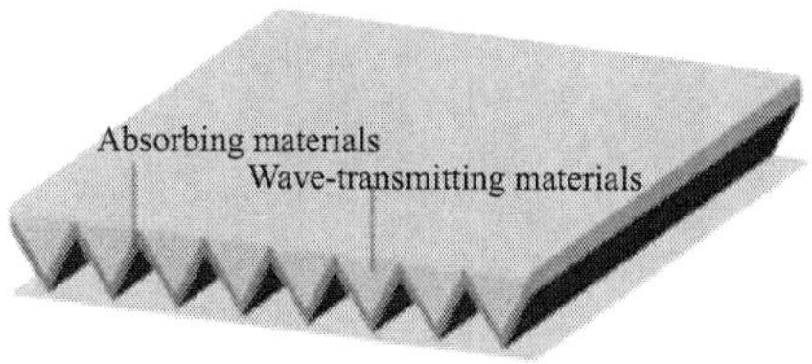

Figure 6.16 Corrugated sandwich structure.

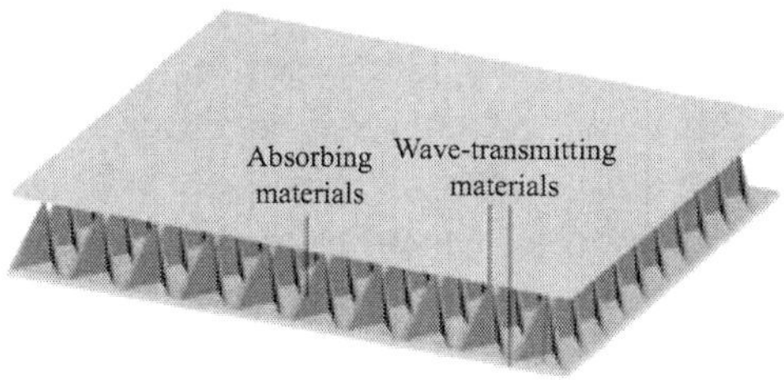

Figure 6.17 Pyramidal structure.

6.3.2.3 Frequency-selective surface absorbing structure

A frequency-selective surface (FSS) is a layered quasi-plane structure composed of a large number of passive units distributed as periodic arrays. An FSS has good selectivity on the transmission and reflection of EM waves. For EM waves in the pass band, an FSS presents all-pass features, while for EM waves in the stop band, an FSS shows total reflection characteristics. In other words, FSSs have a good spatial filter function[35].

An FSS usually has a substrate support and a cover layer, and multilayer FSS structures are often separated by dielectric layers. The units of an FSS can be divided into two basic types: one is constructed by metal patches pasted periodically on a dielectric back, and the other is composed of apertures opened periodically on a conductive screen. The patch and aperture could be of any shape. Figure 6.18 is a structure diagram of a single-screen FSS structure[36].

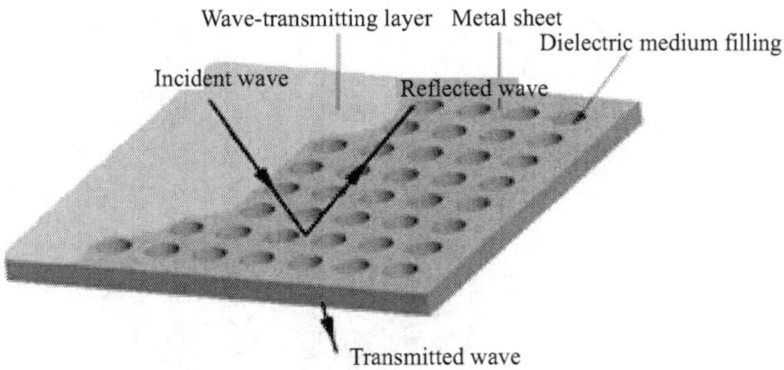

Figure 6.18 Structure diagram of a single FSS screen.

6.3.2.4 Circuit simulation absorber

As described above, the Salisbury absorber and the Jaumann absorbing structure used only the real part of the admittance of the material. If the imaginary part of the admittance is also introduced in the process of material design, there will be more parameters to adjust. Therefore, absorbing materials with a broad band could be obtained. The circuit analog absorber is established on the basis of this concept. This absorbing structure consists of many grid units and the spacer layer structures, as shown in Fig. 6.19. Its role is similar to that of an FSS[37], which reflects in one or more frequencies, while it is transparent to other frequencies. Michielssen et al. have studied the surface of different FSSs, and they have put forward a design of multilayer absorbers[38].

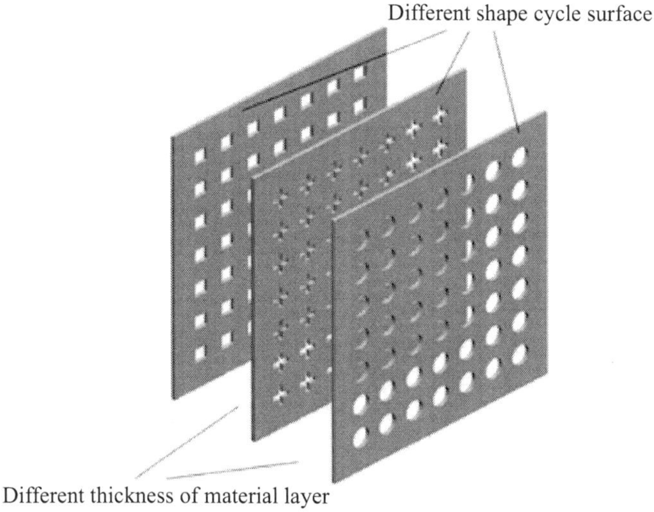

Figure 6.19 Structure of circuit simulation absorbing material.

In the application of absorbing materials, the structure of a circuit analog absorber belongs to the active frequency selection. It usually joins active devices or ferrites on the FSS unit or dielectric substrate; in addition, the chiral medium can be used as a dielectric substrate, etc. Studies by Xing et al. show that there is little influence on the strength properties of the resistance gradient composite system when the circuit analog absorber joined[39,40]. The shearing strength of the composite which contains a fiber circuit analog structure is slightly higher than other systems. The composite contained a fiber circuit analog structure which is prepared after optimized design has excellent absorbing/bearing performances. Because of the filtering action, the structure of circuit analog absorbing is widely used in the EM field, especially in stealth aircraft and satellite antenna systems.

6.3.2.5 Ferrite grid structure

This is a kind of two-dimensional structure with a matrix of ferrite tile, as shown in Fig. 6.20. This kind of composite structure has an advantage over normal ferrite tile in that it has an additional parameter (filling factor g). Through properly adjusting the thickness of the ferrite grid absorber, the parameter of ferrite, and the filling factor, an absorber with a wider frequency band and better absorbing performance can be designed. Meanwhile, it can realize zero reflection in a fixed frequency[41–43]. When the frequency is under 600 MHz, this kind of grid structure has better absorption performance, while as the frequency becomes more than 600 MHz, the absorbing performance gradually decreases with the increase in frequency[1,44].

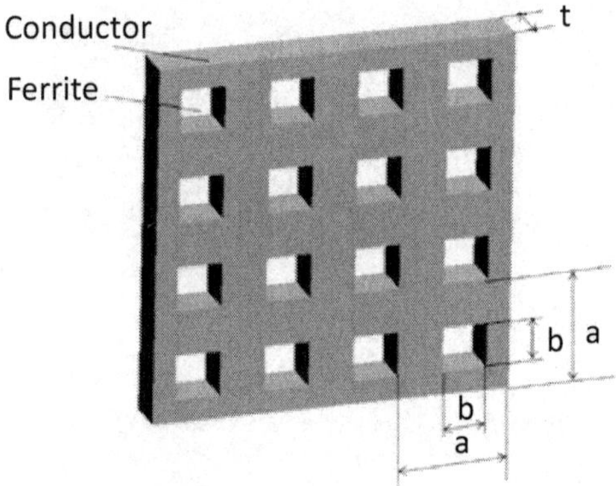

Figure 6.20 Ferrite tile structure absorber.

6.3.2.6 Pyramidal absorbing structure

The pyramidal absorbing structure is the most widely used absorbing material in a microwave darkroom and the design theory of a pyramid absorber is completely consistent with that of stealth material. Pyramid material first appeared during World War II, when Neher first installed pyramidal absorbing materials on the wall of the darkroom to improve the testing capability of the darkroom[45]. Pyramid material has the advantage of its special geometry, which is in favor of multiple reflections between the cones, as shown in Fig. 6.21. Thus, the absorbing performance is greatly increased than other plate materials.

The absorbing properties of pyramid materials have a complex relationship with the height of the pyramid, the angle of the vertex, the incidence angle of the incident wave, and EM parameters of the materials. For a given material, when the EM wave incidents vertically, the reflectivity reduces with an increase in the height of the pyramid firstly and then increases sharply after reaching the critical value. The variation of reflectivity with the vertex has a similar tendency. The reflectivity decreases first with the reduction of the vertex and then increases slowly. Therefore, a reasonable size and appropriate EM parameters are very important for the materials to have higher reflectivity[46–49].

Figure 6.21 Pyramid absorber.

If each pyramid is rotated 45° around its central axis, a twisted pyramidal structure can be obtained[50,51], as shown in Fig. 6.22. This kind of microwave absorbing structure has lower efficiency than a normal pyramid, but it has a lot of advantages, such as saving of raw materials and better durability, and it is not easy to sag when used in the darkroom[52,53]. In addition, a pyramid can also be combined with ferrite to improve its absorbing properties at a low frequency, as shown in Fig. 6.23.

Figure 6.22 Deformation pyramid absorber.

Figure 6.23 Pyramid composite ferrite structure.

6.4 Epoxide Resin Composites

As the matrix material, epoxide (EP) resin has many advantages, such as good thermal stability and high-frequency dielectric properties ($\tan \delta \approx 10^{-3}$–$10^{-4}$). So, EP was selected for the matrix material in this section.

Composite coatings with different absorbents were fabricated by a traditional mechanical mixing method. The preparation process can be briefly described as follows. The absorbents in flexible loadings were mixed with a resin-based solution, and the mixture was stirred for 15 min at 500 rev/min. Then, a hardener was added to the solution and blended with a homemade motor stirrer for 4 min. The uniform mixture was poured on an aluminum plate (200 mm × 200 mm). During the preparation process, the thickness of the coating was controlled by a mold located outside the plate. This operation process is easy to realize, and the obtained absorbers are uniform and flat, which can result in widespread application in the field of preventing EM interference.

The curves of RL versus frequency of the prepared absorbers were studied by an HP8720B vector network analyzer and standard horn antennas (including a transmitter and a receiver) in an anechoic chamber and the samples were backed by a steel board. The testing frequency bands were in the range from 2 to 18 GHz. A schematic illustration of the test system is shown in Fig. 6.24.

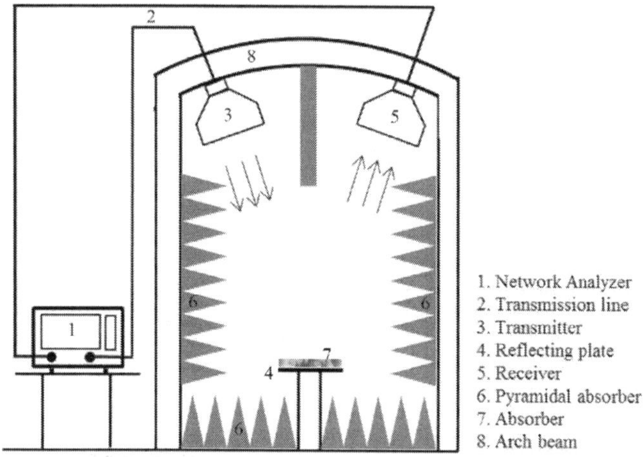

Figure 6.24 Sketch of arch reflection testing in an anechoic chamber. An HP8720B vector network analyzer is used for test analysis, and the testing frequency range is 2–18 GHz.

6.4.1 Epoxide Resin/Barium Titanate Composites

Barium titanate ($BaTiO_3$ or BT) as one of the perovskite-type structures is widely used in multilayer ceramic capacitors (MLCCs), positive temperature coefficient (PTC), ferroelectric nonvolatile memory, dynamic random access memory (DRAM), transducers, thermistors, and electro-optic devices as a dielectric due to its high dielectric constant, good ferroelectric properties, and nonlinear optical properties[54–58]. There have been many investigations concerning the interaction of the temperature, phase transitions, grain size, and dielectric properties of BT and its compounds. It is well known that BT is a kind of dielectric material. The relaxation presence of BT occurred in the gigahertz frequency, which can be characterized by a decrease in the dielectric constant (K) and a peak in the dielectric loss ($\tan\delta$) with increasing frequency. One may speculate whether BT can serve as a lossy filler according to this definition[59–61].

In Fig. 6.25, experimental values of RL are shown as a function of frequency of samples with different contents and thicknesses of the BT/EP composite. The thickness of the composite was increased from 1 to 4 mm, as shown in Fig. 6.25a-c. Figure 6.25a shows that the RL for the BT/EP composite with a thickness of 1 mm was poor throughout the whole frequency range when the volume concentration of BT was lower than 8 vol%. After the thickness of the sample reached 2 mm, the RL of the composite with 6 vol% of filler contents is below −10 dB in the frequency band of 11–14.5 GHz. However, the 8 vol% BT composite exhibits excellent RL performance and a broad frequency absorption peak appears at 14.8 GHz. As the thickness reaches 4 mm, a higher-absorption-frequency region is observed when the volume concentration of the BT is higher than 6 vol%.

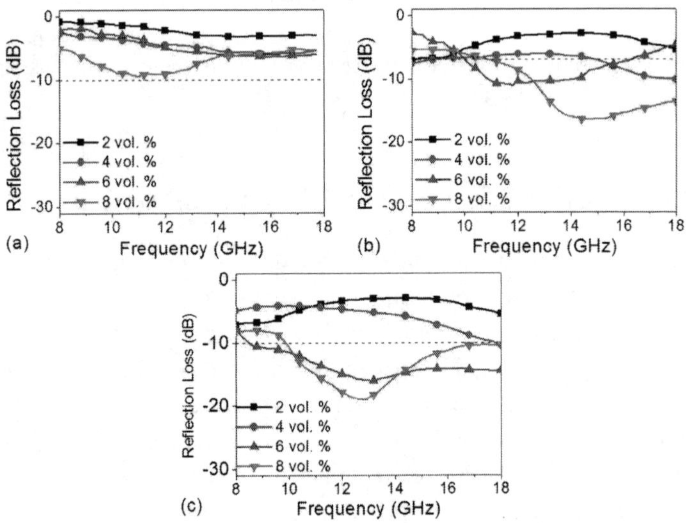

Figure 6.25 Frequency dependence of the RL of the BT/EP composite with different thicknesses: (a) 1 mm, (b) 2 mm, and (c) 4 mm.

As for the absorbing property of the BT/EP composite, the sample thickness has some effects on the EM RL of a wave absorbing material, and the thickness can influence the material's absorbing peak values and bandwidth[62,63]. However, the change in thickness also changes the impedance matching between the material and free space; so for each sample with different components, there are one or more optimum thicknesses (matching thickness)[64].

When the sample thickness is beyond the matching thickness in a certain frequency band, the microwave RL will not increase but drop gradually. So there is an absorbing peak for the optimum thicknesses of the BT/EP composite.

As is known, the dielectric relaxation occurring in the gigahertz-frequency region of BT will lead to dielectric loss. The relaxation of BT came from the inertial boundary–leaded piezoelectric resonance and the presence of a ferroelastic or 90° domain wall in BT[65–67]. From Fig. 6.25, it can also be seen that the trend of the RL values is flat before the filler concentration reaches a certain content. At low filler levels, the effect of BT relaxation is not obvious, and the transmission of microwave radiation through the composite was high. So the RL of the BT/EP composite is small. Conversely, at high filler concentrations, the interaction and wave loss enhance when the EM wave transmits through the material. It is considered that the high loss is attributed to polarization relaxation which forms BT particles in the EP matrix. In addition, the composite can be polarized repeatedly in a high-frequency electric field when the absorber filler content is higher, which causes the electric energy to be changed into other forms of energy, such as heat energy[68].

In conclusion, BT is mainly a kind of dielectric loss material with little magnetic loss according to the dielectric constant and permeability in the 8–12 GHz range. The RL of the BT/EP composites indicates that BT is a promising microwave absorption material in the X and Ku bands due to the high relaxation of BT in the BT/EP composite.

6.4.2 Epoxide Resin/Barium Titanate and Carbon Black Composites

It is well known that carbon black (CB)-filled polyolefines have very interesting scientific and applied aspects. Much research has been done on the EMI properties of carbon powder[69,70]. The CB percolation concentration leading to a rapid increase in composite electroconductivity has been found[71]. When the content of CB in the matrix is too high, it can cause an increase of the amount and size of the CB-contacting aggregates. The result will rapidly decrease the absorption properties of CB.

To improve the EM properties (electroconductivity, absorption, etc.) of BT, a novel composite was prepared in this study.

A current-voltage method was used for the volume resistivity measurements (as shown in Fig. 6.26). A digital multimeter GDM-8055 and a Fluke 8840A were used to measure the electrical current and to provide the voltage. The volume resistivity of samples was calculated by the following formula[72]:

$$\rho_V = R \times \delta \times \frac{d}{L} (\Omega \cdot \text{cm}) \tag{6.6}$$

where R is the volume resistance of the measured sample (Ω), δ is the thickness of the measured sample (cm), d is the width of the sample (cm), and L is the effective length of the sample (cm).

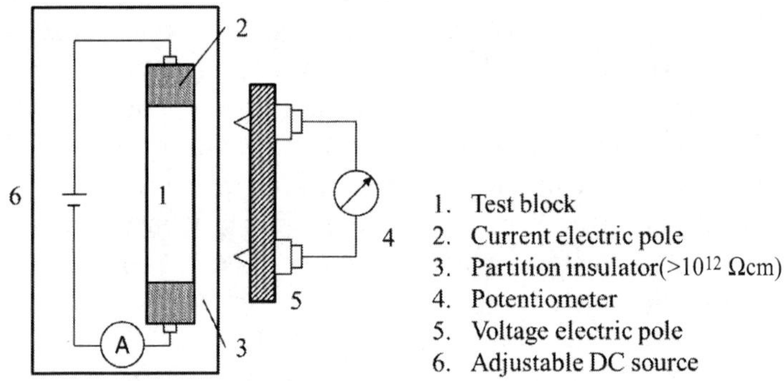

Figure 6.26 Schematic diagram of the volume resistivity test system.

6.4.2.1 Electric conductivity

From Table 6.1, it was indicated that the volume resistivity of BT is very high (6.6526×10^8 Ω·cm). BT belongs to nonconductors. CB exhibited good conductivity with low volume resistivity. The insulated BT was coated by conductive CB, which embedded among BT grains showed up in transmission electron microscopy (TEM). CB can form an electric net in the CB/BT compound particle (CP). Therefore, CPs showed much higher conductivity than BT attributed to the carbon powder which was embedded among the surface of the BT grains.

Table 6.1 Test results of the electrical resistivity of barium titanate and carbon black

Sample	Voltage (V, mV)	Electric current (I, A)	Volume resistivity (ρ_v, Ω·cm)
Barium titanate	2.022×10^3	0.61×10^{-3}	6.6526×10^8
Carbon black	0.124	0.2433	2.0456×10^{-1}
Compound particles	5.049	0.2409	8.4126

6.4.2.2 Microwave absorption properties

Figure 6.5 shows the typical relationship between RL and frequency for EP resin composites with different wt% CP powders. As to the absorbing property of the CP/EP composite, sample thickness has some effects on the EM RL of a wave absorbing material, and thickness can influence the material's absorbing peak values and bandwidth[73,74]. However, the change of thickness also changes the impedance matching between the material and free space, so to each sample with different components, there are one or more optimum thicknesses (matching thickness)[64]. The matching thickness d_m can be expressed as Eq. 6.7:

$$d_m = \frac{c}{2\pi S_0} \tag{6.7}$$

where c is the wave velocity in free space and S_0 is Snoek's value correlating to EM parameters of the material. When the sample thickness is beyond the matching thickness in a certain frequency band, the microwave RL will not increase but drop gradually. So the optimal percentage of CPs will lead to the optimum thickness. Figure 6.27 indicates that when the CP percentage is lower than 15 wt% in the matrix, the CP/EP composite cannot exhibit good EM absorbing properties. After the CP percentage in the matrix exceeded 15 wt%, the RL of CP/EP composite effectiveness improved. In addition, a minimum RL value of −23 dB was obtained at 13.4 GHz with 30 wt% CP percentage.

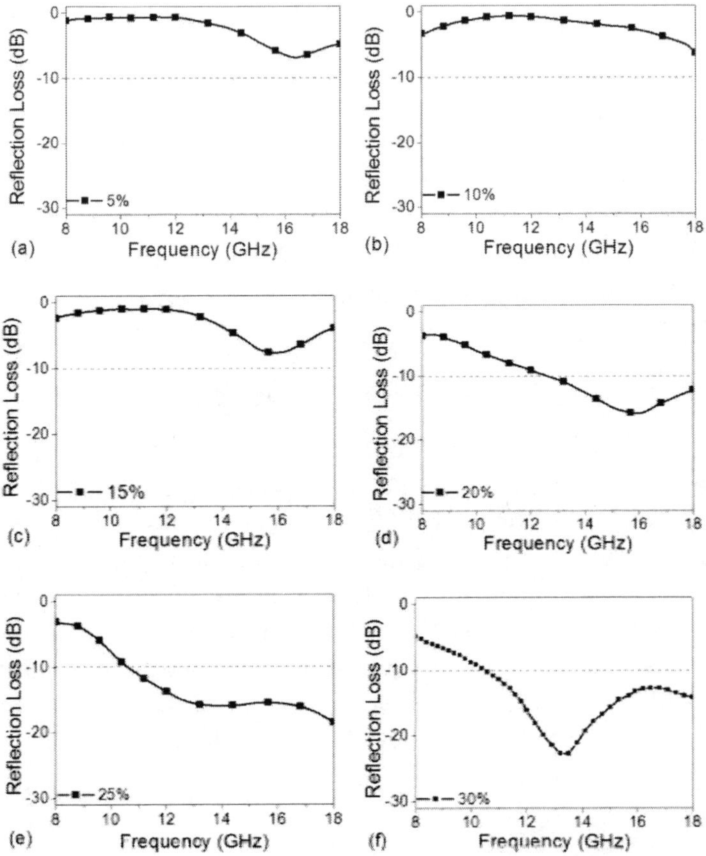

Figure 6.27 The frequency dependence of RL of the CP/EP composite with different percentage lossy fillers: (a) 5 wt%, (b) 10 wt%, (c) 15 wt%, (d) 20 wt%, (e) 25 wt%, and (f) 30 wt%.

EP is an insulator and a nonmagnetic material, so it is transparent to EM waves[75]. In the CP/EP composite, EP only functions as a matrix. Moreover, new substances and chemical bonds, which can provide electrical dipoles for the composite, will be hardly formed from the interactions among EP, CB, and BT. Thus, it is considered that the electric loss of the composite mainly stems from contribution of CB and dielectric loss of the composite primarily due to BT. It is known from the result of the electroconductivity that CB has high electric conductivity, so the addition of CB in the EP matrix can provide electric dipoles for the composite. Therefore, the composite can be polarized repeatedly in a high-frequency electric field, which causes the electric energy to be transformed into other forms of energy, mainly heat energy, and then dissipated. According to the relation $\varepsilon''\ \sigma/\omega$[71], where σ is the electrical conductivity of the material, the ε'' value of the CP/EP composite should increase with the increase of CP percentage. That is, the electric loss of the composite will increase as the CP percentage in the matrix increases. BT is a type of dielectric material, the most prominent characteristics of which are relaxation and internal electric domain wall motion contribution in the EM field. The relaxation of BT comes from the inertial boundary–leaded piezoelectric resonance and presence of a ferroelastic or 90° domain wall in BT[65–67].

In conclusion, the CP is a kind of dielectric loss material with some magnetic loss according to the dielectric constant. The microwave RL of the CP/EP composite indicates that the CP is a kind of good microwave loss material in the X and Ku bands with the increased CP percentage when the CP filler in the matrix is higher than 15 wt%.

6.4.3 Double-Layer Absorber: α-Manganese Dioxide and Carbon Black Composites

Currently, research on manganese dioxide (MnO_2) focuses on its application in batteries, such as the Zn/MnO_2 battery, Mg/MnO_2 battery, alkaline Zn/MnO_2 battery, and rechargeable alkaline Zn/MnO_2 battery[72,73]. Also, MnO_2 has been used as an important material for Mn-Zn ferrites. But there are few research papers on its absorbing properties except those reported by Guan et al., who found that MnO_2 can be used as an absorber with good absorbing properties[75,76].

In this section, double-layer wave absorbing materials, which are composed of a matching layer and an absorbing layer, were devised. The α-manganese dioxide and CB were used as absorbents in the matching layer and the absorbing layer, respectively.

The mixed proportion of double-layer samples is shown in Table 6.2.

Figure 6.28 shows a typical relationship between the RL and the frequency for samples 1–6, of which the mass fraction of CB in the absorbing layer is 30% and the thickness of the absorbing layer is 3 mm. As for the absorbing properties of the CB/α-MnO_2 composite, the sample thickness has some effects on the EM RL and can influence the material's absorbing peak values and bandwidth. However, the change in thickness also changes the impedance matching between the material and free space, so for each sample with different components, there is one or more optimum thicknesses (matching thickness)[77]. In the case of the mass fraction of α-MnO_2 in the matching layer being constant, the microwave absorbing properties will be worse with decreasing matching thickness. For samples 1–6, the effective absorption band (below −10 dB) reaches 8.6 GHz, 5.1 GHz, 7.6 GHz, 8.7 GHz, 5 GHz, and 5.7 GHz, respectively, and samples 1, 3, and 4 have better absorbing properties than the others.

Table 6.2 Mixed proportion of double-layer samples

Sample	Matching layer		Absorbing layer		Areal density ($kg\cdot cm^{-2}$)
	Mass fraction (wt%)	Thickness (mm)	Mass fraction (wt%)	Thickness (mm)	
1	10% MnO_2	2.0	30% CB	3.0	0.76
2	10% MnO_2	1.5	30% CB	3.0	0.71
3	10% MnO_2	1.0	30% CB	3.0	0.67
4	20% MnO_2	2.0	30% CB	3.0	0.84
5	20% MnO_2	1.5	30% CB	3.0	0.81
6	20% MnO_2	1.0	30% CB	3.0	0.79

Note: In addition to the filler of carbon black or manganese dioxide, the others are the matrix of EP resin and its solidified agent in the samples and *w* is the mass fraction.

A comparison of Figs. 6.28a and 6.28b shows that the absorbing properties of the double-layer absorber will be worse as the content of α-MnO_2 is increasing in the matching layer. For example, the absorbing properties of samples 4, 5, and 6 are worse than those of 1, 2, and 3, respectively. The possible reason for this is that the impedance match between the absorbing layer and the matching layer becomes worse with increasing content of α-MnO_2 in the matching layer, which will lead to a strong reflection wave on the sample surface and will act against incidence and absorption of the transmission EM wave on the sample.

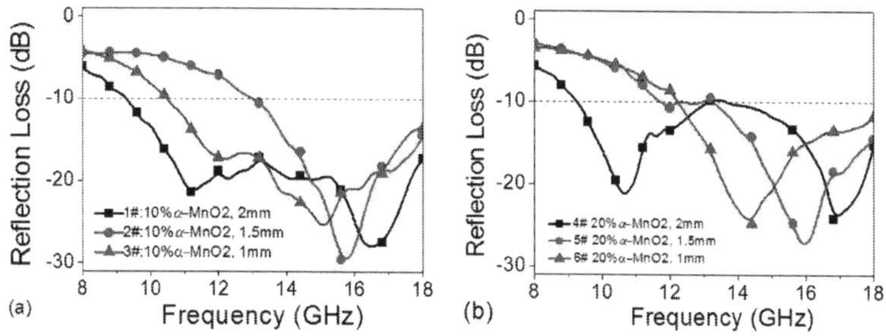

Figure 6.28 Reflection loss curves of double-layer samples. Samples 1–6, of which the mass fraction of CB in the absorbing layer is invariably 30% and the thickness of the absorbing layer is fixed at 3 mm, but the mass fraction of MnO_2 and the thickness of the impedance matching are different.

In this section, when using α-manganese dioxide and CB as absorbents in the matching layer and the absorbing layer, respectively, double-layer absorbers show good absorbing properties. In particular, the double-layer absorbers show an applied effectual absorption band (below −10 dB) when the mass fraction of α-MnO_2 in the matching layer is 10% and the thicknesses of the matching layer are 2 mm and 1 mm, respectively, or when the mass fraction of α-MnO_2 in the matching layer is 20% and the thickness of the matching layer is 2 mm in the case of the mass fraction of CB in the absorbing layer being 30% and the thickness of the absorbing layer being 3 mm.

6.4.4 Double-Layer Absorber: Carbonyl-Iron/Carbon Black

In the present study, a single-layer wave absorbing coating employing carbonyl-iron powder (CIP) as a magnetic loss absorbent and CB as a dielectric loss absorbent with epoxy resin (ER) as a matrix was prepared.

6.4.4.1 Effect of CIP content on microwave absorption properties

Figures 6.29a and 6.29b show the microwave absorption properties of samples with the same CB mass content (20 wt%), different CIP mass ratios (25 wt%, 40 wt%, and 55 wt%), and thicknesses of 1.5 mm (1, 2, and 3) and 2 mm (4, 5, and 6). The detailed data for the microwave absorption properties of the samples are listed in Table 6.3. As one can see, the absorption bands shift toward the lower frequency range with increasing CIP content, but the absorption bandwidths with RL < −4 dB change slightly. So, by adjusting the content of CIP in composites, the wave absorbing coating can be used in different frequency ranges.

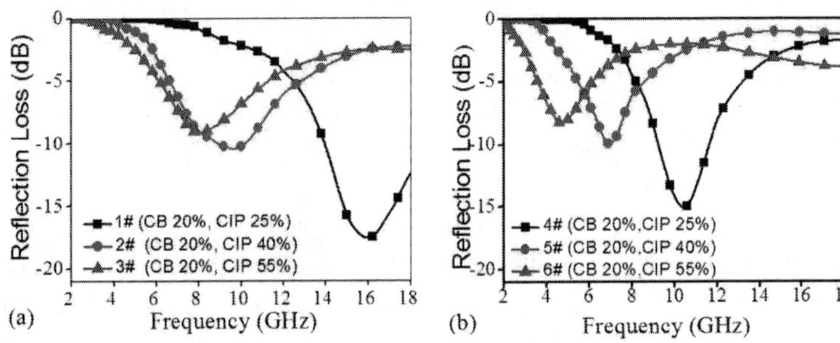

Figure 6.29 Absorption curves for samples with the same CB content, different CIP ratios, and thicknesses of (a) 1.5 mm and (b) 2.0 mm.

The results show that the samples containing higher CIP content have better absorption efficiency in the lower frequency range as a whole. With increasing CIP content, the number of magnetic poles in the nonmagnetic matrix per unit volume rises. As a result, both μ' and μ'' are enhanced. Especially, higher μ'' is very beneficial to improve the microwave absorption properties in the lower frequency range.

6.4.4.2 Effect of CB content on microwave absorption properties

Figures 6.30a and 6.30b show the variations of RL versus frequency in the range of 2–18 GHz for samples with thicknesses of 1.5 mm (7, 1, and 8) and 2 mm (9, 4, and 10). The coatings are composed of the same CIP mass ratio (25 wt%) and different CB mass ratios (15 wt%, 20 wt%, and 25 wt%). The microwave absorption properties of the six samples are shown in Table 6.3. From the data in Table 6.3 and curves in Fig. 6.30a,b, with increasing CB content, the absorption bands of the samples shift toward the lower frequency range, and the absorption bandwidths (RL < −4 dB) are obviously enhanced for the samples with a thickness of 1.5 mm, especially the bandwidth of sample 8 reaches 10.1 GHz (7.9–18 GHz).

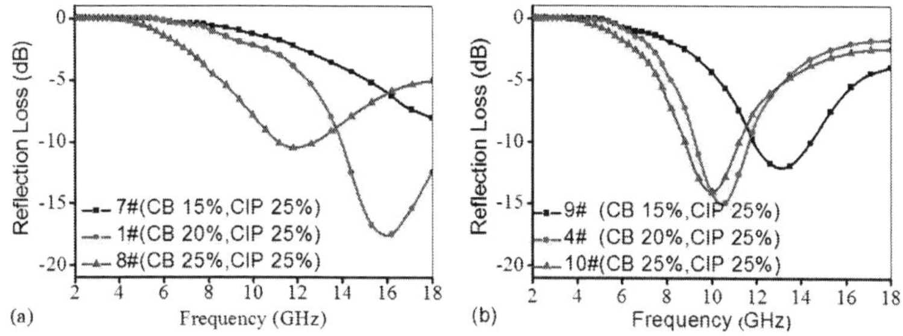

Figure 6.30 Absorption curves for samples with the same CIP content, different CB ratios, and thicknesses of (a) 1.5 mm and (b) 2.0 mm.

Table 6.3 Microwave absorption properties of the samples

	Frequency of absorption peak	**Minimum RL**	**Absorption band (GHz)**
Sample	**Absorption (GHz)**	**Value (dB)**	**RL < −4 dB**
1	15.9	−17.5	11.9–18
2	9.6	−10.5	6.3–13.8
3	7.8	-9.1	5.7–12.3
4	10.4	−15.1	7.9–13.8
5	7	-9.9	5.1–9.2
6	4.7	-8.3	3.3–6.7
7	18	-7.9	14.2–18
8	11.9	−10.6	7.9–18
9	13.1	−12	9.7–18
10	9.9	−14	7.4–14.2
11	18	-5.5	15.8–18
12	10.6	-7.1	7.1–15.4

The results may be from the following factors: Firstly, the dielectric polarization is increased because of higher CB content. Secondly, the more CB particles in the matrix per unit volume, the more surface areas provided to attenuate the EM wave by multiscattering and reflection. Finally, the increasing CB content is helpful to form a conductive network, and thus the EM wave is attenuated by eddy current loss.

As can be seen from Fig. 6.30a,b, the wave absorbing coating used in different frequency ranges could be obtained by tailoring the content of CB. Moreover, CB has low cost and density, which is significant to make the coating apply in practice.

6.4.4.3 Effect of thickness on the microwave absorption properties

Figures 6.31a and 6.31b show RL curves of samples with the same CB ratio (20 wt%), different thicknesses (1 mm, 1.5 mm, and 2 mm), and CIP content of 25 wt% (11, 1, and 4) and 55 wt% (12, 3, and 6). Some key data for the microwave absorption properties of the samples are shown in Table 6.3. According to Fig. 6.31 and Table 6.3, the absorption peaks of samples shift obviously toward the lower frequency range with increasing thickness of the coating.

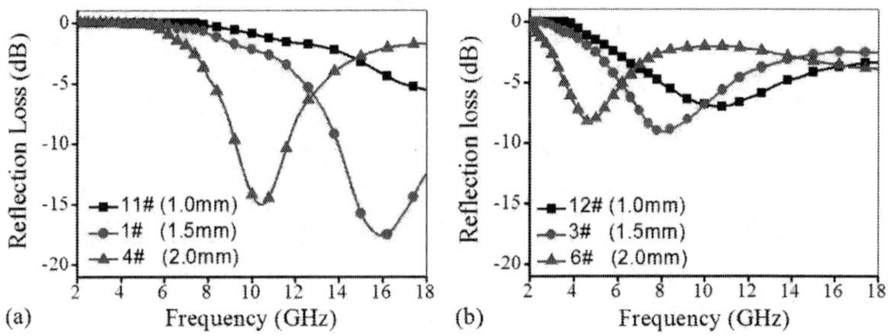

Figure 6.31 Absorption curves for samples with different thicknesses: (a) CIP 25%, CB 20% and (b) CIP 55%, CB 20%.

From the above results, it is found that increasing the thickness of an absorber could improve the microwave absorption properties of the coating in the lower frequency range. The phenomenon can be explained by the equation[78]

$$f_m = c/2\pi\mu'' d \tag{6.8}$$

where f_m, c, and d are the matching frequency (the frequency of absorption peak), the velocity of light, and the sample thickness, respectively[79]. This equation indicates that the matching frequency f_m shifts toward a lower frequency with increasing sample thickness. So the microwave absorption properties of the samples can be affected by thickness.

In this section, the microwave absorption properties of single-layer coatings composed of CIP and CB are studied in the 2–18 GHz range. The effective frequency bands shift toward the lower frequency range with increasing CIP ratio, CB content, or thickness. So the wave absorbing coating could be applied in the desired frequency range through adjusting the content of CIP or CB in the composites. Besides, the absorption bandwidth (RL < −4 dB) can be enhanced by increasing the CB content for the samples with the thickness of 1.5 mm; especially for the sample with a CIP ratio of 25 wt% and a CB content of 25 wt%, the absorption bandwidth reaches 10.1 GHz (7.9–18 GHz).

6.4.5 Double-Layer Absorber Reinforced with Carbon Fiber Powders

The objective of this section is to fabricate microwave absorbing coatings reinforced with carbon fiber powder (CFP) in matching and absorbing layers, respectively.

6.4.5.1 Section morphology

Figure 6.32a,b shows the morphologies of the sectional structure of coatings with absorbents carbonyl-iron (CI)/CB and CFP in each layer. It is clear that CI and CFP are well separated in the ER, as shown in Fig. 6.32a, as well as CB and CFP being uniformly distributed, as shown in Fig. 6.32b. Meanwhile, the scanning electron microscopy (SEM) images indicate that the boundary between the matching layer and the absorbing layer is quite close and flat. The morphology of the absorbent CFP is also characterized by SEM, as shown in Fig. 6.32c. It indicates that the mean diameter of CFP is about 7 μm and the length–diameter ratio is uneven, which is located in a large range.

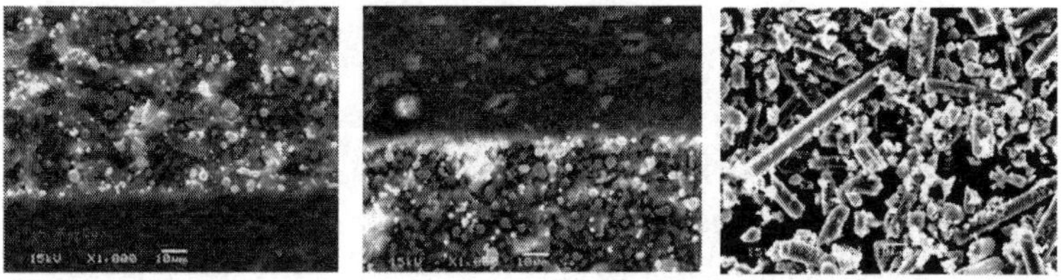

(a) (b) (c)

Figure 6.32 Images (SEM) of fracture of double-layer coatings filled with CI, CB, and CFP: (a) CFP in the matching layer, (b) CFP in the absorbing layer, and (c) morphology of CFP.

6.4.5.2 Microwave absorption properties

The thicknesses and mixed proportions of the samples are shown in Table 6.4.

Table 6.4 Mixed proportions of double-layer samples

Sample	Matching layer (CFP:CI:ER)	Thickness (mm)	Absorbing layer (CFP:CB:ER)	Thickness (mm)
1	0:3:1	1	0:0.3:1	0.6
2	0.2:3:1	1	0:0.3:1	0.6
3	0.3:3:1	1	0:0.3:1	0.6
4	0.5:3:1	1	0:0.3:1	0.6
5	0:4:1	1	0:0.25:1	0.6
6	0:4:1	1	0.17:0.25:1	0.6
7	0:4:1	1	0.25:0.25:1	0.6
8	0:4:1	1	0.33:0.25:1	0.6

The effect of CFP in the matching layer on the microwave absorption properties is explained by the RL, which is shown in Fig. 6.33a, and the detail data are listed in Table 6.5. Combining the analyses of Fig. 6.33a and Table 6.5, it can be seen that there exists an optimal content to obtain the minimum value of RL, and further improvement of CFP content leads to the worse trend of the peak value on the contrary. From the table, one can see that the microwave absorbing performance is not improved by increasing CFP content. The optimal RL is −15.7 dB and the effective bandwidth (RL < −10 dB) is as wide as 3.2 GHz for the CFP-free composite coating.

However, the effective absorption bandwidth (RL < −10 dB) is 3 GHz and the peak value reaches −21.7 dB due to the addition of CFP, which is acting as electric dipoles, when the CFP content ratio rises to 0.3:3:1. It is obvious to see that the additional CFP is not beneficial to expand the absorbing bandwidth. Theoretically, according to transmission-line theory, the incident wave is partially transmitted by the absorber and partially reflected by the absorber. The RL is expressed as follows[80,81]:

$$R = 20\log|\Gamma| = 20\log\left|\frac{Z_{in} - Z_0}{Z_{in} + Z_0}\right| \tag{6.9}$$

$$Z_{in} = Z_1 \frac{Z_1\tanh(\gamma_1 d_1) + Z_2\tanh(\gamma_2 d_2)}{Z_1 + Z_2\tanh(\gamma_1 d_1)\tanh(\gamma_2 d_2)} \tag{6.10}$$

where Γ is the reflection coefficient, Z_0 is the intrinsic impedance of free space, Z_{in} is the input impedance, and Z_1, γ_1, and d_1 are the characteristic impedance, propagation constant, and the coating thickness in the matching layer, respectively. The Z_2, γ_2, and d_2 are the characteristic impedance, propagation constant, and coating thickness in the absorbing layer, respectively. Furthermore, the propagation constant and characteristic impedance can be expressed as a function of complex permeability and complex permittivity:

$$\gamma = \frac{j\omega\sqrt{\mu_0\varepsilon_0}\sqrt{(\mu_r' - j\mu_r'')(\varepsilon_r' - j\varepsilon_r'')}}{c} \tag{6.11}$$

$$Z = \sqrt{\frac{\mu_0}{\varepsilon_0}}\sqrt{\frac{\mu_r' - j\mu_r''}{\varepsilon_r' - j\varepsilon_r''}} \tag{6.12}$$

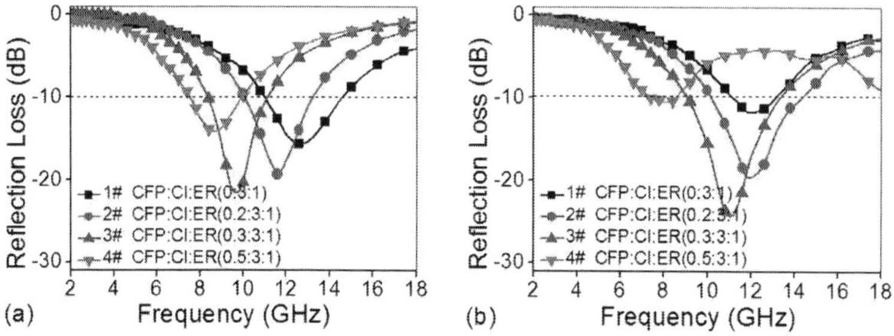

Figure 6.33 Reflection loss curves of double-layer samples for samples 1–8: (a) different CFP content in the matching layer and (b) different CFP content in the absorbing layer.

Table 6.5 Detailed data for coatings with different CFP content in the matching layer

Sample	Optimal RL (dB)	Absorption band for RL < −10 dB (GHz)	Frequency width for RL < −10 dB (GHz)
1	−15.7	11.2–14.4	3.2
2	−19.3	10.2–13.2	3
3	−21.7	8.2–11.2	3
4	−14.2	7.6–9.8	2.2

As deduced from Eqs. 6.9–6.12, the impedance matching condition is determined coherently by the EM parameters ε', ε'', μ', and μ'' and the thickness of the absorber, d. That is to say, the microwave absorbing properties of the absorber can be controlled by modifying the absorbents or designing the structures. All these methods have been verified to be effective to realize controllable absorbing characterization.

The addition of CFP to the matching layer changes the input impedance to some degree. This modification works mainly through altering the EM parameters in the matching layer, as indicated in Eqs. 6.10–6.12. At a certain content, the CFP:CI:ER mass ratio in the matching layer is 0.3:3:1, and the input impedance after modifying is closest to the intrinsic impedance of free space. As a result, the minimum RL value is obtained at its peak frequency. The further increase of CFP content will destroy the impedance matching condition, leading to the increase of the absorption peak. In conclusion, the addition of CFP to the matching layer has almost no effect on the absorbing bandwidth, while the peak values and the peak frequencies are significantly influenced.

Figure 6.33b shows the absorption properties of absorbers with CFP added to the absorbing layer in different content, and Table 6.6 gives the corresponding data in detail. It should be noted that the minimum RL value reaches −25 dB at 11 GHz and the effective absorption bandwidth (RL < −10 dB) reaches 4.5 GHz when the CFP:CB:ER mass ratio is 0.25:0.25:1. In comparison, the optimal RL is only −11.8 dB and the bandwidth (RL < −10 dB) is just 1.7 GHz for the CFP-free composite coating. The only difference between these two samples is the appearance of CFP in the absorbing layer, so the above improvement is attributed to the addition of CFP.

The information obtained from Table 6.6 indicates that a certain amount of CFP in the absorbing layer can obviously enhance the absorption properties both in the bandwidth and the peak value, while too much CFP goes against the improvement of properties compared to the CFP-free sample. The reason can be explained as follows: Both CB and CFP are conductive, which forms a conductive network in the absorbing layer. The conductive network works by producing eddy loss when the microwave enters the absorbing layer. Consequently, more of the EM wave is attenuated by this layer. However, a stronger eddy current is formed if the CFP content exceeds a certain amount, which induces stronger reflection and hence weaker absorption[82]. Sample 8 is in such a condition. The added CFP in the absorbing layer acting as electric dipoles can also resonate with the incident wave besides the original loss mechanisms[83]. This is another reason for the increase of microwave absorption properties. In other words, the addition of CFP to the absorbing layer in an appropriate amount is beneficial to improve the absorption properties both in the bandwidth and the peak values, but the result turns out to be poor once the additional amount exceeds a certain degree.

Table 6.6 Detailed data for coatings with different CFP content in the absorbing layer

Sample	Optimal RL (dB)	Absorption band for RL < -10 dB (GHz)	Frequency width for RL < -10 dB (GHz)
5	-11.8	11.2–12.9	1.7
6	-19.7	10.2–14.1	3.9
7	-25	9.0–13.5	4.5
8	-10.6	7.7–8.4	0.7

6.4.5.3 Comprehensive analysis of additional CFP

Combining the analyses of Figs. 6.33a and 6.33b, it is obvious to see that the absorption peak shifts to a lower frequency with the increase of CFP content in both layers. The effect of adding CFP to the absorbing layer is more obvious than that of the matching layer. Figure 6.34 shows regular changes of the peak frequency in detail. The higher the content of CFP, the lower the peak frequency. From Eqs. 6.9–6.12, it is shown that the direct reason is derived from the changed EM parameters after adding CFP to the coating. According to the Maxwell–Garnett equation, the relationship between the effective permittivity of the composite, ε_{eff}; the permittivity of the matrix, ε_{m}; the permittivity of the infilling, ε_{i}; and the volume fraction of the infilling, f_{i}, can be expressed as follows[84][Shuo, 2015 #53]:

$$\frac{\varepsilon_{\text{eff}}-\varepsilon_{\text{m}}}{\varepsilon_{\text{eff}}+2\varepsilon_{\text{m}}}=f_{\text{i}}\frac{\varepsilon_{\text{i}}-\varepsilon_{\text{m}}}{\varepsilon_{\text{i}}+2\varepsilon_{\text{m}}} \tag{6.13}$$

The above equation can also be transformed as

$$\varepsilon_{\text{eff}}=\frac{3\varepsilon_{\text{m}}}{1-f_{\text{i}}\dfrac{\varepsilon_{\text{i}}-\varepsilon_{\text{m}}}{\varepsilon_{\text{i}}+2\varepsilon_{\text{m}}}}-2\varepsilon_{\text{m}} \tag{6.14}$$

Analogously, the effective permeability of the composite can be indicated using Eqs. 6.13–6.14 by replacing ε with μ. As reported, Eqs. 6.13–6.14 are generally introduced to predict the EM parameters of the composite qualitatively, especially the variation trend after changing the composition of the composite. It can be deduced from Eq. 6.14 that the increased volume fraction of the infilling, f_{i}, leads to increased effective permittivity. This condition is also suitable to the variation of effective permeability. So the mentioned theory explains that the permittivity and permeability of the composite increase with the increased content of CFP. To understand the shift of peak frequency, the quarter-wavelength resonance equation is introduced[81]:

$$f_{\text{m}}=\frac{c}{4d_{\text{m}}\sqrt{\varepsilon\mu}} \tag{6.15}$$

where f_{m} is the matching frequency, c is the velocity of light, d_{m} is the thickness of the absorber, and ε and μ are the permittivity and permeability of the composite, respectively. Therefore, combining the analyses of Eqs. 6.13–6.15, the variation trend could be explained well. The increased content of absorbent CFP in the composite leads to the increase of effective permittivity and permeability[85]. Furthermore, the increased EM parameters result in the decrease of the matching frequency, as shown in Fig. 6.34. In other words, the matching frequency shifts to a lower frequency when the content of CFP increases.

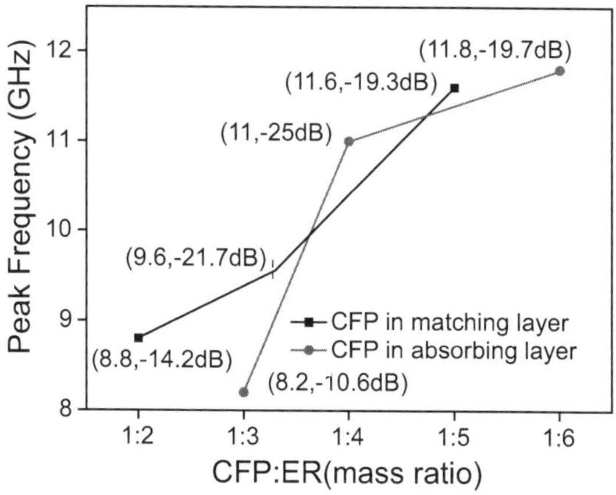

Figure 6.34 Variation of absorption peak frequency versus different CFP content in both the matching layer and the absorbing layer.

In this section, double-layer absorbing coatings were designed with CFP included as an additional absorbent both in the matching layer and the absorbing layer. Results showed that a certain amount of CFP in the absorbing layer was beneficial to broaden the bandwidth as well as minimize the value of RL due to the formation of a conductive network. However, too much amount of CFP induced a strong eddy current, which enhanced the reflection of the microwave. In terms of the bandwidth and peak value, the addition of CFP to the absorbing layer was more helpful in improving absorption properties compared to adding CFP to the matching layer. The optimal content of CFP:CB:ER in the absorbing layer was 0.25:0.25:1, and the corresponding absorbing peak value was −25 dB with the bandwidth of 4.5 GHz (RL < −10 dB). The Maxwell–Garnett equation, together with the quarter-wavelength resonance equation, explained the shift of peak frequency toward a lower frequency when the content of CFP increased. The extra addition of CFP in the absorbing layer was claimed to be an effective way to reinforce the absorption properties.

6.4.6 Influence of Matching Fillers SiO_2

In this section double-layer microwave absorption coatings were fabricated using CI and CB as absorbents in the matching layer and the absorption layer, respectively. SiO_2 was firstly introduced into the matching layer to improve matching impendence and provide tunnels for EM waves. The microwave absorption properties of these samples were then evaluated.

6.4.6.1 Section morphology

Figure 6.35 represents the microstructure of the coating with CB, CI, and SiO_2. The SEM image shows that CI and SiO_2 particles are well separated by ER and do not markedly agglomerate. The matching layer and absorption layer bond with each other closely and the boundary is flat.

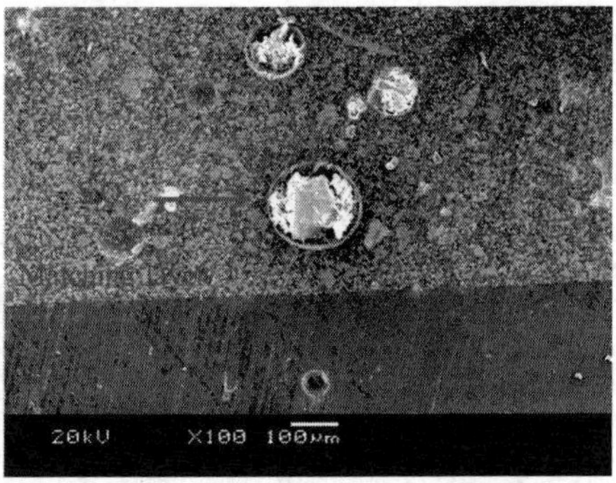

Figure 6.35 SEM micrograph of the fracture surface of CIPs, SiO_2, and CB nanopowder-filled double-layer coating.

6.4.6.2 Microwave absorption properties

Figure 6.36 shows the dependence of the RL for the double-layer coatings on the mass fraction of CI, where the absorption layer and the matching layer are of the same thickness—0.6 mm. As can be seen, with the increase in CI weight concentration (varied from 3:1 to 6:1) in the coatings, the optimal RL is found to increase from −5.9 to −14.8 and the absorption peaks move toward the low-frequency region. It is because with the CI content increasing, the number of magnetic poles in the nonmagnetic matrix per unit volume will rise. As a result, the values of μ' and μ'' are both enhanced with the increasing CI, which could be proved by scale-dependent effective medium theory (SDEMT)[86]. It can be easily concluded that the larger value of μ'' results in good matching properties at a lower frequency.

So increasing the CI content gives rise to a decline of matching frequency.

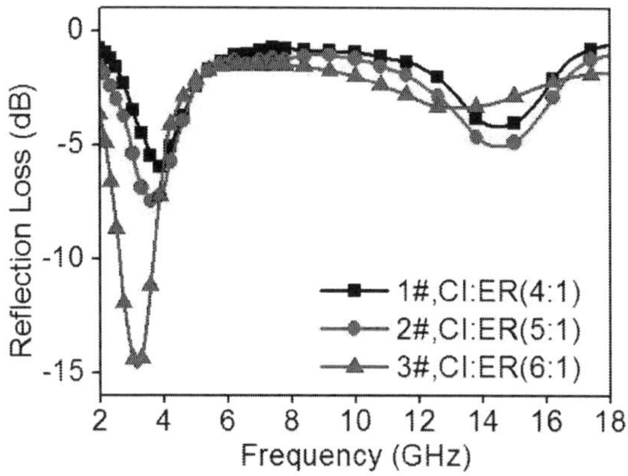

Figure 6.36 Frequency dependence of the RL for coatings with different CI weight concentrations.

It is worth noting that the absorption bandwidths (RL better than −4 dB) of the coatings with the CI weight concentrations (CI:ER) of 3:1, 5:1, and 6:1 are 2.6 GHz (3.1–4.5, 13.9–15.1 GHz), 4.1 GHz (2.7–4.6, 13.4–15.6 GHz), and 2.0 GHz (2.0–4.0 GHz), respectively. It can be found that the bandwidth is enhanced with the weight ratio (CI:ER) increasing from 3:1 to 5:1, but when the content reaches 6:1, the bandwidth becomes narrow. The possible reason for this is that when the mass fraction of CI exceeds the threshold value, the impedance matching between the absorption and the matching layer becomes poor and a large number of EM waves is reflected at the surface of coatings. The strong reflection wave on the matching layer leads to weakening of the absorption properties of the coatings. As shown in Fig. 6.36, the bandwidths of the second peaks which are mainly contributed by CB begin to narrow down when the CI weight ratio reaches 6:1.

The effect of SiO_2 on the microwave absorption properties is shown in Fig. 6.37 and the detailed data are shown in Table 6.7. The microwave properties are improved with increasing SiO_2 content. The minimum RL value reaches −17.3 dB and the effective absorption band (<−4 dB) reaches 5.7 GHz when the SiO_2 weight ratio (SiO_2:CI:ER) rises to 2:5:1. In comparison, the optimal RL is only −5.9 dB and the bandwidth (<−4 dB) is just 4.1 GHz for the SiO_2-free composite coating. It is obviously that the improvements in EM properties are attributed to the additional SiO_2.

Table 6.7 Detailed data for coatings with different SiO_2 content

Sample	Optimal RL (dB)	Absorption band for RL < −4 dB (GHz)	Frequency width for RL < −4 dB (GHz)
2	-5.9	2.7−4.6,13.4−15.6	4.1
4	−10.8	2.6−4.7, 12.5−15.0	4.6
5	−13.0	2.5−4.0, 12.2−15.0	4.3
6	−17.3	2.3−4.5, 11.4−14.9	5.7

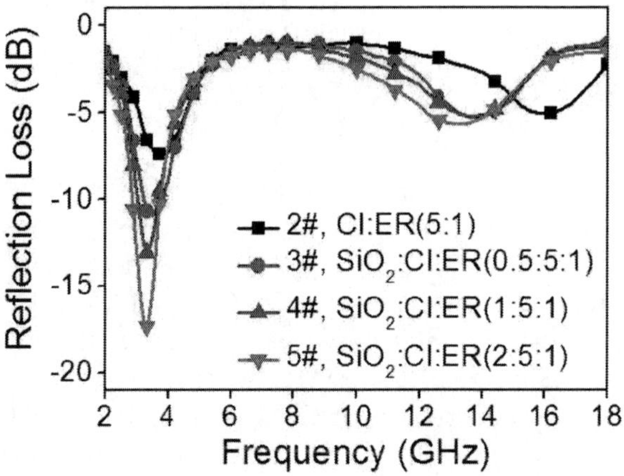

Figure 6.37 Frequency dependence of the RL for coatings with different SiO_2 weight concentrations.

Owning to the low complex permittivity of SiO_2, the input impedance Z_{in} of the coating with SiO_2 is larger than that without SiO_2 and near to the intrinsic impedance Z_0 of free space[87,88]. It can be inferred that the reflection coefficient Γ and the RL value of the coating with SiO_2 are smaller than those of the SiO_2-free composite coating. It is obvious that the additional SiO_2 provides an effective solution to the impendence matching of the coating, ensuring the EM wave transmits into the coating. At the same time, the SiO_2, bringing about channels for the EM wave as wave transmission material, plays an important role in expanding the absorption bandwidth.

On the other hand, the CI powders are separated by SiO_2. Free charges' polarization can occur on CI/SiO_2 interfaces because of the conductivity disparity between interfaces, which then leads to high permittivity[89]. Moreover, the high ε'' caused by the phase lag may benefit the EM losses. High EM parameters, especially the dielectric loss induced by the intense dielectric polarization on conductor/insulator interfaces, are believed to contribute to the high microwave RL performance. Liu[90,91] obtained high RL performance in Fe/ZnO-based coatings and believed that the multipolarization on insulator/conductor interfaces is a critical loss mechanism. In conclusion, the excellent microwave absorption properties of the coatings mainly result from the EM impedance matching and polarization at the CI/SiO_2 interfaces.

In this section, double-layer composite absorbing coatings with a thickness of 1.2 mm were fabricated with CI and CB as absorbents in the matching layer and the absorption layer, respectively. The SiO_2 was employed as a matching filler to improve the matching impendence of the coatings. The microwave absorption properties are improved with the increase of CI concentration. But when the CI content (CI:ER) exceeds the threshold value of 5:1, the bandwidth of the coating begins to narrow down because of the weakened matching impendence. The SiO_2 can improve the matching impendence and has great influence on the absorbing performance of the composite coating. With increasing SiO_2 content in the matching layer, the wave loss is enhanced. When the SiO_2 weight concentration (CI:SiO_2:ER) gets to 2:5:1, the optimal RL reaches −17.3 dB and the effectual absorption band (<−4 dB) is 5.7 GHz.

6.5 Polyurethane Varnish Composites

6.5.1 PU/CIP composites

In this section, the microwave absorbing coatings with a poly(vinyl chloride) (PVC) sheet as a baseplate are fabricated. The coating is composed of CIP as the absorbent and polyurethane (PU) as the matrix. The absorption properties of PVC-based coatings with different CIP content are investigated and compared with the corresponding Al-based coatings. The effects of the thickness of the

PVC sheet and CIP/PU layer on the absorption performance of PVC-based coatings are further discussed.

6.5.1.1 Section morphology

Figure 6.38 shows the SEM microstructures of a composite coating. The image in Fig. 6.38 reveals uniformly dispersed CIP in the PU matrix without significant agglomeration. The addition of a silane coupling agent has made some contribution to particle dispersion and interaction between the particles and the polymer matrix, which provides a "bridge" between the inorganic filler and the organic matrix, and thus improves the interfacial compatibility, finally making the composite combine closely.

Figure 6.38 Morphology of a composite coating.

6.5.1.2 Microwave absorbing properties

To characterize the microwave absorbing properties of composite coatings based on Al or PVC sheets, the RL versus frequency curves for different CIP content are simulated on the basis of EM characteristics and are shown in Fig. 6.39. The content of CIP varies from 1:3 to 1:7 (PU:CIP mass ratio). The thicknesses of the coating and the baseplate are 2 mm and 3 mm, respectively. The sweeping frequency ranges from 2 to 18 GHz. From Fig. 6.39a, it can be found that the allowable RL (<−10 dB for over 90% microwave absorption) can get in the frequency range of 5.5–13 GHz through varying the component content of the coating. It is worth noting that, in Fig. 6.39b, PVC-based coatings display good absorption properties in a lower-frequency region (2–4 GHz, S band), though the overall performance is poor compared to Al-based coatings.

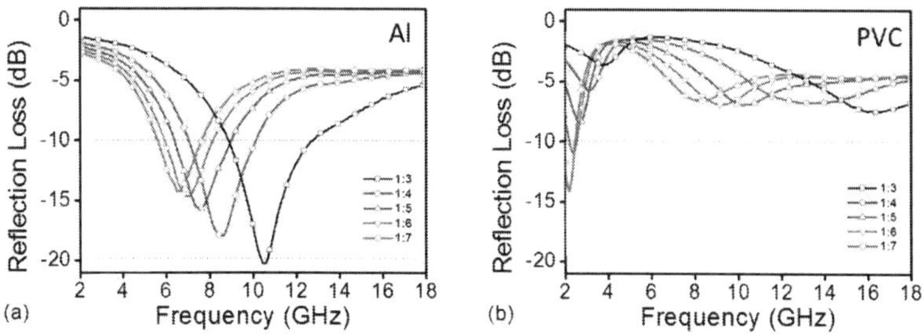

Figure 6.39 Calculated reflection loss versus frequency curves of CIP/PU coatings with different CIP content (PU:CIP mass ratio) based on Al (a) or PVC (b) sheet.

As a validation, the corresponding coatings were fabricated and the properties were measured via the arch method. From Fig. 6.40, one can see that the simulated properties agree well with the measured values except for some individual points, so this method is significant to predict the performance of composite coatings. The detailed comparisons for the absorption peaks of coatings are revealed in Fig. 6.41. On the whole, the peak frequency and minimum RL value of the coating accord with the simulated data. The maximum deviation of the simulated result for the peak frequency and minimum RL value are ca. 1.2 GHz and 7 dB, respectively. These deviations probably result from the errors in the process of preparation and measurement and the idealization of the PVC sheet as a lossless media for microwave radiation.

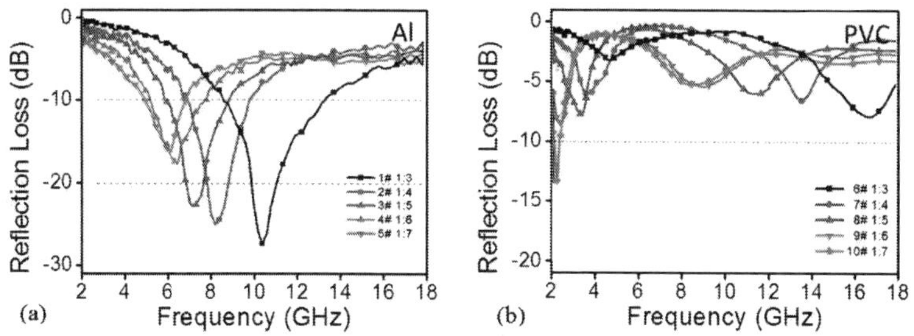

Figure 6.40 Measured reflection loss versus frequency curves of CIP/PU coatings with different CIP content (PU:CIP mass ratio) based on Al (a) or PVC (b) sheet.

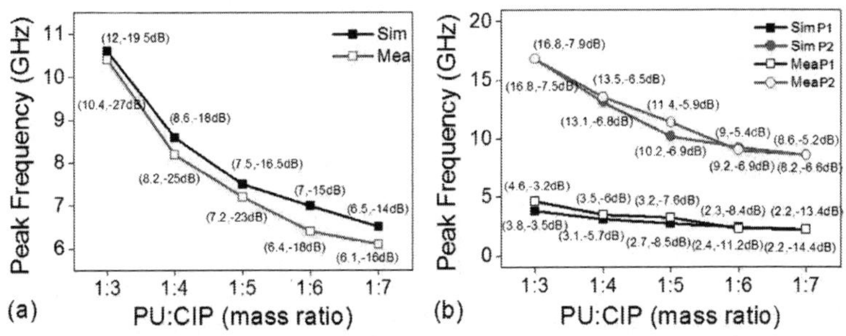

Figure 6.41 Comparison between measured microwave absorbing properties and those simulated of coatings based on Al (a) and PVC (b) sheet.

The classical quarter-wavelength ($\lambda/4$) resonator model can be used to explain the frequency dependence of the peaks in RL curves[81]:

$$f_{\mathrm{m}} = \frac{nc}{4d\sqrt{\varepsilon\mu}} \quad (n = 1,3,5\cdots) \tag{6.16}$$

The peak frequency of the coating is defined as the matching frequency f_{m}, and obviously, f_{m} shifts toward a lower-frequency region with a decreasing mass ratio (PU:CIP), as shown in Fig. 6.41. As can be seen from Eq. 6.16, the peak frequency is relative to the EM parameters and the thickness of the absorber. With increasing CIP content, the EM parameters of the composite coating increase visibly, so f_{m} decreases and the corresponding peak shifts toward a lower-frequency region. It's clearly seen that to obtain high performance at a lower frequency, high concentration of the absorbent or thickness of the coating is needed. However, the exorbitant filling concentration will result in impedance mismatch, namely as being reflected on the surface of the absorber, and the most input EM wave cannot enter the inner part of the composite to be transformed into heat energy. Furthermore, a high filling concentration will also reduce the mechanical properties of the coating, and the application of a coating with a high areal density or thickness is very limited.

Fortunately, a PVC-based coating displays good absorption properties at a lower frequency, which is impossible to achieve for an Al-based coating with the same composite component. The material of PVC is lightweight, economical, and easy to process, so a PVC-based coating is considerably applied in a lower-frequency region. It is remarkable that a PVC-based coating, as a quasi-double-layer absorber with a total thickness of 5 mm, exhibits two peaks simultaneously in the 2–18 GHz range. One peak at a lower frequency is relative to the $\lambda/4$ condition and another at a higher frequency comes from the $3\lambda/4$ condition.

Figure 6.42 shows the predicted RL versus frequency curves of PVC-based coatings with a varied thickness and fixed component content of 1:7 (PU:CIP mass ratio) in the PU/CIP layer. As can be seen in Fig. 6.42a, the absorption peaks of PVC-based coatings with a fixed 2 mm thickness of the PU/CIP layer shift toward a higher frequency with decreasing thickness of the PVC sheet from 3 mm to 1 mm. Similarly in Fig. 6.42b, the absorption peaks of PVC-based coatings with a fixed 1 mm thickness of the PVC sheet move toward a higher frequency with decreasing thickness of th ePU/CIP layer from 3 mm to 1 mm. The permissible RL (<–10 dB) can get in the frequency range of 2–6 GHz for PVC-based coatings through varying the thickness of the PVC sheet or the PU/CIP layer. We believe that the PVC-based coating could be used to make up for the deficiency of current metal-based coatings.

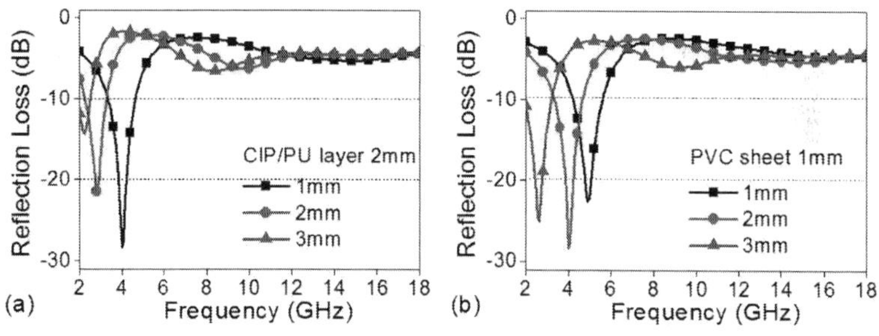

Figure 6.42 Reflection loss versus frequency curves of PVC-based coatings with different thickness of (a) PVC sheet and (b) PU/CIP layer.

In conclusion, the microwave absorbing properties of composite coatings based on Al and PVC sheets in the 2–18 GHz range show that the RL peaks of the coatings shift toward a lower-frequency region by increasing the CIP content or coating thickness. PVC-based coatings with a fixed component content of 1:7 (PU:CIP mass ratio) in the CIP/PU layer exhibit a minimum RL value of –29 dB at 4 GHz and a permissible RL (<–10 dB) band of 2–6 GHz through varying the thicknesses of the PVC sheet and the PU/CIP layer. So PVC-based coatings could be applied in a lower-frequency region, making up for the deficiency of common metal-based coatings.

6.5.2 PU/CB, FSA, and nmCIP Composites

Composite microwave absorbing coatings reinforced with CI nanoparticles were systematically studied. CB, CI, and Fe-Si-Al powders (FSA) are widely used as microwave absorbents, which play important roles in the field of EM protection[91–94]. However, there are several deficiencies that limited the further application of CB and FSA absorbents, for instance, the absorption spectrum is selective, the effective absorption band is too narrow, and the areal density is large. Fortunately, as research focuses on absorbing materials, nanomaterials have made some progress in aspects of a new type of wave absorbing material preparation[95,96]. However, the absorption properties of a single-layer coating are limited. Though the absorption performance can be enhanced by increasing the absorbent content or the thickness of the coating, the effects of these two methods are proven to be very limited. It is an effective way to improve the absorbing properties of the coating using a multilayered impedance-graded structure, but the cumbersome process is also a considerable factor restricting its wide application[97,98]. Therefore, it is the first choice of fabricating highly efficient absorbing materials by employing different loss types of multicomposite absorbents.

In this section, the composite microwave absorbing coatings reinforced with CI nanoparticles (nmCIP) are systematically studied for the first time on the basis of composite coatings that were fabricated with PU as the matrix and CB and FSA as absorbents, respectively. It provides a reference for the design of high-performance wave absorbing coatings by discussing the effects of microwave absorbing properties of composite coatings with different loss mechanisms of absorbents.

6.5.2.1 Section morphology

The SEM images of CB/PU composites with different content are shown in Fig. 6.43. As can be seen, the density of CB particles (small bright spot) increases obviously in the section, and CB is dispersed uniformly without significant agglomeration as the CB content increases from 0.2:1 to 0.6:1 (CB/PU in mass ratio). The addition of a silane coupling agent (M301) is helpful for particle dispersion and interaction between the particles and the polymer matrix, which could provide a bridge between inorganic filler and organic matrix. It not only improves the interfacial compatibility, which is beneficial to increase the filling content of powders, but also reduces the impact on the integrity of the matrix.

Figure 6.43 SEM images of CB/PU composites with different content: (a) 0.2:1, (b) 0.4:1, and (c) 0.6:1.

The SEM images of FSA/PU composites with different content are shown in Fig. 6.44. It can be seen that the density of FSA increases gradually in the section, and FSA is dispersed uniformly without significant agglomeration with increasing FSA content from 3:1 to 7:1 (FSA/PU in mass ratio).

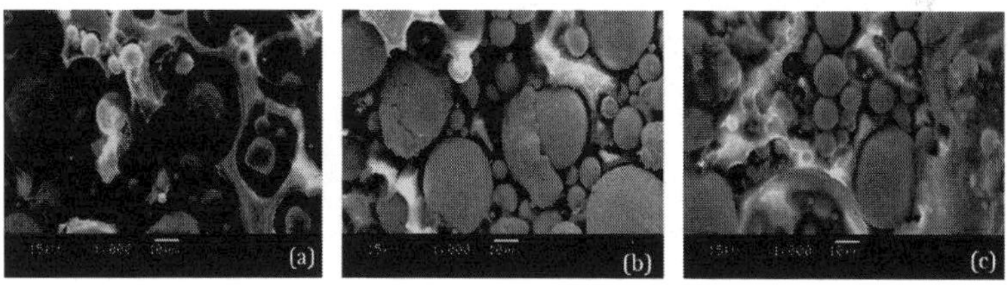

Figure 6.44 SEM images of FSA/PU composites with different content: (a) 3:1, (b) 5:1, and (c) 7:1.

The SEM images of nmCIP-reinforced composite coating sections are shown in Fig. 6.45. The nmCIP disperses uniformly and agglomerate slightly in individual regions of the composite coating; only nmCIP is included as an absorbent in Fig. 6.45a. Agglomeration can be attributed to the strong nanoparticle activity and high surface energy.

Figure 6.45 SEM images of nmCIP-reinforced composite coating sections: (a) nmCIP, (b) CB+nmCIP, and (c) FSA+nmCIP.

Figures 6.45b and 6.45c reveal that the SEM microstructures of CB and FSA coatings were composited with nmCIP separately. But Figs. 6.45b and 6.45c could not exhibit the dispersion of two kinds of absorbents well at the same time. This is because the gap of size between nanoparticles (small bright spot) and metal powders is large and the content of nmCIP is also relatively low.

6.5.2.2 Microwave absorbing properties

Figure 6.46 shows the microwave absorption properties of CB/PU coatings with varied CB content. The result exhibits the absorption peaks of coatings in the range of 2–18 GHz with increasing mass ratio from 0.2:1 to 0.6:1. It can be found that the peak value of RL reaches –1.8, –9.9, and –15.7 dB at 18, 18, and 15.6 GHz, respectively. Moreover, the effective absorption bandwidth (RL < –10 dB for over 90% microwave absorption) is broadened by increasing CB content, which reaches 0, 0, and 3.3 GHz, respectively. Detailed data for the performance of CB/PU coatings are revealed in Table 6.8.

Table 6.8 Detailed data for the performance of CB/PU coatings

Sample	Mass ratio (CB:PU)	Peak frequency (GHz)	Peak value (dB)	Absorption frequency (GHz) RL < -10 dB	Absorption bandwidth (GHz) RL < -10 dB
1	0.2:1	18	-1.8	–	0
2	0.4:1	18	-9.9	–	0
3	0.6:1	15.6	-15.7	14.7–18	3.3

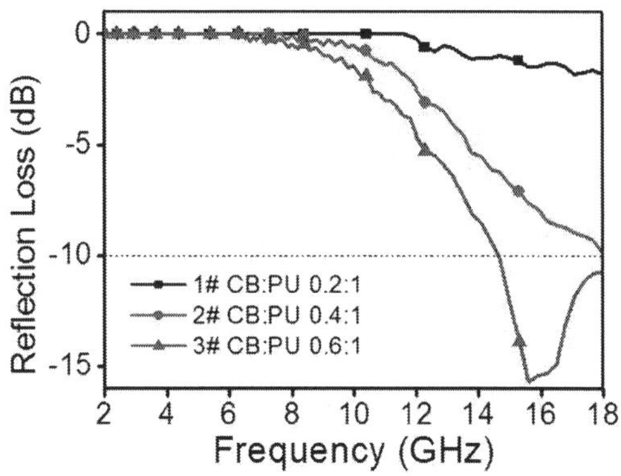

Figure 6.46 Reflection loss versus frequency curves of CB/PU coatings with varied CB content (CB: PU mass ratio).

The microwave absorption properties of FSA/PU coatings with varied FSA content are shown in Fig. 6.47. It can be found that peak frequencies of coatings shift toward a lower frequency range as the FSA content increases. Meanwhile, the peak values gradually weaken with increasing mass ratio from 3:1 to 7:1. The result shows that the peak value of RL reaches −20.6, −14.3, and −10 dB at 14.1, 9.2, and 6.3 GHz, respectively. Moreover, the effective absorption bandwidth also reveals a decreased tendency, which reaches 6.5, 4.7, and 0 GHz, respectively. Table 6.9 gives detailed data for the performance of FSA/PU coatings.

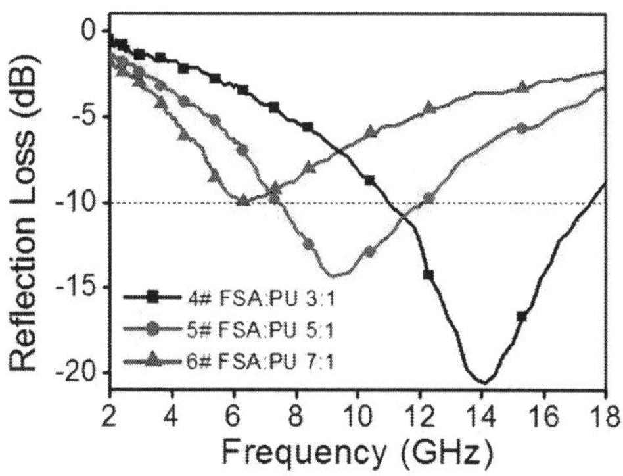

Figure 6.47 Reflection loss versus frequency curves of FSA/PU coatings with varied FSA content (FSA: PU mass ratio).

Table 6.9 Detailed data for the performance of FSA/PU coatings

Sample	Mass ratio (FSA:PU)	Peak frequency (GHz)	Peak value (dB)	Absorption frequency (GHz) RL < –10 dB	Absorption bandwidth (GHz) RL < –10 dB
4	3:1	14.1	–20.6	11–17.5	6.5
5	5:1	9.2	–14.3	7.4–12.1	4.7
6	7:1	6.3	–10	–	0

Therefore, the absorbing properties of coatings are not improved with increasing the absorbent content. When the content of absorbents exceeds a certain threshold, it leads to impedance mismatch between the coating and free space. In this condition, most of the EM wave is reflected on the surface, and the absorbing performance of the coatings also declines greatly.

Figure 6.48 indicates the effects of nmCIP on the microwave absorption properties of composite coatings with PU as the matrix. The results are presented in Figs. 6.48b and 6.48c when CB and FSA coatings were composited with nmCIP, respectively. In addition, detailed data for the performance of nmCIP-reinforced composite coatings can be seen in Table 6.10.

Table 6.10 Detailed data for the performance of nmCIP-reinforced composite coatings

Sample	Absorbent	Peak frequency (GHz)	Peak value (dB)	Absorption frequency (GHz) RL < –10 dB	Absorption bandwidth (GHz) RL < –10 dB
7	nmCIP	18	-8.1	–	0
2	CB	18	-9.9	–	0
8	nmCIP/CB	13.5	-25.8	11.5–15.1	3.6
4	FSA	14.1	-20.6	11–17.5	6.5
9	nmCIP /FSA	11	-15.2	9–15	6

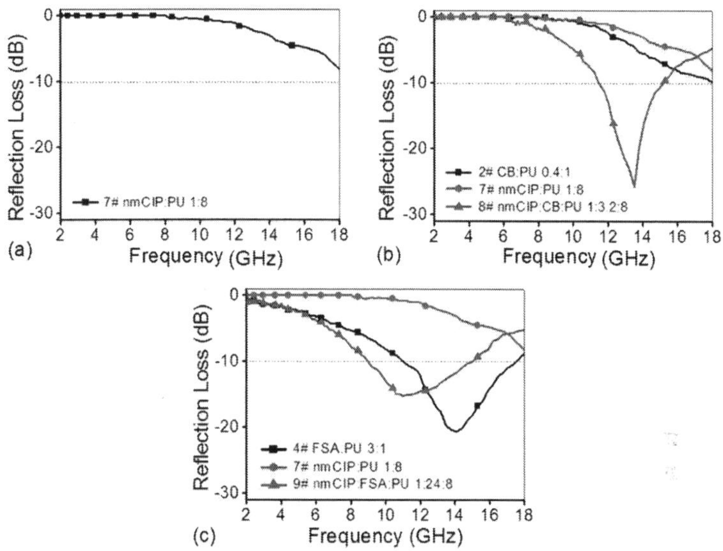

Figure 6.48 Reflection loss versus frequency curves of nmCIP-reinforced composite coatings: (a) nmCIP, (b) CB+nmCIP, and (c) FSA+nmCIP.

It can be seen in Fig. 6.48a that absorption properties are limited when only a low content of nmCIP is included as an absorbent, reaching –8.1 dB at 18 GHz. The conclusion coincides with the preceding analysis of the EM parameters of nmCIP. In this report, nmCIP was composited with other absorbents to improve absorbing performance. Figure 6.48b indicates the absorption properties when nmCIP was composited with CB. The attenuation mechanism of CB is dielectric loss, which is different from nmCIP. It is clear that RL curves shift toward a lower frequency when CB/nmCIP composite absorbents are employed. Compared to of a single CB coating, the absorbing properties of a CB/nmCIP composite absorber are significantly improved: the location of peak frequency moves from 18 to 13.5 GHz, and the peak value shifts from –9.9 dB to –25.8 dB. In addition, the effective absorption bandwidth (RL < –10 dB) also expands from 0 to 3.6 GHz.

Figure 6.48c shows the absorbing performance of an FSA/nmCIP composite absorber. The attenuation mechanism of FSA belongs to magnetic loss, which has the same loss mechanism as nmCIP. It can be found that RL curves shift toward a lower frequency when FSA and nmCIP are employed as absorbents. The peak frequency moves from 14.1 to 11 GHz, and the peak value decreases from −20.6 to −15.2 dB. Additionally, the effective absorption bandwidth (RL < −10 dB) reduces from 6.5 to 6.0 GHz. So the addition of nmCIP does not enhance the absorbing performance of FSA/PU composites. However, a low content of nmCIP can modify the location of the bandwidth.

In this section, results showed that nmCIP was an efficient magnetic loss absorbent with low density. When nmCIP was composited with CB, which has a different loss mechanism than nmCIP, the absorption bandwidth increased effectively: the peak value of the coating reached −15.7 dB at 15.6 GHz, and the effective absorption band (RL < −10 dB) reached 3.3 GHz. In the case of FSA/PU composites, the peak value of the coating reached −20.6 dB at 14.1 GHz, and the effective absorption band (RL < −10 dB) reached 6.5 GHz. When FSA was composited with nmCIP, which had the same loss mechanism, the location of the effective absorption band could be modified with the areal density reduced.

6.5.3 Discrete Slab Absorber: PU/CB/ABS Composites

A slab absorber is an important topic not only for military purposes but also for commercial application[99]. A single-layer compact slab absorber cannot meet the required wave absorption in a broad bandwidth besides some absorption peaks, and it is very necessary to develop a new kind of absorber to meet the need of absorption in a broad bandwidth. In this section, a discrete slab absorber (DSA) has been developed by isolating a conductive medium in an absorbing matrix, and the RL test indicates that the DSA exhibited excellent absorption effectiveness at 8–18 GHz. The discrete absorber was made of isolated conductive particles and was prepared by the way of binding CB/ABS composite particles with a poly(vinyl alcohol) (PVA) binder.

6.5.3.1 Electrical conductivity

Figure 6.49 shows the change of volume resistivity of CB/ABS composites with different CB mass fractions from 5% to 30%. From the chart, it could be seen that the volume resistivity value undergoes a sharp transition over a CB mass fraction range of 10%–15%. When CB loading is 30 wt%, volume resistivity declines to 10^3 Ω·cm.

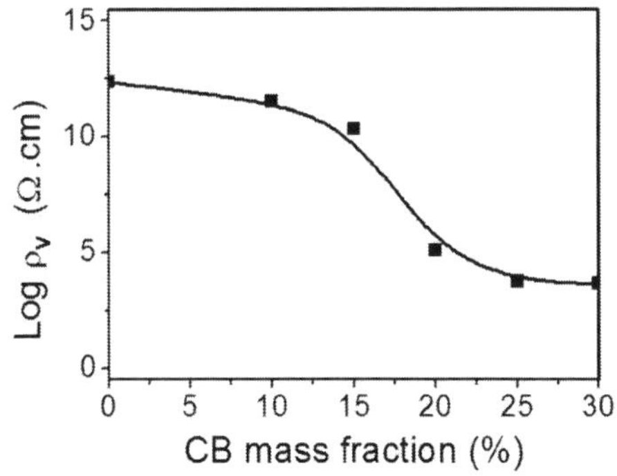

Figure 6.49 Effect of CB mass fraction on the volume resistivity of CB/ABS composites.

6.5.3.2 Microwave absorption properties

Figure 6.50 shows the dependence of RL of the DSA on the mass fraction of CB. As can be seen, with increasing CB concentration (varied from 5 to 30 mass%) in the DSA sample, the EM wave attenuation is increasing in the whole frequency band of 8–18 GHz. It is of interest that when the CB content reaches 20%, a broad characteristic absorption peak of −30.04 dB at 15.9 GHz appears in the frequency loss pattern. The frequency loss pattern of the compact slab sample at various CB concentrations is shown in Fig. 6.51. With the increase of CB concentration in the sample in the range of 5–20 mass%, the RL of the compact slab samples increases in the frequency range of 8–18 GHz. When the CB content obtained was 20 mass%, both characteristic absorption peaks at 10 and 17 GHz reached the maximum.

By comparing the results of Fig. 6.50 and Fig. 6.51, the RLs of the two types of slab samples with the same CB content are different, and wave attenuation of the DSA is greater than that of the compact slab absorber. For the two types of absorbers with a CB content of 30 mass%, the RL of the compact sample (marked with C-4) only appears as two absorption peaks, and then the effective absorption band (below −10 dB) is under 1 GHz. However, the RL of the DSA (marked with D-4) is greater compared to that of the compact sample and is below −10 dB in the 8.5–18 GHz range, with the effective absorption band getting to 9.5 GHz.

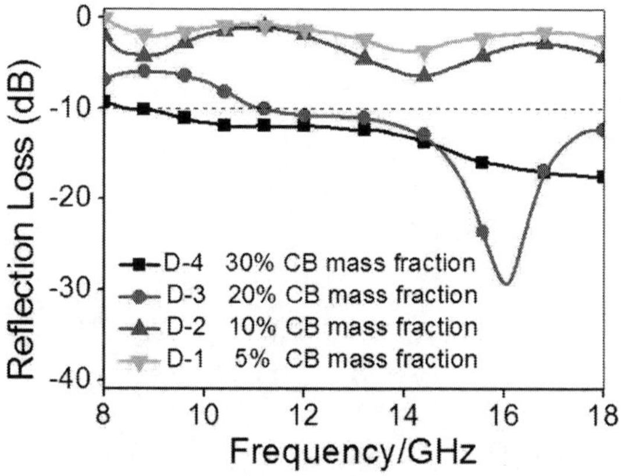

Figure 6.50 Reflection loss of a DSA sample: it was prepared by bonding the CB/ABS cylindrical particles with a PVA binder.

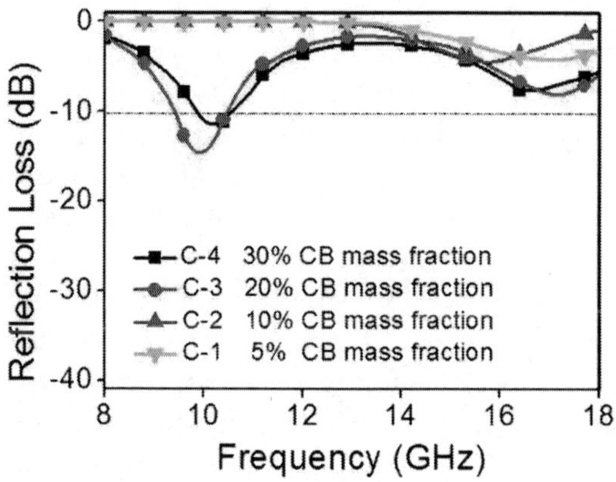

Figure 6.51 Reflection loss of a compact slab sample: it was prepared by hot-compressing a molding with CB/ABS cylindrical particles under 10 MPa at 180°C.

In conclusion, isolated conducting grain mediums can greatly improve the wave loss of a slab absorber in a broad frequency band. Compared to the compact slab absorber, the EM wave loss of the DSA improves greatly. With increasing CB content of the CB/ABS particles, the wave loss of the DSA increases. When the CB mass fraction gets to 30%, the RL is below −10 dB in the 8.5–18 GHz range and the effective absorption bandwidth gets to 9.5 GHz.

6.6 Other Resins

6.6.1 Silicone Rubber/Carbonyl-Iron Composites

As a synthetic rubber, silicone rubber has excellent weathering resistance, resistance to aging, chemical resistance, and insulating properties, and it has good compatibility with many kinds of fillers[100]. Thereby, it is favorable to be the matrix of rubber radar absorbing material.

In this study, the EM parameters and microwave absorbing property of CI were studied using silicone rubber as the base material. The results of the study show that CI is a type of magnetic loss material and performs well in both low and high frequency bands.

6.6.1.1 Influence of carbonyl-iron powder filling ratio

Four types of samples were prepared with CI weight fractions of 50 wt%, 60 wt%, 70 wt%, and 80 wt%. The samples were of 4 mm thickness and were labeled as 1–4.

Figure 6.52 shows the frequency dependence of the reflectivity in the 2–18 GHz range. Results show that with the increase of the CI filling ratio, the reflectivity increases gradually and the absorption peak value shifts to alower frequency band, which can also be seen in Table 6.11.

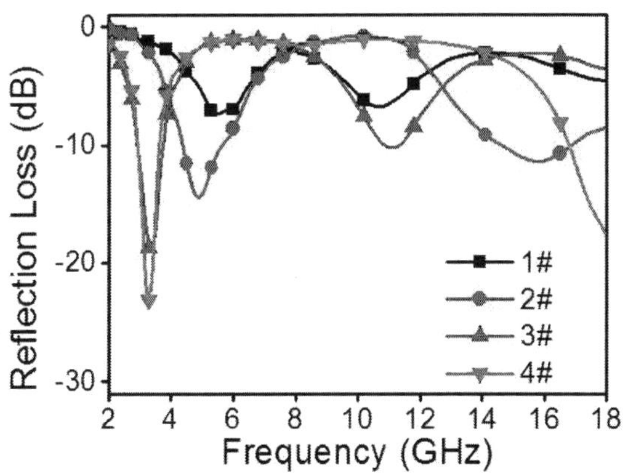

Figure 6.52 Reflectivity of carbonyl-iron at a higher frequency band with a thickness of 4 mm.

Table 6.11 Absorbing peak frequency and values for carbonyl-iron composites

Thickness (mm)	Carbonyl-iron filling proportion			
	50 wt%	**60 wt%**	**70 wt%**	**80 wt%**
2	9.8 GHz (–9.26 dB)	8.4 GHz (–16.26 dB)	6.0 GHz (–20.5 dB)	5.9 GHz (–20.48 dB)
4	5.5 GHz (–7.36 dB)	4.9 GHz (–14.48 dB)	3.4 GHz (–18.92 dB)	3.3 GHz (–23.06 dB)
	–	15.9 GHz (–11.42 dB)	11.2 GHz (–10.28 dB)	10.8 GHz (–6.67 dB)

The reflectivity of sample 1 has only one peak value of –7.36 dB at 5.5 GHz, but it can be deduced from the curve that there must have been another absorption peak at the frequency band above 18 GHz. Sample 2 is better than sample 1, with one peak value of –14.48 dB at 4.9 GHz and a second peak value of –11.42 dB at 15.9 GHz. Sample 3 has shifted its absorption peak to a lower frequency of 3.4 GHz with a value of –18.92 dB, and the second absorption peak also shifts to a lower frequency of –10.28 dB at 11.2 GHz. Sample 4 has the approximate trend with sample 3. The reflectivity peak of sample 4 shifts to a lower frequency than any other samples and it gets a maximum reflectivity of –23.06 dB at 3.3 GHz and the frequency bandwidth at which the RL is less than –10 dB is 0.8 GHz. It has the second peak value of –6.67 dB at 10.8 GHz. With the CI content increasing, the imaginary part and loss tangent of the composite also increased[101] and thus the absorbency of the composite can obviously be raised.

6.6.1.2 Influence of sample thickness

Figure 6.53 indicates the absorbing characteristics of 2 mm thick samples in the 2–18 GHz range with CI loading of 50 wt%, 60 wt%, 70 wt%, and 80 wt%, the samples being noted as 5–8.

Figure 6.53 indicates that there is one absorption peak for each sample in the frequency range of 2–18 GHz, and for samples with the same CI filling proportion, decreasing the thickness shifts the absorbing peak value to a higher frequency band, which is also shown in Table 6.11.

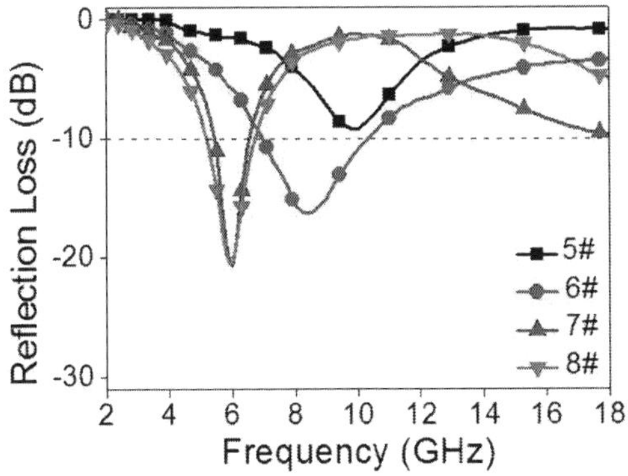

Figure 6.53 Reflectivity of carbonyl-iron with thickness of 2 mm.

Take samples 4 and 8 with the same CI loading of 80 wt%, for example; when the thickness of the sample changes from 4 mm to 2 mm, the absorbing peak values have been changed from 3.3 GHz to 5.9 GHz. From the trend of the absorbing curve of sample 8, it is obvious that the reflectivity decreases drastically with the increasing of frequency over 12 GHz, and we can see that there must have been another peak value at some frequency higher than 18 GHz. It is known that thickness has an obvious influence on EM wave absorbing performance, but their relationship is not simply linear[63]. In a certain range, the number of EM waves scattering and reflection are increased in the composites with an increase in thickness, which also increases the spread optical length, so the thickness increase have some effect on the attenuation of the EM wave.

Comparing samples 3 and 4 in Fig. 6.52 and samples 7 and 8 in Fig. 6.53, we can see that they have almost the same absorbing curve. According to the classical statistical percolation model, when the weight concentration reaches 70%, the volume ratio of the particles reaches the percolation threshold and the conductive network is completed. Under this circumstance, the transmission of the composite is diminished, but the reflectivity is augmented. So the surface reflection of the EM wave is larger and EM wave cannot enter the composite to have effective attenuation. So the increase of the particle content did not have much effect on the absorbing performance.

In conclusion, CI is a kind of magnetic loss absorbent of EM microwaves. With the increase of CI filling and composite thickness, the absorption peak of the CI/silicone rubber composite shifts to a lower-frequency region. When the weight concentration reaches 70%, the increase of the particle content does not have much effect on the absorbing performance.

6.6.2 Double-Layer Absorber: Natural Rubber Composites

Double-layer materials were devised to improve the absorbing properties of EM wave absorbing plates[102]. The double-layer wave absorbing materials are composed of a matching layer and an absorption layer. The matching layer is the surface layer through which most of the incident waves can enter the material, and the absorption layer beneath it plays an important role in wave attenuation. The total thickness of the double layer is the sum of the thicknesses of these two layers. CI and CB were used as absorbents in the matching and the absorption layer, respectively.

The mixed proportion of samples is shown in Table 6.12.

Table 6.12 Mixed proportions of samples

Sample	Matching layer (wt %)	Thickness (mm)	Absorbing layer (wt%)	Thickness (mm)
1	50%CI	2	40%C	2
2	50%CI	2	50%C	2
3	50%CI	2	60%C	2
4	70%CI	2	40%C	2
5	70%CI	2	50%C	2
6	70%CI	2	60%C	2

Figure 6.54 shows a typical relationship between the RL and the frequency for samples 1–6, where the absorption layer and the matching layer are of the same thickness of 2 mm. From the reflectivity curves, one can see that all six samples have two remarkable absorbing peaks, resulting in an additional effective bandwidth in the low-frequency region, across the whole frequency range tested. On the same curve, the value of the first peak at a lower-frequency position is always superior to the second one. For samples

1–3 in Fig. 6.54a, when the mass fraction of CI in the matching layer is fixed at 50%, there are regular changes in the values of the first peak (−23.3, −14.2, and −10.7 dB) and the positions of the second peak (14.5, 13.6, and 11.9 GHz), with CB concentrations from 40% to 60%. The effective absorption bandwidth (below −8 dB) of the three samples reaches 5.5 GHz (4.5–6.5 GHz, 12.7–16.2 GHz), 5.8 GHz (4.5–6.5 GHz, 11.7–15.5 GHz), and 6.5 GHz (4.5–6.5 GHz, 10–14.5 GHz), respectively. In Fig. 6.54b, it can be found that an especially low RL value of −20.6 dB is obtained at 5.3 GHz on the reflectivity curve for sample 4, and an RL less than −4 dB covers almost the whole measuring band (3.3–16.6 GHz) for samples 5 and 6.

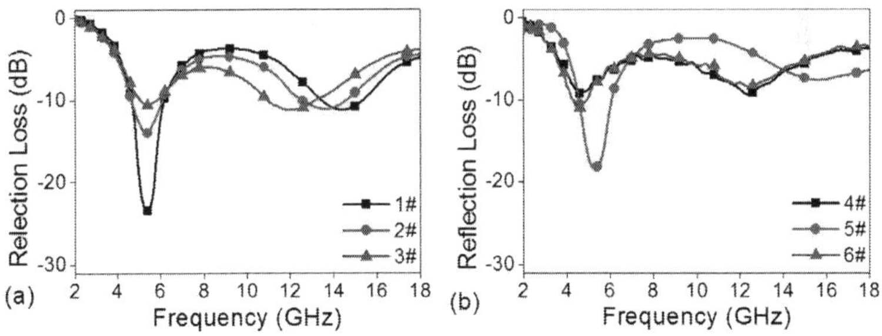

Figure 6.54 RL curves of samples 1–6.

In conclusion, double-layer absorbers with a total thickness of 4 mm, fabricated with CI as the absorbent in the matching layer and CB as the absorbent in the absorption layer, show good absorption properties. There are two clear absorption peaks with significantly different values and positions. In particular, they show an applied effectual absorption band (below −8 dB) when the mass fractions of CB in the absorption layer are 40%, 50%, and 60% and the mass fraction of CI in the matching layer is 50%. The bandwidth of RL values less than −4 dB is larger than 10 GHz for samples composed of 70% CI in the matching layer and 50% or 60% CB in the absorption layer.

6.6.3 Chlorinated Polyethylene/CIP Composites

Chlorinated polyethylene (CPE) (chlorinity: ~35%) is chosen as the polymer matrix due to a series of excellent properties, such as good flexibility, corrosion resistance, etc.[103] What's more, powders can be filled in it easily by the opened refine rubber machine.

To make absorbents disperse in a matrix very uniformly, absorbents and CPE were mixed by extruding with two rolls of the opened refine rubber machine for 10–15 min, then the composites were put into a rubber internal mixer for 30–40 min, and the microwave absorbing sheets were prepared by extruding with two rolls of the opened refine rubber machine ultimately. The thicknesses of microwave absorbing sheets were controlled by adjusting the distance of the two rolls. To measure the microwave absorption properties, the sheets were trimmed into square samples (200 mm × 200 mm).

CIP as the metallic magnetic material is a conventional absorbent and is widely filled in resin or rubber as an EM wave absorber. In a previously published study, Wu et al. discussed that the relative complex magnetic permeability of CIP was influenced by some material properties. But the microwave absorption property of CIP was not concerned[104]. Li et al. prepared one multiband microwave absorption film based on the defective multiwalled carbon nanotube-added CIP/acrylic resin and found that the composites were very promising for fabricating broadband and light microwave absorption materials[105]. However, the microwave absorption films did not show good absorption efficiency in a lower frequency range. Pinho et al. fabricated one CB/CIP/polychloroprene composite absorbing sheet and the microwave absorption properties were discussed only in the 8–16 GHz range[106].

In this section, the EM properties and microwave absorbing mechanism of CIP are described and analyzed, respectively. One single-layer microwave absorbing sheet, which is composed of CIP and CPE, is fabricated.

The mixed proportions and thicknesses of samples are shown in Table 6.13.

Table 6.13 Mixed proportions and thicknesses of samples

Sample	1	2	3	4	5	6	7	8
CIP:CPE (weight ratio)	10:1	12:1	14:1	16:1	16:1	16:1	16:1	17:1
Thickness (mm)	1.0	1.0	1.0	0.5	1.0	1.5	2.0	1.0

According to transmission-line theory, for a single-layer absorbing material backed by a perfect conductor, the input impedance (Z_{in}) at the air–material interface is given by

$$Z_{in} = \sqrt{\frac{\mu_0 \mu}{\varepsilon_0 \varepsilon}} \tanh\left(j2\pi ft\sqrt{\mu_0 \mu \varepsilon_0 \varepsilon} \right) \tag{6.17}$$

where t is the thickness of a single absorber, f is the frequency, μ_0 is the free-space permeability, ε_0 is the free-space dielectric constant, μ is the complex permeability ($\mu' - j\mu''$), and ε is the complex permittivity ($\varepsilon' - j\varepsilon''$).

The RL of the normal incident EM wave at the absorber surface is shown as Eq. 6.9.

Figure 6.55 shows the variation of the RL versus frequency curves in the range of 2–18 GHz for samples 1, 2, 3, 5, and 8 with 1 mm thickness and 10:1, 12:1, 14:1, 16:1, and 17:1 of the weight ratio (CIP:CPE), respectively. From these curves, the minimum RL values for samples 1, 2, 3, 5, and 8 are obtained as −10.7 dB at 10 GHz, -9.6 dB at 8.4 GHz, −9.3 dB at 7.1 GHz, −8.7 dB at 3.9 GHz, and −7.6 dB at 6 GHz, respectively. Meanwhile, the absorption bandwidth, which is the frequency range of RL below −4 dB, is 14 GHz (4–18 GHz), 12.6 GHz (3.8–16.4 GHz), 9.8 GHz (3.2–13.0 GHz), 7.6 GHz (2.0–9.6 GHz), and 8.8 GHz (2.8–11.6 GHz), respectively. It can be seen that the minimum RL values rise with CIP content increasing. Besides, the absorption bandwidth is reduced when the raising weight ratio (CIP:CPE) is less than 16:1, and begins to increase more than the ratio of 16:1.

The frequency of the absorption peak is defined as the matching frequency f_m. It is obvious that f_m shifts toward a lower frequency with increasing weight ratio (CIP:CPE) when it is less than 16:1. However, when the ratio is over 16:1, f_m moves to a higher frequency. Overall, the samples which contain higher CIP content show better absorbing performance in a lower frequency range, except sample 8. The above results could be explained as follows: The magnetic fillers are embedded in a nonmagnetic matrix in the composite material, and the demagnetizing field is generated by the magnetic poles on the surface of the magnetic fillers. Without agglomerative magnetic particles, when the volume loading of the composite material increases, the demagnetizing field nearly keeps unchangeable, but the number of magnetic poles per unit volume will increase. As a result, both μ' and μ'' are enhanced. The higher value of μ'' is very helpful to improve the absorbing property in a low frequency range. However, when the amount of CIP filled into the matrix exceeds a limit, this will result in an impedance mismatch, namely causing wave reflection at the surface of the absorber. Thus the incident EM wave cannot enter the inner part of the composites to be transformed into heat energy. This may explain the result that the absorption characteristic of sample 8 is worse than that of sample 5. Besides, the above result also may be due to the uneven dispersion of CIP in the matrix.

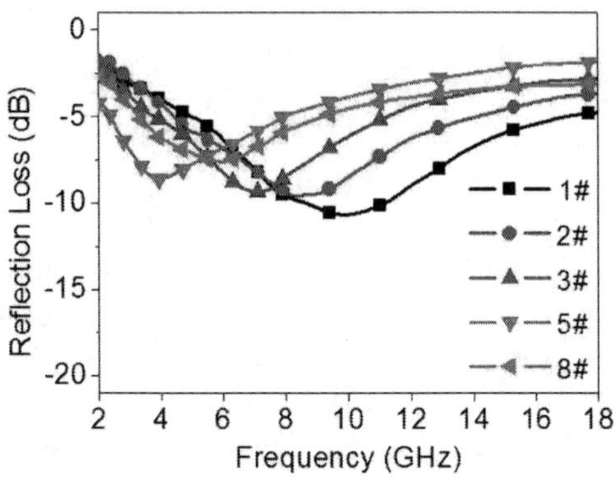

Figure 6.55 Absorption curves for samples with the same thickness (1 mm) and different weight ratios (CIP/CPE).

Figure 6.56 shows the frequency dependence of the RL of samples 4, 5, 6, and 7, which are with thicknesses of 0.5, 1.0, 1.5, and 2.0 mm, respectively, and the same weight ratio (CIP:CPE) of 16:1. As can be seen from these curves, samples 4, 5, 6, and 7 have the minimum RL values of –5.7 dB at 4.9 GHz, –8.7 dB at 3.9 GHz, –13.2 dB at 2.2 GHz, and –16.4 dB at 2 GHz, respectively. Besides, the absorption bandwidths (RL < –4 dB) of the samples are 7.6 GHz (3.1–10.7 GHz), 7.5 GHz (2–9.5 GHz), 4.7 GHz (2–6.7 GHz), and 4.1 GHz (2–6.1 GHz), respectively. It is found that the minimum RL value decreases, the matching frequency f_m moves toward a lower frequency, and the absorption bandwidth becomes narrow with increasing thickness. This phenomenon is consistent with Eq. 6.8[75].

This equation says that the matching frequency f_m shifts toward a lower frequency with increasing thickness. So the microwave absorbing properties of the samples can be affected by thickness. The thicknesses of these samples are smaller than those of the usual ferrite material[107], and so they are good thin MAs. Especially, the bandwidth (RL < −10 dB) of sample 6 reaches 1.1 GHz (2–3.1 GHz), which signifies that sample 6 is very propitious to use in the lower frequency range as one thin absorber.

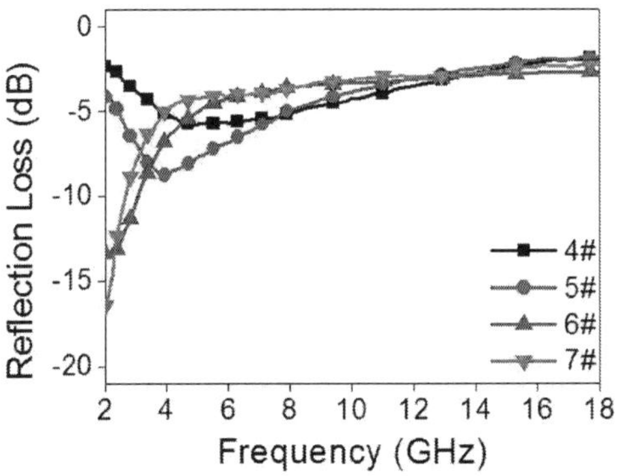

Figure 6.56 Absorption curves for samples with the weight ratio (CIP/CPE) of 16:1 and different thicknesses.

The most significant result of this section is to provide a low-frequency absorber which is composed of CIP and CPE and is thin enough (less than 2 mm). The CIP content in the matrix has an effect on the microwave absorbing properties of the composites. The optimal CIP:CPE weight ratio is 16:1 when the absorber is used at a low frequency. In addition, the absorbing properties at a lower frequency are improved by increasing absorbing sheet thickness. For the sample with a weight ratio (CIP:CPE) of 16:1 and 1.5 mm thickness, the effective bandwidth (RL < −10 dB) reaches 1.1 GHz (2–3.1 GHz).

6.6.4 Water-Based Varnish/Carbonyl-Iron Composites

By analyzing the impedance mismatch, this section evaluates the structural design using an induced current. The intensity of the induced current was reduced using a discrete structure process, which can clearly enhance the absorption ability of a conventional microwave absorber (CMA) due to weakening of the reflection of the EM wave from the surface.

6.6.4.1 Section morphology

Figure 6.57 shows the SEM microstructures of a CIP and composite coatings. Figure 6.57a indicates that CIP is spherical with an average diameter of 1–5 μm, which means the particles can be dispersed uniformly into water-based varnish. Figure 6.57b–d shows that CIP has been uniformly dispersed in water-based varnish without significant agglomeration under different CIP content. As the content of CIP increased, the matrix between neighboring particles became thinner and particles have more possibility of coming into contact.

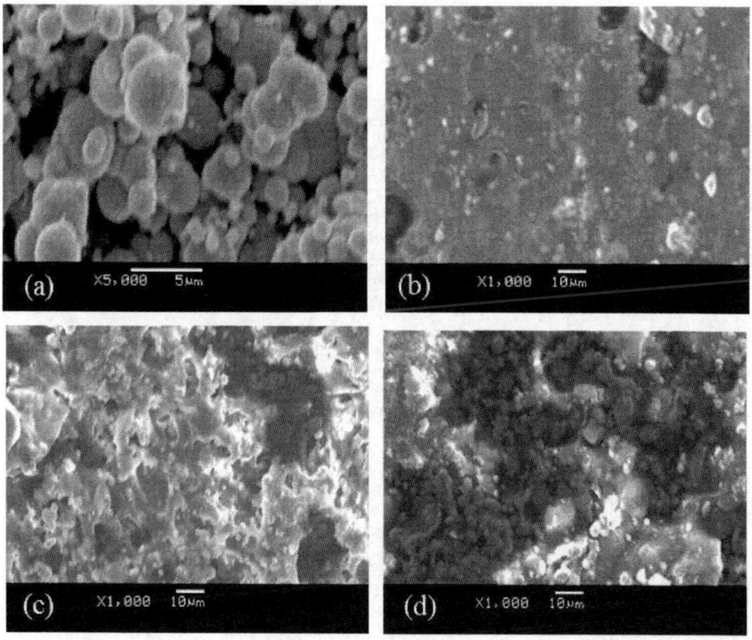

Figure 6.57 Morphologies of (a) carbonyl-iron particles, (b) composite coating (17 wt% CIP), (c) composite coating (33 wt% CIP), and (d) composite coating (50 wt% CIP). As the CIP content increased, the matrix between neighboring particles thinned and particles had greater odds of coming into contact.

6.6.4.2 Microwave absorbing properties

When the EM wave is incident on an absorber, the electric field can induce two different electrical currents, that is, conduction and displacement currents. In general, the displacement current is dominant because most of the charge carriers are bound and cannot participate. However, the situation changed drastically when the content exceeded a percolation concentration, when free charges appeared and the conduction current was produced. The induced currents can create a field that opposes the applied field, as shown in Fig. 6.58.

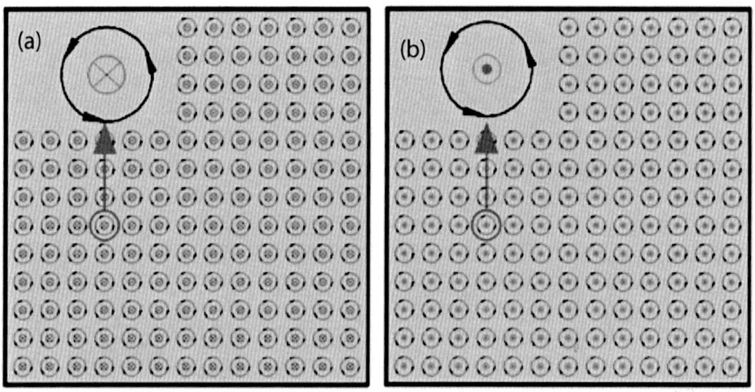

Figure 6.58 Schematic of an incident EM wave radiates to the CMA: (a) incident EM wave creates induced currents and (b) induced currents generate an induced EM field.

The current work proceeds from the angle of impedance match, probing into the formation of an induced current. From the above analysis, it can be speculated that if the formation of an inducted current is discouraged, an overall suitable impedance-matching pattern can be attained and the reflection of the EM wave can be reduced. For this reason, the induced current was weakened by giving the CMA discontinuous processes and then the corresponding RL was tested. The principal diagram is shown in Fig. 6.59.

As shown in Fig. 6.59, the gaps between interfacing isolated regions broke the original routes of the induced current. Under the same conditions, the intensity of the reflected EM wave decreased because the induced current formed from the discrete structure absorber (DMA) was weaker and the impedance match improved. Theoretically, the property of the DMA was more outstanding. The RLs based on the two structures are shown in Fig. 6.60.

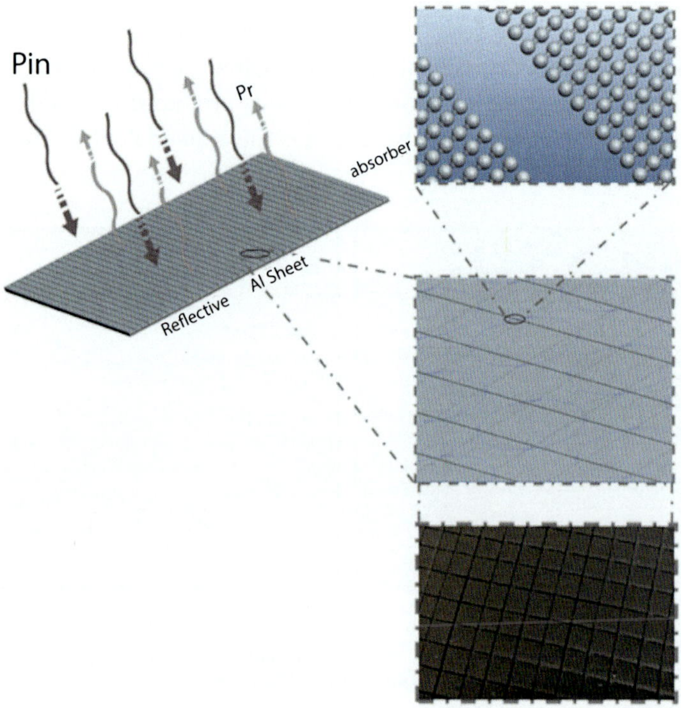

Figure 6.59 Schematic representation of the mechanism for a discrete structure absorber.

As shown in Fig. 6.60, the EM absorbing properties showed a notable increase in most cases as the discrete structure process was adopted. When the thickness reached 2 mm, the minimum RL of the coating with a concentration of 1:0.5 (water-based varnish:CIP) improved from −13 dB to −32 dB. In addition, the frequency region in which the minimum RL was less than −10 dB shifted to a significantly higher frequency, from 5–7 GHz to 7–12 GHz, which visibly broadened the effective frequency region. These are all consistent with expectations.

However, results showed that the property of absorbers at the other two concentrations did not improve visibly. For the low-concentration absorbers, there were two possible causes for this: First, the induced displaced current itself was weak and it had a very little effect on the reflection. Second, the discontinuous process did not decrease the induced currents, because the induced current was formed by the polarization effect. For the high-concentration sample, this situation was found to contribute to the formation of smaller induced current loops. This indicates that smaller isolated regions are needed for higher content absorbers. Hence, the discontinuous process is continued and smaller sections are created on the basis of the existing DMA. The corresponding RL curves are also shown in Fig. 6.61. The final minimum RL value is improved from –5.5 dB at 3.0 GHz to –17.6 dB at 10.6 GHz and a permissible RL (<–10 dB) band of 8.0–15.0 GHz, which is much better than the CMA of the same content and the result is acceptable. This supplements discontinuous process theory.

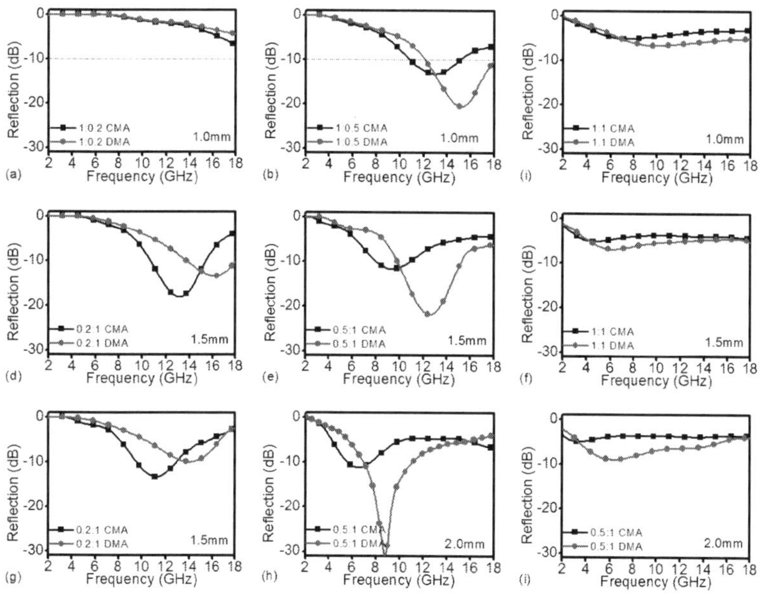

Figure 6.60 Contrasts of the RL based on the CMA and DMA with different thicknesses of samples and component content of CIP.

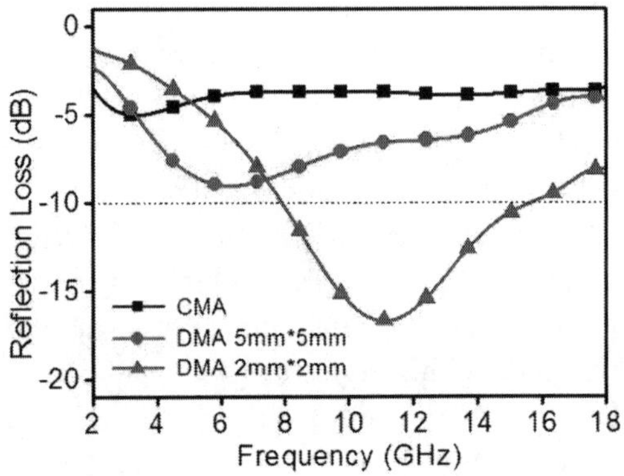

Figure 6.61 Reflection loss curves of DMA with different discrete unit sizes.

In addition, as the thickness and concentration increased, the absorbing peak shifted toward the lower frequency range. This shows that the absorbing properties of RL in high-frequency regions cannot be enhanced only by increasing the thickness or concentration. Fortunately, the discontinuous process can move the absorbing peak to a high-frequency region effectively. This makes it possible to adjust the effective absorption region.

In conclusion, the RL peaks of the coatings can be improved substantially by a discontinuous process. The technology is simpler than most other techniques. The DMA with a fixed component content of 33 wt% in the CIP/water varnish (WV) layer exhibited a minimum RL value of –32 dB at 9 GHz and a permissible RL (–10 dB) band of 7–12 GHz, which is much better than the performance of the corresponding CMA. This shows that discontinuous processes are an effective means to improve the properties of absorbers without increasing the technology difficulty.

6.6.5 Acrylonitrile-Butadiene-Styrene/Carbon Black Composites

A titanate coupling agent (TCA) is widely used as a plasticizer in filled polymer processes[108]. In this section, the effect of a TCA with different content (2 and 10 wt%) on the electrical conductivity and wave absorption of CB/acrylonitrile-butadiene-styrene (ABS) composites are investigated.

6.6.5.1 Section morphology

To investigate this effect and to relate the TCA mechanism to the morphology, the fracture surfaces of the composites were examined with a scanning electronic microscope, as illustrated in Fig. 6.62a–c.

Figure 6.62a–c shows the fracture morphology of 20 wt% CB-filled ABS composite specimens containing no TCA and containing a 2 and 10 wt% TCA to the filled CB. Figure 6.62a shows poor CB particle dispersion in the ABS matrix and poor adhesion between the CB particles and the ABS resin when no TCA was added to the system. CB particles agglomerated as large white spots under microscopic examination. A significant improvement in CB particle dispersion in the matrix and improved adhesion between CB and the ABS resin are shown in Fig. 6.62b,c, where 2 and 10 wt% of TCA were added to filled CB, respectively. From Fig. 6.62c, although the CB particle dispersion and the adhesion between CB and ABS resin were improved greatly compared to that of no TCA, some CB particle aggregation was still observed in the micrograph. This was because some CB particles were embraced by the TCA and were bound each other, which leads to the fact that the embraced CB particles did not disperse well in the matrix and formed an isolated conducting island. Other well-dispersed CB particles in the matrix were also coated by the TCA and were isolated from each other, as shown in Fig. 6.62c.

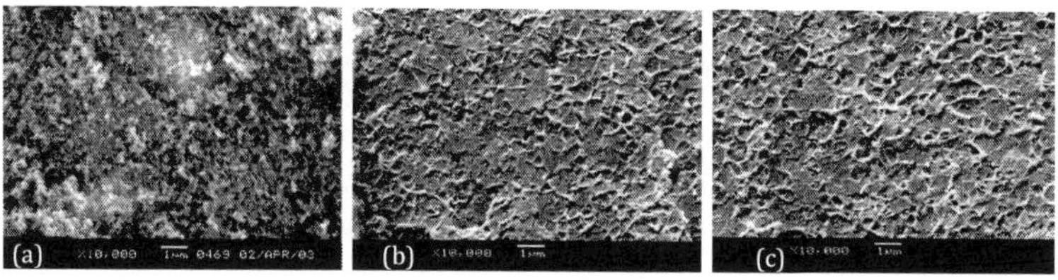

Figure 6.62 Scanning electron micrographs of fractured surfaces of CB/ABS resin composites containing 20 wt% CB filler: (a) without TCA, (b) 2 wt% TCA, and (c) 10 wt% TCA.

6.6.5.2 Electrical conductivity

Figure 6.63 shows electronic conductivity plots of various composites for different TCA content as a function of CB mass fraction, wherein the points represent the measured values and lines are only to guide the eye. As indicated, the electrical conductivity increased with the CB mass fraction, but it showed interesting variation with TCA incorporation. When the addition of a TCA to the filled CB is 2 wt%, the electrical conductivity of the CB/ABS composites improved greatly in comparison to that with no TCA, and the improved tendency by the TCA was more and more obvious with the content of the CB filler. However, with the incorporation of a 10 wt% TCA, the composites were poor in electrical conductivity compared to those having no TCA, as illustrated in Fig. 6.63.

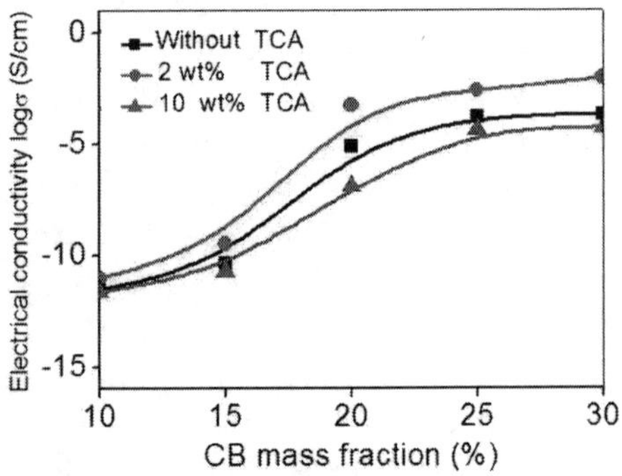

Figure 6.63 Effect of a TCA on electrical conductivity of a CB/ABS composite. The curves are only a guide to the eye.

It has been postulated that a TCA will improve the compatibility of a CB filler with an ABS polymer by the enhancement of its interfacial adhesion[109]. The mechanism of the TCA reaction on the inorganic CB filler surface is represented in Fig. 6.64. The TCA reacts with the hydroxyl groups at the inorganic CB filler surface, which results in the formation of a monomolecular layer on the inorganic surface to increase the compatibility of the filler–matrix interface. When the content of the TCA was 2 wt%, the dispersion of CB particles in the ABS phase was enhanced by the replacement of the water of hydration at the inorganic surface of the CB with the organofunctional titanate, making the CB/ABS interface compatible and thereby eliminating air voids in the system[110,111]. Consequently, this resulted in de-agglomeration and more uniform dispersion in melt blending, which contributed to a decrease in the net resistance ($R_C + R_A$; see Fig. 6.65), where R_A is the resistance within the conducting particles' aggregate (Ω) and R_C is the contact resistance between the conducting particles' aggregate (Ω)][112]. The gaps between the conducting particles agglomerated at high filler loadings became very small or negligible, and the net resistance became practically equal to R_A. So the conducting channel and the electrical conductivity of the filled composites improved greatly.

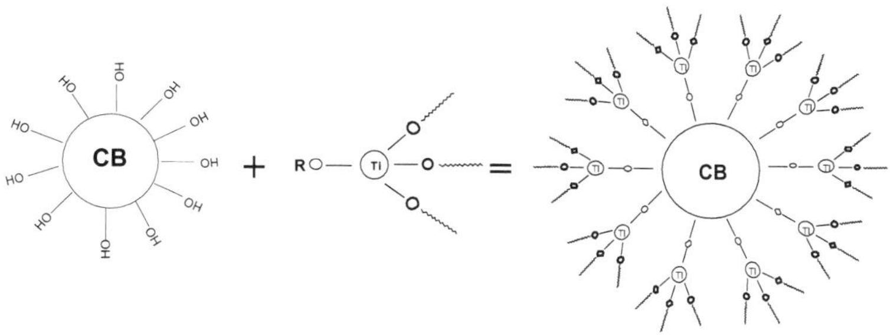

Figure 6.64 Mechanism of a titanate reaction in a CB filler surface.

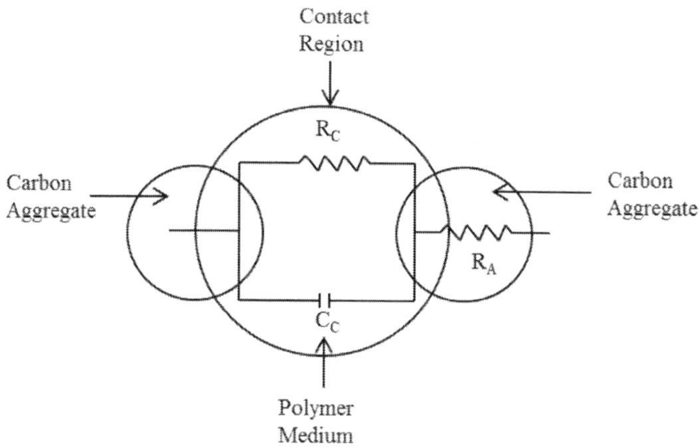

Figure 6.65 Equivalent resistor-capacitor circuit in the contact region of CB filler aggregates in a polymer compound. R_A is the resistance within the conducting particles' aggregate, R_C is the contact resistance between the conducting particles' aggregate, and C_c is the capacitance on the interface of the insulator between the conducting particles' aggregate.

However, when the content of the TCA reached 10 wt%, the filled CB particles had a chemical reaction not only with the TCA (as mentioned previously) but also with the physical coating by the

TCA (as shown in Fig. 6.66), which had an adverse influence on the formation of conducting chains among the CB particles, although the TCA also improved the compatibility of the CB filler with the ABS polymer and uniform dispersion in the ABS matrix. This was due to the fact that there were barriers to the flow of electrons through the TCA coating. So R_C between the CB particle agglomerate was enhanced; this resulted in an increase in the net resistance ($R_C + R_A$). Moreover, the interface of the TCA coating or ABS matrix between the CB particles' aggregate brought on the presence of the capacitance (C_C in Fig. 6.65). R_C and C_C played an important role in the electrical conductivity of low CB loading (≤15 wt%) or coated CB loading (by a 10% TCA) composites, which is generally called hopping or tunneling conduction[98].

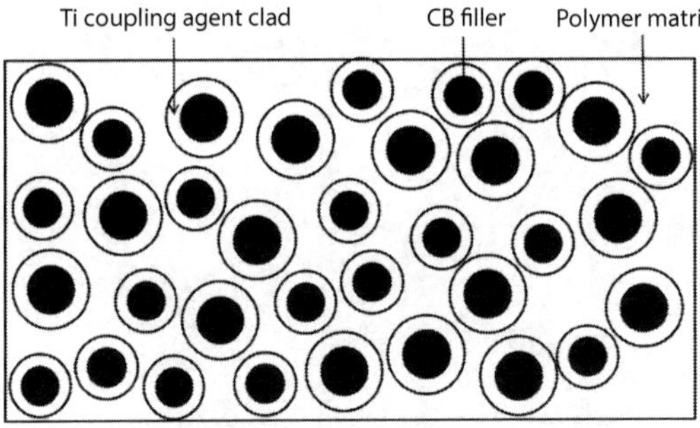

Figure 6.66 Coated CB particles by a TCA. When the content of the TCA gets to 10 wt%, CB particles will be coated by the TCA, for which the conduction channel is blocked.

6.6.5.3 Microwave absorption

Figure 6.67 shows the RL for 20 wt% CB/ABS composites with different TCA content as a function of frequency, which represents the wave absorption performance of the composites versus frequency. As demonstrated, the RL of the CB/ABS composites depended on their electric conductivity and filler particle dispersion in the matrix. With the addition of a 2 wt% TCA to the filled CB, the electrical conductivity of the CB/ABS composites was greatly improved compared to those having no TCA, so the incident EM waves were mostly reflected on the sample surface on account of a continuous conductive network in the sample, and the RL was poor. However, with the incorporation of a 10 wt% TCA in the filled CB particles, the RL of the sample was improved dramatically, as shown in Fig. 6.67. The characteristic absorption peak was −21.4 dB at 12.3 GHz, and the effective frequency bandwidth of wave absorption surpassing −10 dB reached 2.3 GHz. Correspondingly, the characteristic absorption peak of the sample without the TCA was −15.38 dB at 10 GHz, and its effective frequency bandwidth was under 1 GHz. This phenomenon can be due to the fact that the dispersed CB particles in the matrix were coated by the TCA and were isolated from each other. Furthermore, some CB particles were embraced by the TCA and formed isolated conducting aggregates. The contactless (isolated) distribution of the CB aggregates in the ABS matrix led to the improvement of the absorbing and reflection activities. The probable reason for the improved wave absorption was that the plane wave could transmit into the sample containing contactless CB particles and was mostly decayed because of the resistance and interfacial polarization attenuation of the isolated CB particles (aggregates) and multiple scattering losses caused by different discrete CB particles (aggregates)[112].

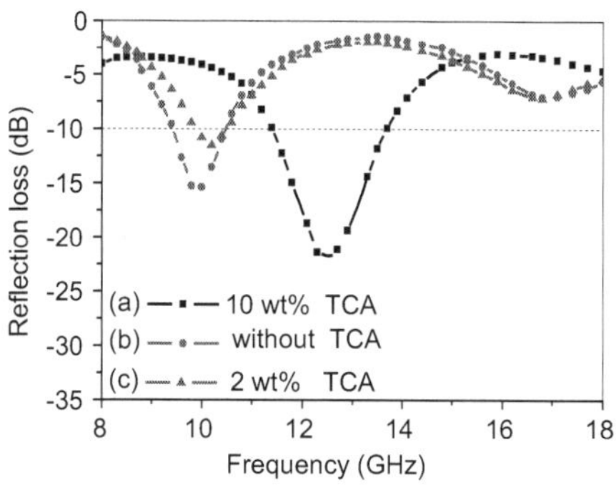

Figure 6.67 Reflection loss of 20 wt% CB/ABS composites for different TCA content versus frequency. Added amount of the TCA in filled CB: (a) 10 wt%, (b) 0 wt%, and (c) 2 wt%. Thickness of all samples is 6 mm. The curves are only a guide to the eye.

In this section, the addition of a TCA to CB-filled ABS had a notable influence on the electrical conductivity and EM wave absorption of CB/ABS composites. With the addition of a 2 wt% TCA to the filled CB, the electrical conductivity of the CB/ABS composites was greatly improved, but their wave absorption performance was reduced. In contrast, the addition of a 10 wt% TCA to the filled CB improved the wave absorption performance of the CB/ABS composites but led to poor electrical conductivity. However, regardless of the content of the TCA (2 or 10 wt%), its addition greatly improved the mechanical properties of the composites. The reasons for this are that the TCA improved adhesion between the CB and the ABS resin through the formation of a chemical band between the filling and resin and the dispersion of CB in the ABS resin during compression molding.

References

1. Emerson WH (1973). Electromagnetic wave absorbers and anechoic chambers through years, *IEEE Trans Antennas Propag*, **21**(4), 484–490.
2. Chen YJ, Cao MS, Wang TH, Wan Q (2004). Microwave absorption properties of the ZnO nanowire-polyester composites, *Appl Phys Lett*, **84**(17), 3367–3369.
3. Knott EF (1979). Thickness criterion for single-layer radar absorbents, *IEEE Trans Antennas Propag*, 27(5), 698–701.
4. Meshram MR, Agrawal NK, Sinha B, Misra PS (2004). Characterization of m-type barium hexagonal ferrite-based wide band microwave absorber, *J. Magn Magn Mater*, **271**(2–3), 207–214.
5. Singh P, Babbar VK, Razdan A, Srivastava SL, Goel TC (2000). Microwave absorption studies of ca-nitihexaferrite composites in X-band, *Mater Sci Eng B*, **78**(2–3), 70–74.
6. Nohara EL, Miacci MAS, Peixoto GG (2003). Radar cross section reduction of dihedral and trihedral corner reflectors coated with radar absorbing materials (8–12 GHz), *Proc SBMO/IEEE MTT-S IMOC*, **1**, 479–484.
7. Tkalich NV, Mokeev YuG, Onipko AF (2003). Absorbing-and-diffusing coating, *Int Conf Antenna Theory Techn*, 677–678.
8. Gao HF, Chen YL (2003). Research status of antiradar coatings, *Surf Technol*, **32**(5), 1–3.
9. Zhao DL, Zhou WC (1997). Structural radar absorbing materials and their structural model design, *Ordnance Mater Sci Eng*, **20**(6), 53–57.
10. Feng L, Lu CX (1992). A new type of wide-band radar absorbing coating, *J Electron*, **14**(6), 618–623.

11. Terracher F, Berginc G (2000). Thin electromagnetic absorber using frequency selective surfaces, *IEEE Antennas and Propagation Society International Symposium*, **2**, 846–869.

12. Jaggard DL, Liu JC, Sun X (1991). Spherical chiroshield, *Electron Lett*, **27**(1), 77–79.

13. Ruck GT, Barrick DE, Stuart WD (1970). *Radar Cross Section Handbook*, Plenum Press, vol. 2, 616–622.

14. Fante RL, Mccormack MT (1988). Reflection properties of the Salisbury screen, *IEEE Trans. Antennas Propag*, **36**(10), 1443–52.

15. Lederer PG (1992). Modeling of practical Salisbury screen absorbers, *IEEE Colloquium on Low Profile Absorbers and Scatterers*, 1/1–1/4.

16. Liu JC, Ho SS, Bor SS (1993). Tchebyshev approximation method for Salisbury screen design, *Microwaves, Antennas and Propagation, IEE Proceedings H*, **140**(5), 414–416.

17. Chambers B (1994). Optimum design of a Salisbury radar absorber, *Electron Lett*, **30**(16), 1353–1354.

18. Chambers B, Tennant A (1994). Characteristics of a Salisbury radar absorber covered by a dielectric skin, *Electron Lett*, **30**(21), 1797–1799.

19. Chambers B (1995). Symmetrical radar absorbing structures, *Electron Lett*, **31**(5), 404–405.

20. Zhao DL, Zhou WC (1998). Radar absorbing coating materials and coating structure design, *Ordnance Mater Sci Eng*, **21**(4), 58–62.

21. Li QL, Jiao PN, Ge DB (1999). HF scattering cross-section of the stealth warplane, *Chinese J Radio Sci*, **14**(1), 31–35.

22. Jaggard DL, Engheta N (1989). Chirosorb™ as an invisible medium, *Electron Lett*, **25**(3), 173–174.

23. Yu RG, Qiao XJ, Zhang TL (2004). Research progress of novel radar wave absorbing materials, *Ordnance Mater Sci Eng*, **27**(3), 63–66.

24. Liu L, Zhang MX, Hu LC (1994). Present situation and trends of radar absorbing material coating, *Aerospace Mater Technol*, (1), 1–9.

25. Zhao HY, Xiang JZ (2002). Wu X H, electromagnetic wave pollution and electromagnetic wave absorbing materials, *J Yunnan Univ (Nat Sci)*, **22**(1A), 58–62.

26. Wang WL (2001). Developments in stealth technology, *Winged Missiles J*, (1), 26–29.

27. Sung C, Ro R, Chan Y (1999). Scattering characteristics of the chiral slab for normally incident longitudinal elastic waves, *Wave Motion,* **30**(2), 135–142.

28. Varadan (1990). Electromagnetic shielding and absorptive materials. US Patent 4948922.

29. Xu YQ, Duan YF (2001). Research progress and prospects of new absorbing materials, *J Air Force Radar Acad,* **15**(1), 45.

30. Liao SB, Yin GJ (1993). Reflection of a chiral plate absorber, *Appl Phys Lett,* **62**(20), 2480–2482.

31. Lakhtakia A, Varadan VV, Varadan VK (1986). A parametric study of microwave reflection characteristics of a planar Achiral-Chiral interface, *IEEE Trans Electromag Compat,* **28**(2), 90–95.

32. Wan MX, Li SZ, Li CJ (1989). Research of a new type of conductive polymer microwave absorbing materials, *Aerospace Mater Technol,* (4–5), 28–32.

33. Shi DM, Deng H, Du SG (2002). Development of radarwave absorbing material technology, *Ordnance Mater Sci Eng,* **25**(1), 64–67.

34. Courric S (2000). Electromagnetic properties of poly (p-phenylene-vinylene) derivatives, *Polym Adv Technol,* **11**(6), 273–79.

35. Mittra R, Chan CC, Cwik T (1988). Techniques for analyzing frequency selective surfaces: a review, *Proc IEEE,* **76**(12), 1593–1615.

36. Lu J, Gao JS, Sun LJ (2003). Frequency-selective surface and its application to stealthy technology, *OME Inform,* (9), 1–8.

37. Puscasu I, Schaich W, Boreman GD (2002). Resonant enhancement of emission and absorption using frequency selective surfaces in the infrared, *Infrared Phys Technol,* **43**(2), 101–107.

38. Weile DS, Michielssen E (2000). The use of domain decomposition genetic algorithms exploiting model reduction for the design of frequency selective surfaces, *Comput Meth Appl Mech Eng,* **186**(2–4), 439–458.

39. Xing LY, Jiang SC, Li BT (2004). Microwave absorbing composite with circuit, *Acta Mater Compos Sin,* **21**(6), 27–33.

40. Xing LY, Jiang SC, Li BT (2004). Study on the mechanical properties of microwave-absorbing composite with circuit analogue, *J Aeronaut Mater,* **24**(2), 22–26.

41. Williams TDN (1992). A new ferrite grid absorber for screened rooms, *Eighth International Conference on Electromagnetic Compatibility,* 220–226.

42. Kim DY, Chung YC (1997). Electromagnetic wave absorbing characteristics if Ni-Zn ferrite grid absorber, *IEEE Trans Electromagn Compat*, **39**(4), 356–361.

43. Naito Y, Anzai H, Mizumoto T (1993). Ferrite grid electromagnetic wave absorbers, *IEEE International Symposium on Electromagnetic Compatibility*, **2**, 254–259.

44. Holloway CL, Delyser RR, German RF (1997). Comparison of electromagnetic absorber used in anechoic and semi-anechoic chambers for emissions and immunity testing of digital devices, *IEEE Trans Electromag Compat*, 39(1), 33–46.

45. Neher LK (1953). Nonreflecting background for testing microwave equipment. US Patent 2656535.

46. Janaswamy R (1992). Oblique scattering from lossy periodic surfaces with application to anechoic chamber absorbers, *IEEE Trans Antennas Propag*, **40**(2), 162–169.

47. Wang XY, Zhao HF, Qian J (2000). Analysis and design for absorbing materials used in microwave anechoic chamber, *J Microw*, **16**(4), 389–398.

48. Wang XY, Qian J (2001). Pyramidal absorbers made of hard foam polystyrene shell, *Chinese J Radio Sci*, **16**(1), 41–44.

49. Qian J, Wang XY, Zhu HF (2000). Novel pyramidal microwave absorbers for out-door using. *The 5th International Symposium on Antenna, Propagation and EM Theory*, 347–349.

50. Holloway CL, Johansson M (1997). Effective electromagnetic material properties for alternating wedges and hollow pyramidal absorbers. *IEEE Antennas and Propagation Society International Symposium*, **4**, 2292–2295.

51. Holloway CL, Johansson M, Kuester EF (1999). A model for predicting the reflection coefficient for hollow pyramidal absorbers, *IEEE International Symposium on Electromagnetic Compatibility*, **2**, 861–866.

52. Kuester EF, Holloway CL (1990). Comparison of approximations for effective parameters of artificial dielectrics, *IEEE Trans Microw Theory Techn*, **38**(11), 1752–1755.

53. Gibbons HT (1990). Design of backing layers for pyramid absorbers to minimize low frequency reflection, MS Thesis University of Colorado at Boulder.

54. Sayer M, Sreenivas K (1990). Ceramic thin films: fabrication and applications, *Science*, **247**(4946), 1056–1060.

55. Janas VF, Safari A (1995). Overview offine-scale piezoelectric ceramic/polymer composite processing, *J Am Ceram Soc*, **78**(11), 2945–2955.
56. Cohen RE (1992). Origin of ferroelectricity in perovskite oxides, *Nature*, **358**(6382), 136–138.
57. Lee MB, Kawasaki M, Yoshimoto M, Koinuma H (1995). Heteroepitaxial growth of $BaTiO_3$ films on Si by pulsed-laser deposition, *Appl Phys Lett*, **66**(11), 1331–1333.
58. Roy D, Krupanidhi SB (1992). Pulsed excimer laser ablated barium-titanate thin-films, *Appl Phys Lett*, **61**(17), 2057–2059.
59. Powles JG (1948). Dielectric properties of titanates at ultra-high frequencies, *Nature*, **162**(4120), 614–614.
60. Davis L, Rubin LG (1953). Some dielectric properties of barium-strontium titanate ceramics at 3000 megacycles, *J Appl Phys*, **24**(9), 1194–1197.
61. Lurio A, Stern E (1960). Measurements of the dielectric constant of batio3 single crystals in the paraelectric region at X-band, *J Appl Phys*, **31**(10), 1805–1809.
62. Truong VT, Riddell SZ, Muscat RF (1998). Polypyrrole based microwave absorbers, *J Mater Sci*, **33**(20), 4971–4976.
63. Oh JH, Oh KS, Kim CG, Hong CS (2004). Design of radar absorbing structures using glass/epoxy composite containing carbon black in X-band frequency ranges, *Compos Part B*, **35**(1), 49–56.
64. Naito Y, Suetake K (1971). Application of ferrite to electromagnetic wave absorber and its characteristics, *IEEE Trans Microw Theory Techn*, **MT19**(1), 65.
65. Kittel C (1951). Domain boundary motion in ferroelectric crystals and the dielectric constant at high frequency, *Phys Rev*, **83**(2), 458–458.
66. Devonshire AF (1951). Theory of barium titanate- part II, *Philos Mag*, **42**(333), 1065–1079.
67. Arlt G, Bottger U, Witte S (1993). Emission of GHz shear-waves by ferroelastic domain-walls in ferroelectrics, *Appl Phys Lett*, **63**(5), 602–604.
68. Chen XD, Wang GQ, Duan YP (2007). Microwave absorption properties of barium titanate/epoxide resin composites, *J Phys D*, **40**(6), 1827–1830.
69. Motojima S, Hoshiya S, Hishikawa Y (2003). Electromagnetic wave absorption properties of carbon microcoils/PMMA composite beads in W bands, *Carbon*, **41**(13), 2658–2660.

70. Kaynak A, Polat A, Yilmazer U (1996). Some microwave and mechanical properties of carbon fiber-polypropylene and carbon black-polypropylene composites, *Mater Res Bull*, **31**(10), 1195–1206.

71. Miyauchi S, Togashi E (1985). The conduction mechanism of polymer-filler particles, *J Appl Polym Sci*, **30**(7), 2743–2751.

72. Neelakanta PS, Subramaniam K (1992). Controlling the properties of electromagnetic composites, *Adv Mater Process*, **141**(3), 20–25.

73. Frysz CA, Shui XP, Chung DDL (1996). Carbon filaments and carbon black as a conductive additive to the manganese dioxide cathode of a lithium electrolytic cell, *J Power Sources*, **58**(1), 41–54.

74. Guan HT, Liu SH, Zhao YB, Duan YP (2006). Electromagnetic characteristics of nanometer manganese dioxide composite materials, *J Electron Mater*, **35**(5), 892–896.

75. Wang GQ, Chen XD, Duan YP (2008). Electromagnetic properties of carbon black and barium titanate composite materials, *J Alloy Compd*, **454**(1–2), 340–346.

76. Guan H, Zhao Y, Liu S, Lv S (2006). Application of manganese dioxide to electromagnetic wave absorber: effective permittivity and absorbing property, *Eur Phys J Appl Phys*, **36**(3), 235–239.

77. Duan YP, Yang Y, He M, Liu SH, Gui XD, Chen HF (2008). Absorbing properties of α-manganese dioxide/carbon black double-layer composites, *J phys D*, **41**(12), 125403–125408.

78. Maeda T, Sugimoto S, Kagotani T, Tezuka N, Inomata K (2004). Effect of the soft/hard exchange interaction on natural resonance frequency and electromagnetic wave absorption of the rare earth-iron-boron compounds, *J Magn Magn Mater*, **281**(2–3), 195–205.

79. Liu LD, Duan YP, Ma LX, Liu SH, Yu Z (2010). Microwave absorption properties of a wave-absorbing coating employing carbonyl-iron powder and carbon black, *Appl Surf Sci*, **257**(3), 842–846.

80. Feng YB, Qiu T, Shen CY (2007). Absorbing properties and structural design of microwave absorbers based on carbonyl iron and barium ferrite, *J Magn Magn Mater*, **318**(1–2), 8–13.

81. Duan Y, Wu G, Gu S, Li S, Ma G (2012). Study on microwave absorbing properties of carbonyl-iron composite coating based on PVC and Al sheet, *Appl Surf Sci*, **258**(15), 5746–5752.

82. Honghuan WU, Xiaoyan W, Ling Z, Dongmei Z, Wancheng Z (2007). Present development of absorbing composites containing carbon fibers, *Mater Rev*, **21**(5), 115–117, 124.

83. Shen G, Xu M, Xu Z (2007). Double-layer microwave absorber based on ferrite and short carbon fiber composites, *Mater Chem Phys*, **105**(2–3), 268–272.

84. Chen W, Jialin GU, Feiyu K (2009). Research progress in theoretical designs of microwave absorbing materials, *Mater Rev*, **23**(5), 5–8.

85. Duan YP, Wang L, Liu Z, Gu SC (2013). Microwave properties of double layer absorber reinforced with carbon fibre powders, *Plast Rubber Compos*, **42**(2), 82–87.

86. Yu M, Li X, Gong R, He Y, He H, Lu P (2008). Magnetic properties of carbonyl iron fibers and their microwave absorbing characterization as the filer in polymer foams, *J Alloy Compd*, **456**(1–2), 452–455.

87. Peng ZH, Cao MS, Yuan J, Xiao G (2004). Strong fluctuation theory for effective electromagnetic parameters of fiber fabric radar absorbing materials, *Mater Des*, **25**(5), 379–384.

88. Cao MS, Zhu J, Yuan J, Peng ZH, Xiao G (2002). Simulation of multiple composite coatings based on conducting plate and investigation of microwave reflectivity, *Microw Opt Technol Lett*, **34**(6), 442–445.

89. Bowler N (2006). Designing dielectric loss at microwave frequencies using multi-layered filler particles in a composite, *IEEE Trans Dielectr Electr Insul*, **13**(4), 703–711.

90. Liu XG, Geng DY, Meng H, Shang PJ, Zhang ZD (2008). Microwave-absorption properties of ZnO-coated iron nanocapsules, *Appl Phys Lett*, **92**(17), 173117.

91. Chen LY, Duan YP, Liu LD, Guo JB, Liu SH (2010). Influence of SiO_2 fillers on microwave absorption properties of carbonyl iron/carbon black double-layer coatings, *Mater Design*, 32(2), 570–574.

92. Liu L, Duan Y, Ma L, Liu S, Yu Z (2010). Microwave absorption properties of a wave-absorbing coating employing carbonyl-iron powder and carbon black, *Appl Surf Sci*, **257**(3), 842–846.

93. Duan Y, Li G, Liu L, Liu S (2010). Electromagnetic properties of carbonyl iron and their microwave absorbing characterization as filler in silicone rubber, *Bull Mater Sci*, **33**(5), 633–636.

94. Liu L, Duan Y, Guo J, Chen L, Liu S (2011). Influence of particle size on the electromagnetic and microwave absorption properties of FeSi/paraffin composites, *Physica B*, **406**(11), 2261–2265.

95. He Y-F, Gong R-Z, Li X-C, Wang X, He H-H (2006). Preparation and microwave absorbing properties of multi-layered radar (microwave) absorbing materials composites, *J Inorg Mater*, **21**(6), 1449–1453.

96. Li X, Zhang YH, Chen JL, Duan YP, Wu GL, Ma GJ (2013). Composite coatings reinforced with carbonyl iron nanoparticles: preparation and microwave absorbing properties, *Mater Technol*, **29**(1), 57–64.

97. Duan YP, Wang L, Liu Z, Gu SC (2013). Microwave properties of double layer absorber reinforced with carbon fibre powders, *Plast Rubber Compos*, **42**(2), 82–87.

98. Matitsine SM, Hock KM, Liu L, Gan YB, Lagarkov AN, Rozanov KN (2003). Shift of resonance frequency of long conducting fibers embedded in a composite, *J Appl Phys*, **94**(2), 1146–1154.

99. Duan YP, Liu SH, Wen B, Guan HT, Wang GQ (2006). A discrete slab absorber: absorption efficiency and theory analysis, *J Compos Mater*, **40**(20), 1841–1851.

100. Duan YP, Li GF, Liu LD, Liu SH (2010). Electromagnetic properties of carbonyl iron and their microwave absorbing characterization as filler in silicone rubber, *Bull Mater Sci*, **33**(5), 633–636.

101. Qing Y, Zhou W, Luo F, Zhu D (2009). Microwave-absorbing and mechanical properties of carbonyl-iron/epoxy-silicone resin coatings, *J Magn Magn Mater*, **321**(1), 25–28.

102. Wang M, Duan YP, Liu SH, Li XG, Ji ZJ (2009). Absorption properties of carbonyl-iron/carbon black double-layer microwave absorbers, *J Magn Magn Mater*, **321**(20), 3442–3446.

103. Liu LD, Duan YP, Liu SH, Chen LY, Guo JB (2009). Microwave absorption properties of one thin sheet employing carbonyl iron powder and chlorinated polyethylene, *J Magn Magn Mater*, **322**(13), 1736–1740.

104. Wu LZ, Ding J, Jiang HB, Chen LF, Ong CK (2005). Particle size influence to the microwave properties of iron based magnetic particulate composites, *J Magn Magn Mater*, **285**(1–2), 233–239.

105. Li Y, Chen C, Pan X, Ni Y, Zhang S, Huang J, Chen D, Zhang Y (2009). Multiband microwave absorption films based on defective multiwalled carbon nanotubes added carbonyl iron/acrylic resin, *Physica B*, **404**(8–11), 1343–1346.

106. Pinho MS, Gregori ML, Nunes RCR, Soares BG (2001). Aging effect on the reflectivity measurements of polychloroprene matrices containing carbon black and carbonyl-iron powder, *Polym Degrad Stabil*, **73**(1), 1–5.

107. Matsumoto M, Miyata Y (1997). Thin electromagnetic wave absorber for quasi-microwave band containing aligned thin magnetic metal particles, *IEEE Trans Magn*, **33**(6), 4459–4464.

108. Duan YP, Liu SH, Wang GQ, Guan HT, Wen B (2006). Effect of a coupling agent on the electromagnetic and mechanical properties of carbon black/acrylonitrile-butadiene-styrene composites, *J Appl Polym Sci*, **102**(2), 1839–1843.

109. Wah CA, Choong LY, Neon GS (2000). Effects of titanate coupling agent on rheological behaviour, dispersion characteristics and mechanical properties of talc filled polypropylene, *Eur Polym J*, **36**(4), 789–801.

110. Miller AC, Berg JC (2003). Effect of silane coupling agent adsorbate structure on adhesion performance with a polymeric matrix, *Compos Part A*, **34**(4), 327–332.

111. Ghosh P, Chakrabarti A (2000). Conducting carbon black filled EDPM vulcanizates: assessment of dependence of physical and mechanical properties and conducting character on variation of filler loading, *Eur Polym J*, **36**(5), 1043–1054.

112. Biwa S, Idekoba S, Ohno N (2002). Wave attenuation in particulate polymer composites: independent scattering/absorption analysis and comparison to measurements, *Mech Mater*, **34**(10), 671–682.

Chapter 7

Cement-Based Electromagnetic Functional Materials

The electromagnetic wave (EMW) absorbing materials are previously applied in military fields, such as moving targets, and seldom in civil usage. With the increasing threat of electromagnetic interference (EMI) pollution to the environments, it has been brought to the attention of everyone all over the world that electromagnetic shielding and anechoic chambers should be built to weaken or eliminate the effects of EMI caused to people or the environment people are living in. Moreover, according to modern building design theory, electric and electronic equipment that is relevant to electric power and telecommunication is gradually assembled underground. Thus shielding is particularly needed for the vaults containing this equipment. Electromagnetic shielding is also needed for deterring spying and bugging in electromagnetic forms[1]. Cement, as a kind of construction materials essentially, can hold both EMI shielding and EMW absorbing properties through modification with some other additives. In this sense, cement-based composite materials have attracted more and more attention from all over the world[2].

The cement or cement paste we discussed here refers to the powdery hydraulic gelation material which can exist in the form of plastic slurry after mixing with water and can harden in air or water after cementing with moderate sand and stone. Cement or cement paste has no aggregates, and when some aggregates are added into

Microwave Absorbing Materials
Yuping Duan and Hongtao Guan

ISBN 978-981-4745-10-9 (Hardcover), 978-981-4745-11-6 (eBook)
www.panstanford.com

the cement matrix, they are called cement mortar or concrete with fine or coarse aggregates, respectively[3].

Cement is the most commonly used engineering building material due to its abundant resources, low energy consumption, and environmental friendliness. However, the electrical or electromagnetic properties of cement materials are mainly attributed to the metal oxides in the cement components, such as Fe_2O_3, Al_2O_3, FeO, and CaO, which leads to a much worse electromagnetic shielding and absorbing performance. To improve its electromagnetic characteristics, appropriate additives must be introduced to the cement. For the cement-based materials with high performance, ordinary Portland cement (OPC) and high-alumina cement are often chosen as the matrix and the additive components are mainly focused on three types, which are conductive polymer, carbon material, and metal material[2].

Due to the complexity of cement components, there may be complicated physical and chemical interactions between the components when mixed with other additive fillings, so when the electromagnetic absorbing fillings are chosen, special attention must be paid to such aspects as the interaction between the cement components and aggregates, the EMW absorption properties of the fillings, and both types and addition ratios of the fillings. Moreover, the EMW absorbing bandwidth, chemical activity, filling grain composition, and mechanical properties also need to be taken into consideration[4]. Considering comprehensively the economy, practicability, and processability, the most commonly used fillings are carbon materials and metal materials (mainly metal powder, metal fiber, and metal plate). What needs to be pointed out is that the carbon-based fillings include graphite, carbon black, and carbon fiber, and graphene is often applied in cement-based composites at present to improve their electrical properties due to its much higher electrical conductivity than other carbon materials[5–8]. In recent few years, ferrites also have gained great popularity in the application fields of cement-based composites. Their electromagnetic properties can be tuned conveniently through designing the components, contents, and calcination temperature of the ferrite materials[9].

Until now, there have been various fillings with different crystal structures, morphologies, and electrical properties applied in cement-based composite materials. Through adjusting the morphology, content, and filling patterns of the additive fillings, the composites can present different electromagnetic shielding and absorbing properties and can be applied in various practical engineering fields.

7.1 Electrical Properties of Cement Materials

Cement paste is the hydration product of unhydrated cement and water. Once in contact with water, the hydrolysis and hydration reactions will take place immediately in the cement clinker on the surface of the cement particles. This reaction is the so-called hydration process. Cement is a kind of complicated compound, and during the hydration process, there will be series of physical and chemical reactions. During the hydration process, the water and cement molecules chemically combine into a binder, transforming the initial free water into bound water, and thus the dielectric properties of cement also change[10,11]. Both the electrical and dielectric properties of cement material are dependent on frequency, temperature, hydration time, and its composition. The water in the cement also plays a great role in the electrical properties of the cement material, and in a very long time after hydration, the electrical and dielectric properties vary dramatically with the hydration time. Thus, there is the attractive possibility that electrical and dielectric measurements may provide useful information regarding the state of curing of the cement at any given time. For example, such measurements can give us the information of what process technique we can take after hydration to meet different demands for various properties and what water–cement (w/c) ratios we should take to meet the demands for different application fields[12].

The simplest electrical parameter to measure for cement materials is the direct current (DC) conductance. However, conductance measurements often turn out to be very difficult because the electrode polarization always influences the measurement results. To minimize the effects of electrode polarization, it is customary to use low-frequency alternating current (AC) measurements or four-probe techniques. More information about electrical conductivity may be also obtained through the measurements of dielectric constant and conductivity in a frequency range.

As cement hardens, the water molecules in paste changes from free water to bound states of hydration or crystallization, which will cause a decrease of ion conductivity and thus influence the dielectric characteristics of cement[12]. Moreover, when the water molecule changes from one state to another, its ability to orient in an applied electric field also changes. Thus, through the measurements of dielectric and electrical property changes in the hydration process, changes in the bonding states of cement paste and variations of hydration products can be obtained[10].

The investigation of the different mechanism involved in cement hydration is a challenging problem because the early cement hydration has complicated chemical and physical processes. The common techniques used to study this problem are X-ray diffraction, electron microscopy, thermal analysis, and conduction calorimetry[10]. These methods can provide detailed information of the paste at a given hydration time, but they are noncontinuous methods and not suitable for investigating the hydration process in situ. Since the 1970s, there has been increased interest in using electrical methods to study the physical and chemical changes at the early stage of cement hydration, because the electrical methods are continuous methods and can study the cement hydration in situ[12–15]. Electrical measurement is also a convenient and nondestructive method for determining the bond strength between cement paste and the aggregates. Reports have shown that the bond strength between a steel fiber and cement paste is linearly related to the contact electrical resistivity[16]. The measurements of reflection and transmission coefficients of an incident wave on the surface of a cement paste sample can also be used to determine the components or water content in the paste.

7.1.1 Electrical Property Measurements

The first application of electrical resistivity measurements to monitor the properties of cement-based materials dates back to the 1920s. Bogue[17] reported about experiments conducted by Shimizu in 1928/1929 showing that the setting of cement can be more accurately monitored by electrical rather than by mechanical or thermal methods. It was found that the end of setting corresponds to an inflection point in the conductivity development versus hydration time.

Cement consists mainly of hydrated calcium silicates, aluminates, and aluminoferrites, together with calcium hydroxide and some unhydrated cement clinker. It is generally considered that the cement consists of amorphous or microcrystalline porous ranging in size from a few nanometers to a few micrometers. Wet cement contains a solution of sodium, potassium, and calcium salts in the gel pores. With the process of hydration, the water in the larger pores evaporates and the ions migrate in the pore water. Thus, when an electric field is applied to the cement, a current will flow in the paste and one or more of the following electrochemical reactions will take place[18]:

$$2H_2O + 2e^- = H_2 + 2OH^- \quad (7.1)$$

$$4OH^- = O_2 + 2H_2O + 4e^- \quad (7.2)$$

$$3Fe + 4H_2O = Fe_3O_4 + 8H^+ + 8e^- \quad (7.3)$$

$$2Fe + 3H_2O = Fe_2O_3 + 6H^+ + 6e^- \quad (7.4)$$

It the early 1980s, there were reports on the DC resistivity measurements of cement paste and mortar by Hansson et al.[18] The electrodes for resistivity measurements consisted of two perforated mild steel sheets with a surface of 3×3 cm^2. The electrodes were embedded in $9 \times 7 \times 5$ cm^3 samples of cement or mortar and were compacted by vibration. This method was successfully applied to measure the DC resistivities at the early hydration state. However, the measurement precision is restricted by the contact areas between the electrodes and samples. With the progress of the hydration process, parts of the samples harden gradually and then the electrodes may adhere to the samples and thus corrosion appears. Subsequently, cracks are formed in the surfaces of the electrodes and samples, and the measurements turn out to be no longer receivable. So DC resistivity measurement can only be used at the early stage of hydration.

Some years later, coaxial electrode measurement was developed to examine the electrical and dielectric properties of cement paste. McCarter studied the electrical resistivity and dielectric constants of cement paste and mortar at the early stage of hydration using the coaxial electrode method[19]. It is displayed that when the cement was mixed with water, a series of complicated chemical reactions began to take place. The cement clinker and water reacted at different rates, for there are various mineral phases, and both hydrolysis and hydration processes were involved in the reaction procedures. When the cement begins to harden, there will be an abrupt change for both the electrical resistivity and dielectric constant, which can be used to monitoring the hydration process of cement paste. With the proceed of hydration process, the water in the cement paste evaporates gradually and thus the ionic conduction paths through the paste are becoming more and more tortuous as the gel on the cement grains extends to form a fibrous rigid structure. So when there is a great change rate of resistivity and dielectric constant, it means that the period of hardening has started[20]. Figure 7.1 shows the electrical response of ordinary Portland cement (OPC) with a w/c ratio of 0.27 in the early stage of hydration. It can be seen that there is a peak for the dielectric constant at about 150–180 min after hydration. It is believed that this peak is caused by the tricalcium silicate (C_3S) hydration, which results in a sudden release of charges into the diffuse electrical double layer around the gel coating in the cement grains[21].

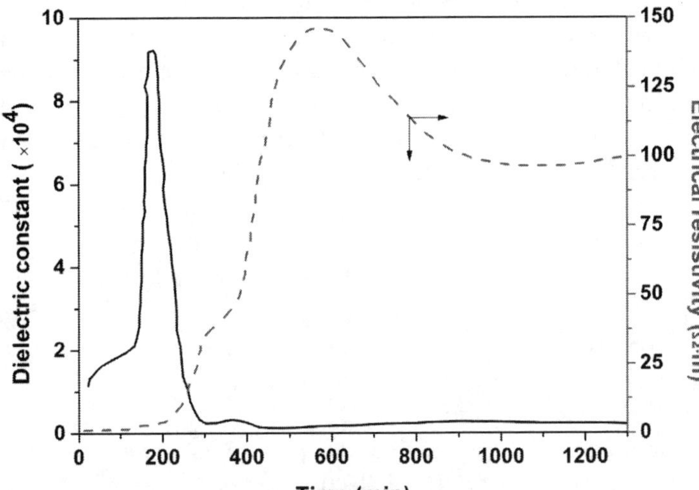

Figure 7.1 Electrical response of OPC for w/c = 0.27 at the early stage of hydration.

Abo-El-Enein et al. discussed the influences of the w/c ratio and silica fume on the electrical conductivity of Portland and slag cement paste. They found that there arise two peak values for electrical conductivity with the hydration process[22]. The first peak in the early stage of hydration is due to the increase in the ion concentration and ionic mobility. The second peak can be attributed to two reasons. The first is due to an increase in the osmotic pressure, which leads to the breaking of the wrapper out of the gel on the cement, and thus makes the contact areas between the cement grains and water turn larger to improve the ionic mobility. The second cause is the transformation of the mineral phases, which makes the ions of Ca^{2+} and $SO_4{}^{2-}$ in the minerals be released into the solution and thus improves the ion concentration.

7.1.2 Relationship between Electrical Property and Hydration

There are various factors influencing the electrical conductivity and dielectric properties of cement material during its hydration process, such as the physical states of the water molecules and ionic concentrations in the paste, the dipoles in the paste and their migration mobility, the degree of association of charges in the paste and the cement grain interfaces, and the temperature of the whole cementitious system. All the factors change more or less in the process of the hydration procedure, and thus cause the changes of the electrical conductivity and dielectric properties[10].

The study on the electromagnetic properties of cement paste during its hydration process can be traced to 1928, when Shimuzu carried out a series of studies on the electrical conductivity of cement paste and tried to clarify the relationship between the DC conductivity and hydration time. The Spainish scientist Calleja also found problems related to the DC conductivity measurements of cement paste and published six academic papers during the years from 1950 to 1953. In these papers, he stated his opinion of substituting DC conductivity with AC conductivity at low frequency to improve measurement precision[23,24]. It was found that distinct points in the resistance development matched the occurrence of marked changes in the temperature development. McCarter conducted AC electrical resistivity measurements at low frequency during the hydration and wet maintenance process of cement paste and found that the electrical conductivity had a close relationship with the porosity of the paste[19–21]. The pores are full of water and the water molecules affect the electrical properties greatly; thus through the testing of electrical resistivity of the paste, the information of porosity in the paste can be obtained.

Most of the early publications mainly focused their attention on the low-frequency electrical resistivity of cement pastes during the first 24 h after casting. These studies all found that the electrical responses of the cement pastes are sensitive to physical and chemical changes within the pastes and can be used to monitor the setting and hardening process. Since then, until 1971, Paquit realized that dielectric properties at high frequency may be more useful and practical for cement paste and conducted the dielectric measurements at 50 MHz for cement paste under hydration for continuous 24 days. Similar reports were found by Taylor, Arulanandan, and Camp et al., who measured the changes of dielectric properties in the megahertz range with the frequency and hydration time[25]. They found that DC conductivity values of the cement samples had a direct relationship with the number of unbound water molecules in the cement paste. The electrical conductivity at the early stage of hydration is decided by the frequency-dependent ionic conduction until the free water molecules lose their ability to migrate due to the decrease of their numbers at the later stage of hydration.

In the frequency range of 1 kHz to 1 MHz, the electrical conductivity of cement paste at the early hydration stage is independent of frequency due to the fact that in this frequency the conductivity is dominated by ionic conduction and there exist large amounts of ions in the cement paste in this stage of hydration. With increasing frequency, the dependence of electromagnetic parameters on frequency gradually turns out to be significant and there will be obvious dispersion effects at a high frequency range.

A comprehensive study of the dielectric constant and electrical conductivity of OPC and slag cement with different w/c ratios was presented by Zhang et al.[10] Emphasis was placed on the influence of the w/c ratio and hydration time on the early hydration of cementitious systems. It was found, for example, that after the mixing of cement clinker and water, there will be a series of chemical reactions taking place in the cementitious system, such as hydrolysis and hydration. The reaction rates are very different at different hydration stages, for there form various mineral phases. They also found that the dielectric and electrical properties are very sensitive to the hydration time and w/c values. Generally speaking, the higher the w/c ratios, the higher the dielectric constants and electrical conductivity. However, a higher w/c ratio may retard the hydration process of the cement paste and prolong the hydration time, thus causing adverse effects on the cement strength.

In their research, the frequency they adopted was in the X-band (8.2 to 12.4 GHz), with w/c ratios keeping at 0.30, 0.35, and 0.40. The chemical process of cement hydration can be categorized into four stages[10,26].

The first stage begins from the contact of water with cement and lasts for about 15 min of hydration. Because of the little solubility of the hydration products, the supersaturated ions of Ca^{2+} and OH^- leach from C_3S, tricalcium aluminate (C_3A), and the grain surfaces, which causes an increase in the ion concentration. The hydration of C_3S will have the following reaction during this stage:

$$C_3S \xrightarrow{H} C_2S + Ca^{2+} + OH^- \quad (7.5)$$

In this stage, most ions are unbound charges and can move easily in the paste. Under an external electrical field, the ions are readily to be polarized, resulting in a large dielectric and electrical conductivity, as shown in Fig. 7.2.

The second stage starts at about 1–3 h after mixing with water and its lasting time is dependent on the cement composition and w/c ratio. In this stage, the Ca^{2+} and OH^- ions leach from the cement grain surfaces rapidly and leave behind a surface layer rich in hydrosilicate ions on the C_3S and C_3A phase, giving the cement grain a negative charge[10]. A few minutes later, an amorphous semipermeable gel membrane with the phase of calcium silicate hydrate (C–S–H) forms outside the surface layer. All these phases in this stage result in an electrical double layer, leading to a slow hydration rate. In this stage, the viscosity of the cement paste is increasing continuously and thus leads to a decrease in the dielectric constant and electrical conductivity, which also can be seen clearly in Fig. 7.2.

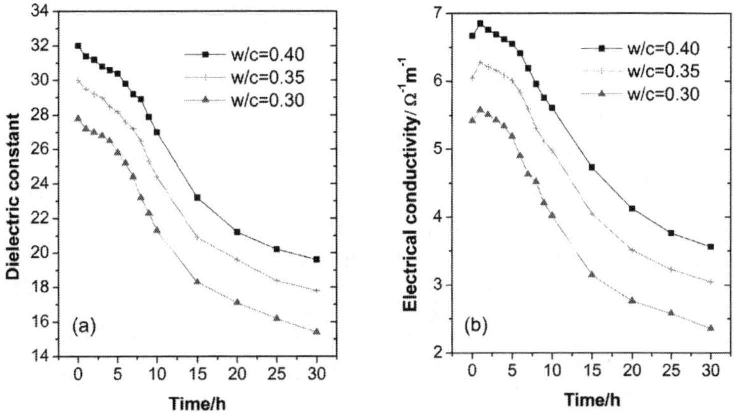

Figure 7.2 Variations of dielectric constant (a) and electrical conductivity (b) with hydration time for OPC with w/c = 0.40 measured at a frequency of 9.5 GHz. The data are from Ref. [10].

The third state starts at about 3–4 h and ends at about 15–17 h after mixing with water. The rupturing of the electrical double layer allows water to reach the cement grains, which results in the accelerated dissolution of the grains[10]. In this stage, Ca^{2+} ions are removed from the solution and form the hydration products C–S–H and $Ca(OH)_2$, leading to increased viscosity of the paste. The cement paste begins to harden and its porosity increases. The formation of porosity and increase of hydration products hinder the free movement of molecules and ions and thus results in decreased electrical conductivity. The formation of hydration products proceeds at an accelerated rate, leading to a sharp decrease in the dielectric constant and electrical conductivity.

The fourth stage starts about 15–17 h after the mixing of cement with water and may be lasting for as long as 1 y. This is the last stage of hydration, and during this stage the porosity decreases gradually with the continuous formation of C–S–H and $Ca(OH)_2$. The changes of paste turn out to be slow in this stage, and both dielectric and electrical conductivities decrease at a very slow rate. With the total solidification of the cement paste, the electrical conductivity and dielectric constant will tend to stable values.

Several subsequent electrical studies of curing cement slurries conducted by Lyomov et al. found a similar phenomenon; only in their studies the hydration of the cement was divided into five periods: preinduction, induction, acceleration, deceleration, and diffusion. With a w/c ratio of 0.5, Lyomov measured the electrical conductivity of a type I cement slurry at 25°C. To avoid the influence of electrolysis on the electrical measurements, an AC was applied through changing the voltage from 1 to 6 V. The findings show that the ion concentrations and porosities have a great influence on the equivalent conductivity. In each period, due to the diversity of the ions and their mobility, there present different electrical performances. Moreover, the slurry temperature also changed with the hydration time due to the formation of different hydration products during the process, as illustrated in Fig. 7.3.

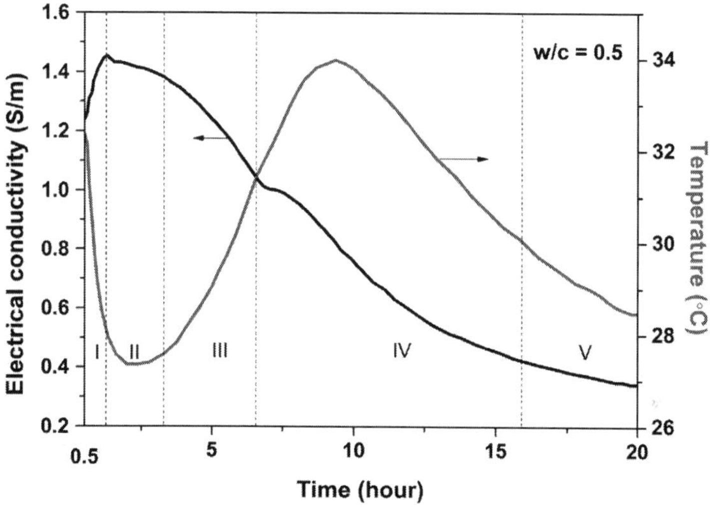

Figure 7.3 Electrical conductivity and temperature changes during the hydration of cement slurry at 25°C.

As to the effects of porosity on the electrical conductivity of the cement slurry, Lyomov found that it is in approximate agreement with Archie's law, which is usually applied to describe the relationship between the conductivity and porosity of rocks saturated with conducting water:

$$F = \frac{\sigma_f}{\sigma_c} = \frac{a}{\phi^m} \tag{7.6}$$

where F is the formation factor; σ_f and σ_c are the conductivities of the cement and the pore fluid, respectively; ϕ is the porosity; and both a and m are constants.

After examining the porosity and conductivity of 33 tested slurries from 25°C to 195°C, σ_f is found to be constant as equaling 4 S/m and F accords with a new version of Archie's law after a least squares fitting to all the experimental data:

$$F = \frac{0.126}{\phi^{5.77}} \tag{7.7}$$

From Fig. 7.2, it can also be seen that the dielectric constant and electrical conductivity are very sensitive to the w/c ratio and with the increase in the w/c ratio, there is also a trend of increase for both these parameters. The cement paste with a w/c ratio of 0.35 has a larger dielectric constant and electrical conductivity, as high as 8% and 11%, respectively, than those of the paste with a w/c ratio of 0.30. Compared to the paste with w/c = 0.35, the paste with a w/c value of 0.40 also has a 7% and 10% larger dielectric constant and electrical conductivity. It can also be seen from Fig. 7.2 that the profiles of the dielectric constant and electrical conductivity have a clear shift to larger values for the cement samples with larger w/c values than those with smaller ones. It thus illustrates that the samples need more hydration time to reach the same electrical parameters due to the prolonging of the hydration process.

Besides electric conductivity analysis, impedance spectroscopy has also been conducted to investigate the electrical properties of hydrated cement pastes, since the impedance technique could provide new information related to the microstructure and hydration mechanism. In the AC impedance analysis of cement pastes, it was always postulated that there existed three basic elements in the cement paste, which were solid, liquid, and solid–liquid interfacial phases. The solid phase included unhydrated cement and hydration products, whereas the liquid phase was mainly the solution in the micropores or capillaries. Thus the hydrated cement paste can be considered as solid–liquid interface unit cells, as illustrated in Fig. 7.4. The terms of R_s, R_l, and R_{int} are the resistances of solid, liquid, and interface phases, respectively. C_{int} refers to the capacitance of the solid–liquid interface[27].

In the AC impedance spectrum, a single arc could be observed in the high frequency range with a small part of the second arc in the low-frequency region. The high-frequency arc was due to the bulk cement–related effect, whereas the arc in the low frequency could be attributed to the electrode–cement interface effect[27].

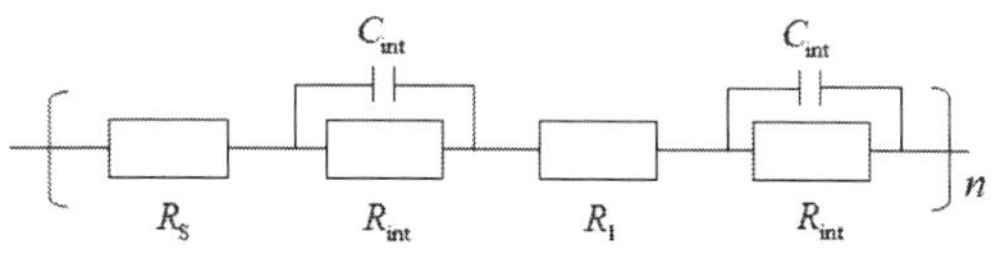

Figure 7.4 The equivalent circuit of the unit cell model for the cement paste.

The AC impedance of a kind of Portland cement with a *w*/*c* ratio of 0.35 was discussed by Gu et al. in the frequency range from 20 MHz to 1 Hz. The samples were kept for 24 h with 100% relative humidity and measured at various hydration times from 48 to 380 h. The impedance spectra exhibit a typical high-frequency arc caused by the solid–liquid interface capacitance after hydration for 169 h and continue to grow in diameter with time, as shown in Fig. 7.5. At the early stage of hydration, the rigidity of the microstructure is too low to detect capacitive behavior and thus only the electrode arc is visible. According to the equivalent circuit illustrated in Fig. 7.4, the total resistance of solid and liquid ($R_{t(int)}$) is about 80 Ω.

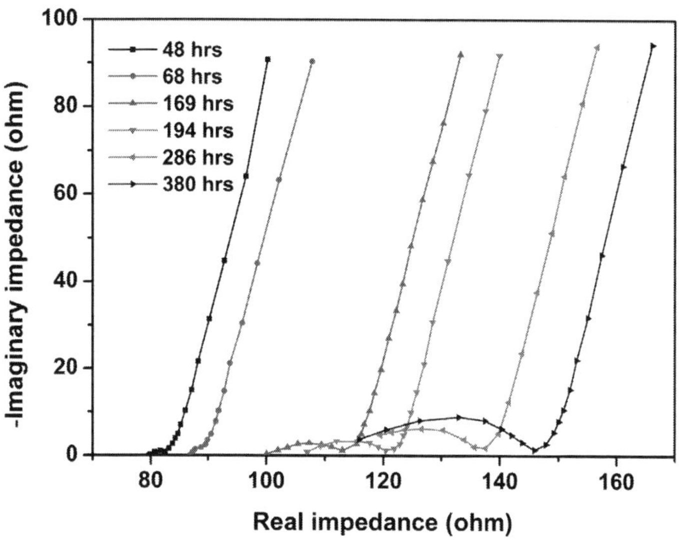

Figure 7.5 Real and imaginary impedance of the cement pastes hydrated for various times (*w*/*c* = 0.35).

The delayed appearance of the high-frequency arc is related to the microstructure changes during the cement hydration process. The arc diameter represents the solid–liquid interfacial resistance. Once hydration proceeds, these hydration reactions generate a huge amount of interfacial areas which has impedance properties different from the bulk solid and liquid phases. The interfacial area is manifested by the increasing values of the solid-liquid interfacial resistance, $R_{t(int)}$, as shown in Fig. 7.6.

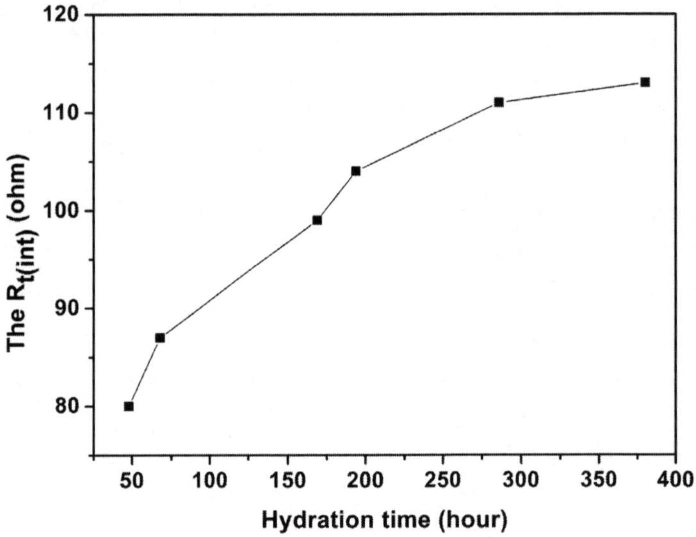

Figure 7.6 Electrical parameters vs. hydration time.

7.1.3 Applications of Electrical Properties of Cement-Based Materials

7.1.3.1 Monitoring of nondestructive features

The electrical property of cement material has a close relationship with the cement chemistry, ion content, and porosity. A little variation of these parameters can be responded to and reflected directly by the change of electrical properties. Thus the intrinsic electrical properties of cementitious systems can be taken as an effective methodology to study and evaluate these materials at both the micro- and macroscale. Moreover, electrical property measurements can also be taken as a nondestructive technique for monitoring civil engineering structures as a technique for surveillance. For example, it has been commonly well known that during the hydration process of cement, the strength of the cement mortar or concrete material increases with the curing age. However, due to the increase in the interfacial void contents between the cement paste and the aggregates, the strength may decrease with the curing time in some cases. To give a nondestructive monitoring in real time, electrical measurements sometimes may play great roles. The electrical property measurements of a steel concrete with a w/c ratio of 0.45 found that the greater the curing age, the lower the bond strength, and the higher the contact resistivity[28]. And also, for the material with the same aging time, the change of contact resistivity approximately follows a linear relationship with the bond strength. Thus through the measurement of the contact electrical resistivity, the bonding conditions between the rebar and the cement paste can be monitored conveniently.

Electrical conductivity of cement-based materials is dependent on the porosity, pore connectivity, and conductivity of the pore solution, all of which are significant in dictating the time-dependent behavior of cement-based materials. Extensive research on the application of electrical methods has shown that many important parameters of cement-based composites can be monitored effectively using these methods, such as time-dependent changes in pore solution and pore connectivity, effects of mineral and chemical admixtures, diffusion, permeability, and determination of setting time[29]. Especially pore structure features have close and direct relationships with the conductivity, durability, and hydration process of the cement-based materials, so there is increasing emphasis on a detailed understanding of the evolution of pore structures.

For a cement-based material, once the hydration and pozzolanic activity have ceased, the conductivity of the cementitious system will be a function of both the level of pore saturation and the pore–fluid conductivity, so an electrical study can provide much information for long-term durability monitoring[30]. A marine exposure test for concrete blocks has shown that after being immersed in seawater for a long enough time, the conductivity of the blocks will increase with the testing days due to increasing ionic concentration within the pore fluid caused by the penetration of NaCl into the covercrete of the blocks. Thus through the investigation of the conductivity values, the salt ion concentration can be obtained, which also has a close relationship with the compressive strength of the concrete. An electrical study on the effect of salt concentration on the cement-stabilized clay has found that the electrical resistivity correlates closely with the cement content, porosity, and salt ion concentration via the following expression:

$$\rho = 41.934\exp\left(-4.964\cdot\frac{n\cdot C_s}{a_w}\right) \tag{7.8}$$

where ρ is the resistivity of the cement-treated clay, n is the porosity, C_s is the ratio of salt weight to the weight of the dry clay, and a_w is the ratio of cement weight to the weight of dry clay.

Equation 7.8 shows that a unique exponential function adapts electrical resistivity values of the composite with cement content and salts concentration, which can also be seen clearly in Fig. 7.7. As shown in Fig. 7.7, the electrical resistivity of cement-stabilized clay increases with the increasing cement content. This tendency can be attributed to the hydration reactions. As is known, a higher cement content yields a greater amount of hydration compounds, which will fill in the pore spaces and intersect each other to form a solid network and thus result in a denser structure. Meanwhile, with the proceeding of hydration, the free water and porosity decrease, so the resistivity increases accordingly[31].

In engineering applications, electrical measurement has also been shown to be an effective tool for characterizing the mechanical properties of soils and rocks due to its economical, nondestructive, and relatively noninvasive advantages[32]. The investigation of the unconfined compression strength (UCS) of the cement-stabilized clay shows that the strength increases with the increasing of cement content and decreases with the salt content, as shown in Fig. 7.8. The increasing of the cement content results in a greater amount of new compounds such as calcium silicate hydrate and calcium aluminate hydrate gels, which are formed as a result of hydration processes. Meanwhile, the continuous pozzolanic reactions caused by the cement increment leads to subsequent crystallization to bind the structure together. All these results lead to an improved strength of the cement composites[31]. It is also indicated that the strength decreases with an increase in salt concentration; however, it can relieve the detrimental effect of salt concentration on the hydration process of the cement-treated clay through increasing the cement content in practice, as shown in Fig. 7.8.

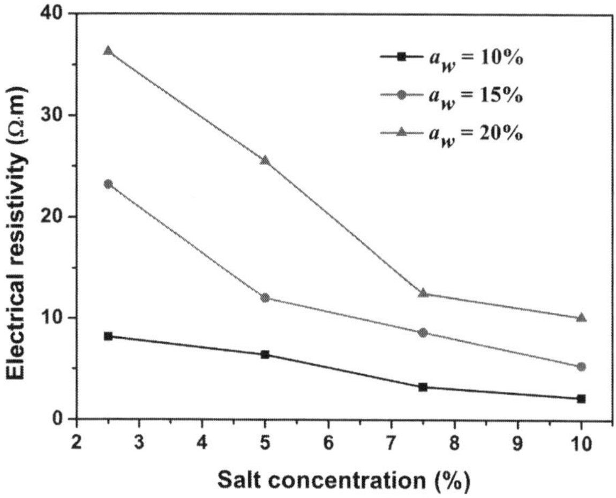

Figure 7.7 Variations of electrical resistivity of cement-treated clay with salt concentration.

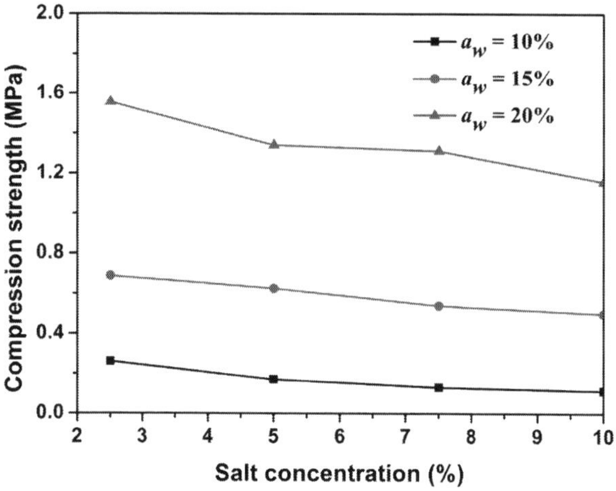

Figure 7.8 Variations of compression strength of cement-treated clay with salt concentration.

The measurement of time-dependent conductivity and porosity of plain cement (w/c = 0.30) and fly ash cement (with 20 wt% cement replaced by fly ash) has been conducted immediately after mixing. The conductivity and porosity appear to have a similar tendency with time at the early stage of hydration, as shown in Fig. 7.9, which demonstrates clearly the relationship between porosity and electrical conductivity of cement composites[29].

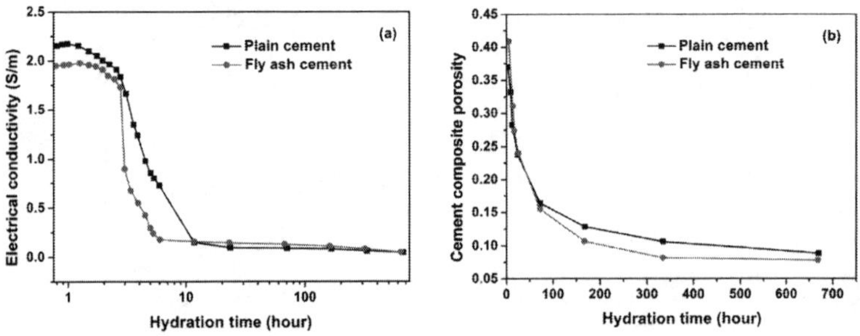

Figure 7.9 Electrical conductivity (a) and porosity (b) changes with hydration time of cement composites.

To predict the porosity features of cement composites from electrical conductivity values, three prediction models have been proposed, which are the modified Archie's law, the Bruggeman–Hanay model, and generalized effective media (GEM) theory. The first two models can be expressed as Eqs. 7.9 and Eq. 7.10, respectively:

$$\sigma_{\text{eff}} = \sigma_{\text{p}}\phi_{\text{p}}^{m} + \sigma_{\text{s}} \tag{7.9}$$

$$\sigma_{\text{eff}} = \sigma_{\text{p}}\phi_{\text{p}}^{m}\left(\frac{\sigma_{\text{p}} - \sigma_{\text{s}}}{\sigma_{\text{eff}} - \sigma_{\text{s}}} \cdot \frac{\sigma_{\text{eff}}}{\sigma_{\text{p}}}\right)^{m} \tag{7.10}$$

In Eqs. 7.9 and 7.10, σ_{eff} is the effective conductivity of the cement composite and σ_{p} and σ_{s} are the conductivity of the pore solution and the solid phase, respectively. ϕ_{p} is the volume fraction of the pore solution and m represents the degree of connectivity of the pore phase.

Experimental results show that both the two models with the parameter m = 4 are capable of predicting the porosity of cement composites with low porosities effectively from the electrical conductivity values. However, there are discrepancies between the measured and calculated values at initial stages with higher porosities, as shown in Fig. 7.10. This can be attributed to the fact that at the initial hydration stage, the connectivity of the pores (m) with high porosities is much different from that at a later stage with low porosities. Thus the prediction with the same m value will result in great discrepancies between the calculated and measured porosity values.

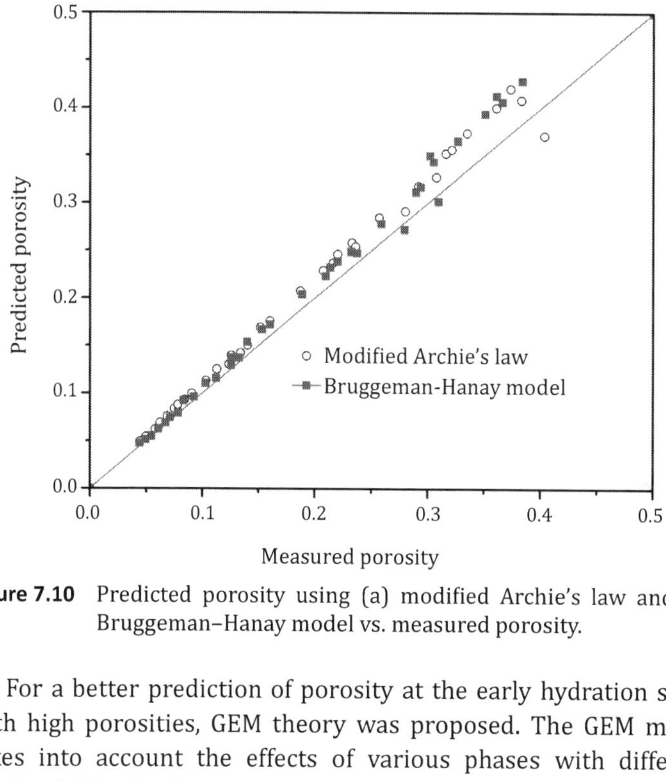

Figure 7.10 Predicted porosity using (a) modified Archie's law and (b) Bruggeman–Hanay model vs. measured porosity.

For a better prediction of porosity at the early hydration stage with high porosities, GEM theory was proposed. The GEM model takes into account the effects of various phases with different conductivities in the cement composite material, and thus it can predict the early-stage porosities with higher precision. According to this model, porosity can be expressed as[33]

$$\phi = \frac{\left[\left(\frac{\sigma_p}{\sigma_{eff}}\right)^{1/t} + \left(\frac{1-\phi_c}{\phi_c}\right)\right] \times \left[\left(\frac{\sigma_s}{\sigma_{eff}}\right)^{1/t} - 1\right]}{\left[\left(\frac{\sigma_p}{\sigma_{eff}}\right)^{1/t} + \left(\frac{1-\phi_c}{\phi_c}\right)\right] \times \left[\left(\frac{\sigma_s}{\sigma_{eff}}\right)^{1/t} - 1\right] - \left[\left(\frac{\sigma_s}{\sigma_{eff}}\right)^{1/t} + \left(\frac{1-\phi_c}{\phi_c}\right)\right] \times \left[\left(\frac{\sigma_p}{\sigma_{eff}}\right)^{1/t} - 1\right]} \tag{7.11}$$

where ϕ and σ_{eff} are the porosity and effective conductivity of the cement composite, respectively; and σ_p and σ_s are the conductivity of the pore solution and the solid phase, respectively. ϕ_c is the percolation threshold and t is an exponent with a value of 0.22, representing the conductivity of the cement system.

The GEM model is proved to more effective to predict the porosity through electrical conductivity values compared to the modified Archie's law and the Bruggeman–Hanay approach during early hydration stages. Nevertheless, in later stages, these three models provide almost comparable results, with a maximum deviation between the experimental and predicted values within 10%[31].

7.1.3.2 Structural health monitoring

In civil engineering, the cement-based infrastructure often needs to evaluate its health, which is just like the experience of the medical profession. The health monitoring of civil engineering structures has been defined as the continuous or regular monitoring of the condition of a structure or a system using in situ and nondestructive testing. In recent years, civil health monitoring has shown significant merits by enabling engineers to predict and prevent structural failure and damage, thereby saving human lives and reducing the structures' costs of maintenance and repair[34]. The term "structural health monitoring" is often used to refer to the assessment of a structure's load-carrying and serviceability parameters, such as the measurements of loads, stresses, strains, deformations, and vibrations. For health monitoring and assessment of cement-based structures, a range of nondestructive sensors may be useful and a variety of material properties and performance parameters can be measured. However, self-sensing by cement materials is advantageous over other embedded or attached sensors, such as strain gages and optical fibers, since it has low cost, high durability, and large sensing precision[35].

Electrical measurements have been shown to be valuable as noninvasive methods to evaluate the material properties of cement or concrete-based structures. For structural health monitoring application, electrical methods have been applied to monitor the damage development inside structures. The previous studies conducted by Chung et al. have shown that the interior damage of carbon fiber–reinforced cement composites can be observed through the gradually decreased electrical resistivity of the composites, since the damage increases the opportunity for adjacent fibers to touch one other and thus decreases the resistivity of the composites[36]. However, the strain-sensing ability may turn out to be less repeatable with the curing age beyond 14 days, and a surface treatment of the carbon fiber with ozone can diminish this undesirable effect. Experimental results show that after pretreating the carbon fibers with ozone through exposing the fibers to O_3 gas at 160°C for 5 min, the fractional DC resistance increase ($\Delta R/R_0$) gives a stable trend during the cyclic tensile loading. The downward trend in $\Delta R/R_0$ for pristine carbon fiber is absent for the ozone-treated composite due to the improved mechanical properties of the carbon fiber/cement composite.

Compared to carbon fiber, carbon nanotubes (CNTs) and carbon nanofibers (CNFs) show much more advantages in health monitoring of cement- or concrete-based structures, such as stress sensing, damage detecting, and traffic monitoring in highway structures, due to their higher electrical conductivity and sensitivity. When subject to stress or strain, their electrical properties change in a linear and reversible piezoresistive response. Hitherto, great efforts have been concentrated on the piezoresistive behavior and sensing ability of cementitious composites embedded with CNTs and nanofibers through electrical property studies. It has been found that cement-based composites with CNTs as reinforcement displayed greatly reduced electrical resistivity and the resistivity changed synchronously with the compressive stress levels[37]. Road experiments also show that the CNT/cement composite can detect vehicular loads through remarkable changes in electrical resistance[38]. The compressive stress sensitivity of cement-based composites reinforced with CNTs and CNFs performed at 28-day-old specimens show that composites with 0.1 wt% loading yield great electrical performances. The average change in resistivity under cyclic stress loading is about 5%, with resistivity decreasing during the stress loading and increasing during unloading, as shown in Fig. 7.11[39]. The results demonstrate that cement composites reinforced with CNTs and CNFs are much more sensitive in monitoring the change in the applied stress and the induced mechanical deformation, which also indicates that these composites are excellent candidates for strain sensing.

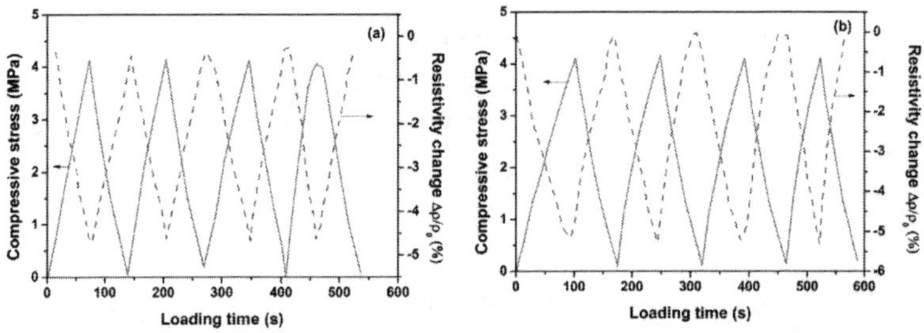

Figure 7.11 Piezoresistive behavior of cement composites reinforced with (a) CNTs and (b) CNFs at an amount of 0.1 wt% of cement.

Besides CNTs and CNFs, graphene nanoplatelets (GNPs) can also be an ideal type of filler for cement-based composites to assess the damage extent and overall structural behavior. As a kind of low-cost sheet-like conductive filler, GNPs can enhance the electrical conductivity of the cement greatly, and moreover, they can also diminish the fiber-bridging effect of carbon fibers when applied as damage self-sensing materials[40].

The damage assessment of the GNP/cement composite was investigated in detail using electrical methods. GNPs were added at a content of 15 wt% by mass of cement, which corresponds to 3.6 vol% of the mortar mixture[40]. The dimension of the specimen is 40 × 40 × 160 ($W \times D \times L$) mm^3 and the spacings of the inner probe are L = 40 and 80 mm. The electrical resistance was tested with a four-probe technique, as shown in Fig. 7.12a. Using the multiphysics software COMSOL, the electrical current flowing through the composite material can be simulated, as shown in Fig. 7.12b. The experimental results show that the conductivity of cement is greatly improved by the addition of GNPs and the change of electrical resistance is much sensitive to the damage.

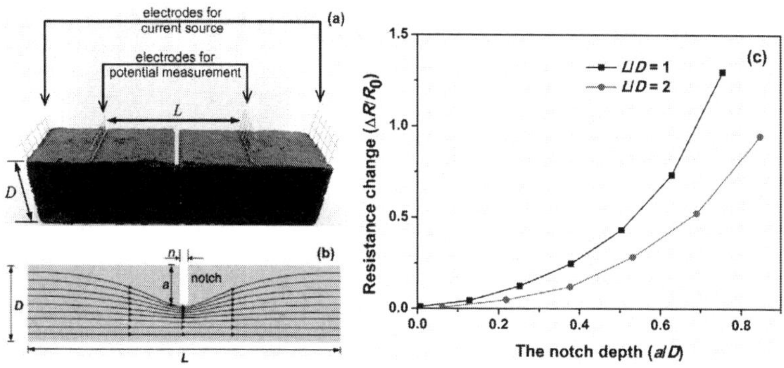

Figure 7.12 Sketch map of the electrical testing technique (a), current flow across the specimen (b), and the electrical resistance changes of the GNP/cement composites (c).

The electrical resistance changes for GNP/cement composites with different notch depths and specimen sizes are shown in Fig. 7.12c. It is found that the electrical resistance of the GNP/cement composite is much sensitive to the notch depth, which means that evolution of cracks inside the composite can be detected conveniently through the observation of electrical resistance. On the basis of the experimental results and the classic elastic fracture mechanics, a relationship between the crack length and the overall electric resistance can be established as follows:

$$\frac{R(\alpha)}{R_0} = 1 + \frac{\int_0^{\alpha} k^2(\alpha')d\alpha'}{bGC_{\tau 0}} \tag{7.12}$$

where $\alpha = a/D$ is the relative crack length and $k(\alpha)$ is a dimensionless stress intensity factor. $R(\alpha)$ is the electrical resistance of the composite with crack length a. G is the energy release rate, C_τ is the elastic compliance of the specimen, and b is the specimen width in the transverse direction[40].

This relationship is based on the assumption that the electrical resistance of the composite is analogous to the elastic compliance of the composite under antiplane shear loading. However, experimental results have shown that this equation is also applicable to composites with other common structural geometries and under different loading conditions.

7.2 Cement-Based Electromagnetic Shielding Materials

Cement is slightly conductive, but its electromagnetic interference (EMI) shielding effectiveness (SE) is much weak, and thus improving the cement materials' EMI shielding performances by introducing conductive and magnetic filling and loadings turns out to be a simple and practical method. So far, extensive and intensive research has been made on shielding properties and shielding mechanisms for cement-based materials. Generally, there are mainly two purposes for a shielding material—to prevent the EMI emissions of the electronic radiation outside the restricted boundaries and to prevent externally radiated emissions to the product's electronics. So, in this sense, a shield is conceptually a barrier to the transmission of electromagnetic fields[41].

According to Schelkunoff's theory, the SE of a shield can be expressed as the product of three terms, with each presenting one of the phenomena of reflection loss, absorption loss, and multiple reflections. The SE can be written as follows[41]:

$$\mathrm{SE} = 20\ \log\left|\frac{E_{\mathrm{tr}}}{E_{\mathrm{in}}}\right| = \mathrm{SE_A} + \mathrm{SE_R} + \mathrm{SE_M} \tag{7.13}$$

where E_{in} and E_{tr} are the electric fields incident on and transmitted through the shield materials, respectively. $\mathrm{SE_A}$ is the absorption loss of the electromagnetic wave (EMW) as it is transmitting through the shield. $\mathrm{SE_R}$ is the reflection loss caused by the reflection at the surface of the shield due to impedance mismatching between the air and the shield, and $\mathrm{SE_M}$ is the multiple reflection loss caused by the multiple reflections interior to the material. The three terms can be expressed in detail by the following equations:

$$\mathrm{SE_A} = 20\ \log(e^{-t/\delta}) \tag{7.14}$$

$$\mathrm{SE_R} = 106 + 10\ \log\left(\frac{\sigma_{\mathrm{r}}}{f\mu_{\mathrm{r}}}\right) \tag{7.15}$$

$$\mathrm{SE_M} = 20\ \log(1 - e^{-2t/\delta}) \tag{7.16}$$

In Eqs. 7.14–7.16, t is the thickness of the shield and σ_r and μ_r are the relative electrical conductivity and magnetic permeability of the shield, respectively, compared to pure copper. δ is the skin depth of the shield, a distance that the incident wave propagates in the shield when its amplitude attenuates to its e^{-142}. The skin depth can be expressed as Eq. 7.17.

$$\delta = \sqrt{\frac{1}{\pi \mu \sigma f}} \tag{7.17}$$

The reflection loss (SE_R) is generally the primary mechanism of EMI shielding. To achieve this goal, the shield must have mobile charge carriers which can react with the EMW in the radiation and attenuate the electromagnetic energy. From Eqs. 7.14–7.17, it can be seen that SE_R has a close relationship with the conductivity of the shielding material and a high conductivity is beneficial to its shielding performance. However, a much high conductivity is not always required, and a volume resistivity of the order of 1 Ω·m is typically sufficient[43].

The absorption loss (SE_A) is usually the second mechanism of EMI shielding. For significant absorption by a shielding material, electric or magnetic dipoles are often needed. The dipoles react with the EMW and turn the incident electromagnetic energy into other energy types, such as heat. SE_A is a function of the product $\sigma_r \mu_r$, and a material with appropriate electrical conductivity and magnetic permeability will be very helpful to its absorption loss.

Besides reflection and absorption, another main mechanism of shielding is the multiple reflection loss, which refers to the reflections at various surfaces or interfaces in the shielding material[43]. This mechanism requires the presence of a large surface area or interface area in the material. Thus materials with a large number of porosities or conductive fillings may be beneficial to their multiple reflections.

Cement is slightly conducting, so the use of a cement matrix allows the conductive filler units in the cement-based composite to be electrically connected, even when the filler ratio is not high enough to let them touch one another. And also, the fillers used in a cement matrix needn't be as conductive as those used in a nonconductive matrix. Generally, there are mainly three types of fillers used in cement matrix composites, which are conductive polymers, carbon materials, and metal materials. Conductive polymers are usually attractive in their low density, but they are not common and their addition often affects the processability and mechanical properties of the composites. Carbon materials, due to their diversity and high conductivity and low percolation threshold, are often taken as the optimal choice in cement-based composite materials.

7.2.1 Carbon Filling Cement-Based Materials

7.2.1.1 Graphite

Graphite is a kind of conductive filler with high conductivity, but its conduction is not very stable, so it is very late when graphite is used as a conductive introduction[2,44]. Introducing graphite powder (GP) into a cement matrix finds that the graphite's threshold is about 17.5 vol%. A very low loading of GP barely affects the shielding capacity of cement paste. Zornoza's research finds that a 0.5 wt% addition only shows a shielding property of about 5 dB at 1 GHz for a 4 mm thickness sample[45]. However, increasing the aspect ratio will improve the electrical conductivity and thus the SE greatly. When the filling ratio is increased to 30 vol%, a specimen with about 3 mm thickness can give an SE of 10–40 dB in the frequency range of 200–1600 MHz. Studies also have shown that the conductivity of a graphite/cement composite confirmed to the power-law percolation model:

$$\sigma = C(\phi - \phi_c)^t \tag{7.18}$$

where, σ, ϕ, and ϕ_c are the conductivity of the composite, the filler volume fraction and the critical volume fraction, respectively. And t is the critical index of conductivity. In Bhattacharya's study, the ϕ_c is as small as 2 wt% and t is 8.2 by experimental fitting[46].

To improve the stability of GP, nickel-clad GP has been developed and has shown great EMI shielding properties and long-life stabilities. It has been reported that a nickel-graphite-filled elastomer showed an SE of at least 80 dB over a wide frequency range of 10 MHz to 10 GHz. The nickel cladding covers the whole surface of each graphite particle and has a stable and oxide-free surface. The particles are rough and irregular in shape to provide good electric contact between the particles. The nickel graphite offers the synergistic effect of reflection loss by the nickel cladding and absorbing loss by the graphite core. It has been demonstrated that nickel graphite exhibited an excellent SE similar to silver, although the volume resistivity of nickel graphite is 20–500 times higher[47].

Suspending ultrafine GP in some water or alcohol, colloidal graphite can be obtained with the addition of a small amount of poly(vinyl alcohol) (PVA). When applying the colloidal graphite to a surface, the water or alcohol will evaporate and thus the graphite particles adhere to each other tightly. Colloidal graphite filling cement composites have shown excellent EMI shielding performances. For instance, in a Portland cement matrix composite with a thickness of 4.4 mm, a filling ratio of 0.92 vol% gives an SE of 22.3 dB at 1.0 GHz and 25.6 dB at 1.5 GHz, respectively[48]. Colloidal graphite conventionally uses submicron graphite particles. However, its SE can be greatly improved after combining with carbon filaments with a diameter of 0.1 μm.

In the graphite family, there is a special kind of graphite, namely flexible graphite. It is a flexible sheet made by compressing a collection of exfoliated graphite flakes without a binder. Flexible graphite is essentially a kind of pure graphite with a large specific surface area and high electrical conductivity[49,50]. Studies show that flexible graphite has as high a skin depth as 44 μm at 1.0–2.0 GHz. With a thickness of 0.79 mm and 3.1 mm, it can have an EMI shielding capacity as high as 101.9 dB and 129.4 dB[51].

Coke is another kind of carbon material with a graphite-like structure. Coke is usually used as an electrode material for the production of aluminum and a sorbent for bitumen and naphtha, and it is also applied as a raw material for making graphite and steel. Coke is not as conductive as graphite, but it is much less expensive and superior in mechanical strength. Studies by Chung et al. have shown that coke is also good filler in cement-based composites to improve its electromagnetic properties. With a granularity of 200 meshes and a filling ratio of 0.51 vol%, coke powder can have an SE of 41–45 dB at the frequency range of 1.0–1.5 GHz. By further increasing the coke content to 9.18 vol%, a coke/cement composite sample with a thickness of 4 mm can get a shielding property of about 50 dB at the same frequency tested[52].

7.2.1.2 Carbon fibers

Carbon fiber–cement matrix composites are gaining attraction and importance rapidly due to the decrease in carbon fiber cost and the increasing demand for superior structural and functional properties[3]. As to the electrical conductivity and EMI SE of cement composites, carbon fibers are more effective than carbon-based particulates due to their small unit size and large aspect ratio, which is helpful to form more conductive networks through intercalating[44]. Moreover, fibers play an important role in increasing the strength, preventing cracks, and resisting impacting for cement-based composite materials. With a *w*/*c* ratio of 0.3 and 0.5 wt% polycaboxylate-based superplasticizer, a carbon fiber–reinforced cement composite shows great EMI shielding properties. With the Portland P·II 52.5 cement as the cementitious material, a filling ratio of 1.0 vol% can gain SE values over 25 dB in a wide frequency range from 200 MHz to 10 GHz. When the frequency reaches 10 GHz, the shielding property can be as high as 45 dB[53].

In the carbon fiber–reinforced cement composites, the carbon fiber with a diameter less than 0.1 μm is often called carbon filament. Due to its low diameter and high aspect ratio, carbon filament is superior to fiber in cement-based composites for EMI shielding[54]. Since Chung et al. reported the first application of carbon filament in cement matrix composites in 1996[55], it has gained popularity in application in cement and concrete functional composites. With Portland cement as the starting material, an introduction of 0.51 vol% carbon filament can give an SE of about 26 dB for a 4 mm thick sample at a frequency of 1.0, 1.5, and 2.0 GHz. The filament/cement composite has an electrical resistivity of 1.93×10^4 Ω·cm, which is only 10% of the plain cement paste[56,57]. Though the filling ratio is far below the percolation threshold, the large aspect ratio of the filament and the conducting nature of the cement matrix provide conductive connectivity between the filaments and the matrix, which is enough for enhancing the EMI shielding performances.

Carbon filaments also show great advantages for EMI shielding in carbon fiber/epoxy and graphite/cement composites to improve the direct contact between the adjacent fillings and the matrix. It is shown that the addition of carbon filament in the epoxy matrix can improve the dielectric loss tangents (tan δ) of carbon fiber/

epoxy composites in the transverse direction of the composite at a frequency of 0.2, 1.0, and 4.0 Hz[57]. A 20 wt% filament addition in the graphite/cement composite can also improve the SE from 24 dB to 37 dB at 1.0 GHz[58]. The improvement can be attributed to the increase in the reflectivity due to the high electrical conductivity of carbon filaments.

7.2.1.3 Carbon nanotubes

The unique structure and excellent properties of CNTs has prompted extensive and intensive study for potential engineering applications[59]. According to classical electromagnetic shielding theory, a thinner diameter, greater aspect ratio, higher conductivity, and mechanical strength of the filler are much beneficial to improve the EMI SE. In carbon-based materials, CNTs can meet all the demands for a better filling and make them an excellent option for creating conductive composites for high-performance EMI shielding materials at a lower filling than other carbon variants[60,61].

The electromagnetic performances of single-walled carbon nanotube (SWCNT)-filled epoxy composites showed that the electrical conductivity threshold in the epoxy matrix is as low as 0.062–0.342 wt%. With a filling ratio of 15 wt%, the composites show an EMI SE as high as 16–30 dB in the frequency of the X-band (8.2–12.4 GHz), and the shielding is found to be dominated by reflection[60,62]. It is also found that the percolation threshold, electrical conductivity, and SE of the composites are highly correlated with the aspect ratio and wall integrity of the CNTs. SWCNTs with a higher aspect ratio and better surface integrity show high conductivity and shielding performances. After annealing at 1100°C for 3 h in a tube furnace under a N_2 atmosphere, the SE of CNTs increased significantly due to their improved wall integrity caused by the annealed process[62].

CNTs have showed the same excellent SE in cement matrix composites. Multiwalled carbon nanotube (MWCNT)-filled Portland cement composites have been fabricated and their EMI SE in the X-band illustrated that the incorporation of 15 wt% MWCNTs in the cement matrix gives a shielding property of more than 27 dB in the frequency range of 8.2–12.4 GHz[61]. The shielding properties increase with the MWCNT contents due to the improved electrical conductivity and decrease with the hydration ages caused by the loss of water molecules, as shown in Fig. 7.13. From Fig. 7.13 it is clearly shown that the CNT/cement composites with a filling ratio of 15 wt% MWCNTs give an SE of about 43 dB in the whole X-band for the first day of hydration (sample CNT-1 day). However, when the hydration proceeds, the shielding properties decrease gradually to about 28 dB after hydration for 28 days (sample CNT-28 days). The cement paste shows the same trend for the shielding performances with hydration time. After hydration for 28 days, the SE properties decrease from 8–10 dB (sample OPC-1 day) to 3–6 dB (sample OPC 28 days)[61]. The EMI shielding properties are proved to be dominated by absorption caused by ionic, electric, orientational, and space charge polarization[61–64].

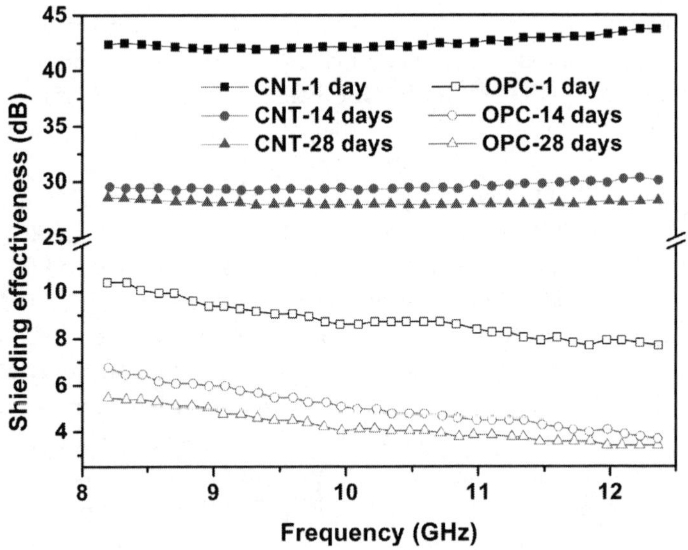

Figure 7.13 The shielding effectiveness of CNT/cement composite and the cement paste (OPC) in the X-band.

To improve the EMI shielding performance of CNTs, it is applicable to attach a magnetic or dielectric layer to the surface of the CNTs. Fe-, Co-, and Ni-decorated CNTs have been successfully prepared and good EMI shielding or absorption properties are displayed[65–69]. For example, Singh et al. carried out a study on the EMI shielding properties of Co- and Ni-decorated SWCNT composites (Co/Ni@SWCNTs) and found that an SE of 24 dB for a

1.5 mm thick sample could be obtained in the frequency range of 12.4–18 GHz. The improved shielding performance can be attributed to the extra magnetic loss caused by the eddy currents and natural resonance attenuation after the incorporation of Co and Ni magnetic nanoparticles[67–70].

7.2.1.4 Graphene and reduced graphene oxide

Graphene is a kind of two-dimensional sheet of sp^2-hybridized carbon and its extended honeycomb-like network constitutes the basic building block of other important carbon allotropes. The long-range π-conjugation in graphene gives it an extraordinary thermal, mechanical, and electrical properties, which have gained great interest in electronics, energy storage, and biotechnology fields[7]. Graphene has great electrical conductivity, high carrier mobility, and high cutoff frequency in the microwave frequency band. Thus it can be used in various technologies such as waveguides, filters, or isolators and can be interconnected with existing microwave devices[6].

Compared to graphene, graphene oxide (GO) introduces some oxygen-containing functional groups, which makes it hydrophilic, and can combine with other functional components to obtain new graphene-based composites. However, the electrical conductivity of GO is much less than graphene due to the damage of the perfect conjugated structures of graphene during the oxidation process. Reducing GO with chemical reduction, photoreduction, or thermal reduction can restore parts of the conjugated structures of GO and thus partly recover its conductivity. Moreover, there are residual defects and oxygen-containing chemical bonds such as C–O and C=O in the reduced graphene oxide (RGO). There was a traditional opinion that the defects and chemical groups in RGO would degrade its performance[7]; however, the defects can act as polarization centers under an external electromagnetic field, which would generate polarization relaxation and attenuate the EMW[71]. Experimental results also show that compared to graphite and other carbon counterparts, RGO demonstrates totally different Debye relaxation performances and it has several relaxation process in the Cole–Cole semicircles. These relaxations, on the one hand, arise from the lag of induced charges and, on the other hand, can also arise from the defects and functional groups[72]. Under these circumstances,

RGO displays much better EMI SE and microwave attenuation properties[73].

Due to their excellent electrical conductivity and microwave attenuation performances, graphene and RGO have been studied comprehensively in polymer and cement matrices for EMI shielding and microwave absorption. For example, RGO-filling polymer matrix composites, such as epoxy[5], polyaniline (PANI)[74,75], polystyrene (PS)[76], polydimethyl siloxane (PDMS)[77], poly(methyl methacrylate) (PMMA)[78], and nitrile butadiene rubber (NBR)[79], all exhibit an EMI SE more than 20 dB in the frequency range tested. Wen et al. discussed the frequency- and temperature-dependent EMI shielding performances of RGO/SiO_2 composites in the X-band under the temperature range of 323–473 K[73]. It is found that when the RGO weight is over 10 wt%, the composite can achieve a dielectric loss tangent more than 1.0 in the whole frequency range. The EMI SE of the composite with 20 wt% RGO reaches a maximum of about 38 dB and the shielding property increases with the temperature. This excellent SE can be attributed to the cooperation of dipole polarization and hopping conductivity of the RGO/SiO_2 composite[73].

Graphene also has been extensive studied in cement matrix composites for its excellent properties such as high conductivity and thermal resistance. GO has been used as a filler in ferrofluid Portland cement–based composites by a planetary ball-milling method. It is found that the presence of GO embedded in the cement matrix enhances the interfacial polarization and effective anisotropy energy of the composite, which leads to higher EMI shielding. Due to its high conductivity, the improvement of the SE after GO addition is mainly attributed to better microwave absorption and strongly depends on the volume fraction of the GO content[8]. With a filling ratio of 30 wt%, the composite presents an SE as high as 46 dB with a thickness of 2.5 mm in the frequency range of the X-band. Moreover, with a filling ratio of 20 wt% GO, the hardness of the composite reaches 200% times higher than that of the cement paste, which indicates that the addition of GO can provide sufficient mechanical properties for the cement paste[8]. GO can also be used to enhance the SE of carbon fiber/cement composites. GO-deposited carbon fiber (GO-CF) was successfully obtained through electrophoretic deposition method and the GO-CF was proved to be more effective than pure carbon fiber in EMI shielding for cement matrix composites[80]. The experimental results show that with a filling ratio of 0.4 wt%, the GO-CF/cement composite (sample GO-CF 0.4%) exhibits an SE as high as 30 dB in the frequency range of 8.2–12.4 GHz (X-band), which is 20% higher than that of the carbon fiber/cement composites (sample CF 0.4%), as shown in Fig. 7.14[80].

The studies on the transport properties of GO/cement composites show that a trace addition (0.03 wt%) of GO to the cement matrix can improve the amount of gel pores and enhance the microstructure of cement paste, thus enhancing the water sorptivity and chloride penetration values, which subsequently improves its durability[81].

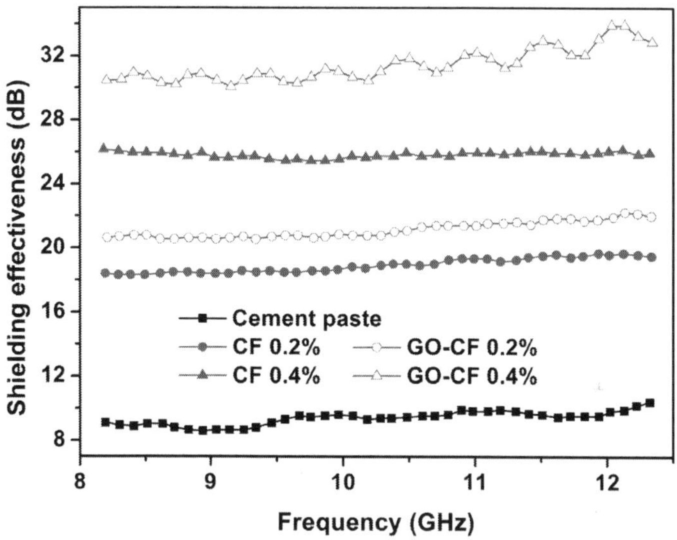

Figure 7.14 Shielding performances of carbon fiber (CF)/cement composites and GO-CF/cement composites with a filling ratio of 0.2 wt% and 0.4 wt%.

7.2.2 Metal Filling Cement-Based Materials

Metal-based fillers, which are used in cement matrices for EMI shielding applications, include metal powder, metal fiber, and metal alloys.

Due to their high electrical conductivity, silver, copper, iron, and nickel powders have been used for a long time as conductive fillers in polymer matrices[2]. Silver has very good conductivity but it is a type of noble metal and is only used on special occasions. Copper and iron also have good conductivity than most other metals; however, they are prone to be oxidized in the normal atmosphere. Nickel is relatively stable and resistant to corrosion, but the electrical conductivity of nickel is much lower than silver, copper, and iron. So it seldom applied solely in cement materials as a conductive filler.

Lead powder filling Portland cement composites have been prepared as shielding materials for gamma radiation emitted from a ^{137}Cs-γ-source. The results show that the attenuation coefficient increases with the lead concentration and curing time. When the weight ratio of lead powder increases to 5 wt%, the attenuation coefficients get two peak values with the curing time at 15 and 28 days, respectively. It is also shown that the shielding property improves with the increase in the shield thickness and the curing process[82]. An extra use of 5% silica fume can reduce the curing time from 12 h to 6 h with the same shielding property[83].

Compared to metal powder, metal fiber has a lower density and higher mechanical strength. And also, metal fiber has a higher aspect ratio and is liable to be intercalated when introduced into the cement matrix. So metal fiber is more prone to form a conductive network and thus provides a better shielding effect than metal powder in the same filling ratio. However, due to their disadvantageous tendency to be oxidized in cement matrices, there are only few studies on metal fiber–filled cement matrix shielding materials[2]. And moreover, most of the studies are focused on the effects of cement or concrete structure and fiber distribution on the shielding properties of the composites and the shielding calculation with mathematical methods.

Chung et al. have studied the magnetic SE of cement pastes filled with steel paper clips, and found that the composites with a filling ratio of 5 vol% can provide a magnetic SE comparable with that of a steel mesh with a diameter of 0.6 mm[84]. In this low-volume fraction, the clips are not continuous, but their intertwining tendency and the conductive nature of cement paste allow the formation of a conductive network continuity required for magnetic shielding.

The proliferation of cellular communication systems around manmade structures has resulted in a growing need to evaluate the SE of various materials used in buildings. This is very useful in radio base station planning and in the evaluation of exposure effects of nearby radiation sources. In the last few years, many simulation models have been built and calculations have been carried out with finite difference time domain (FDTD) methods to examine the shielding performances and influences of concrete structures and both the shapes and distribution configurations of the steel fibers/bars on SE. The results are demonstrated that at the lower frequency bands, the transmitted radio signal is attenuated mainly by the steel structure. As the frequency increases, the effects of the wall structure turn out to be more and more pronounced and result in a larger transmission coefficient. Any changes of the wall structure, steel fiber distribution, electrical properties, and transmission frequencies also influence the shielding properties of the composites[85–87].

7.3 Cement-Based Electromagnetic Absorbing Materials

There are mainly two methods to prevent EMI and radiation, namely EMI shielding and microwave absorption. EMI shielding essentially forms an effective enclosed area with high conductive or magnetic materials, in which the external electromagnetic radiation cannot penetrate and internal radiation cannot be easily leaked out[88]. However, shielding cannot eliminate EMI radiation, and moreover, the reflected wave may interact with the incident one and cause more complicated disturbance. Only by microwave absorbing materials can the EMI radiation be attenuated by transferring the electromagnetic energy to other forms[5].

The electromagnetic absorption property of a microwave absorbing material is denoted with the term "reflectivity" or "reflection loss," which is named by its testing methods and can be expressed as

$$R = 20\log\left|\frac{E_{\text{re}}}{E_{\text{in}}}\right| \tag{7.19}$$

where E_{re} and E_{in} refer to the electric field strength of the incident and reflected EMWs, respectively. A reflectivity of –20 dB for a plate absorber is sufficient for civil use and it means that the incident wave has been reduced by 90%.

According to the classical microwave absorption mechanism,

traditional microwave absorbers can be divided into three types as electric loss, dielectric loss, and magnetic loss materials according to their attenuation mechanisms of the incident microwave. Electric loss media, such as carbon-based materials and conductive polymers, have higher electrical conductivity and higher electric loss tangents (tan δ_e), and the electromagnetic energy is mainly attenuated to heat energy as a resistor. Many metal oxides and ceramic materials, such as TiO_2, MnO_2, and $BaTiO_3$, are dielectric loss absorbents, which mainly attenuate electromagnetic energy by interfacial and surface polarization, space charge polarization, electronic and ionic polarization, and the associated polarization relaxations[89]. This kind of material needs an appropriate conductivity and high dielectric loss tangents. The third type of microwave absorbent is the magnetic loss media, which include ferrites, fine metal powders, and polycrystalline iron fibers. This type of material usually has higher magnetic loss tangents (tan δ_m) and mainly attenuates the incident EMW energy by hysteresis loss and magnetic domain resonance[5].

To obtain a good electromagnetic absorbing property for cement matrix composites, there are many factors to be considered, such as various physical and chemical reactions between the fillings and cement matrix, the physical and electric properties of the fillings, and the electromagnetic characteristics of the mixtures, so as to determine the types and contents of the fillings. In practical applications, the commonly used fillers in cement-based electromagnetic absorbing materials are conductive and magnetic powders and fibers.

7.3.1 Electric Loss Cement-Based Materials

All sorts of carbon materials, especially carbon blacks, are important raw materials which have been widely used in rubber, plastic, and other industries. There are many types of carbon blacks, such as acetylene black, furnace black, and conductive black, and each has its own electrical conductivity and is used in different fields[90]. The electrical conductivity of carbon black is closely correlated with its structure, specific area, and surface chemical activity.

A carbon black structure means the aggregation state of carbon particles caused by fusion and gathering and is usually denoted with the abbreviation DBP, which refers to the dosage of dibutyl phthalate dropped into unit mass carbon black when there is an abrupt change in the viscosity of carbon black. The DBP value is dependent on the size of the aggregates, the amount of particles in each aggregate, and the particle sizes. Carbon blacks with a DBP value less than 0.9 ml/g are often called low-structure carbon, in which the carbon particles are dispersed separately, as shown in Fig. 7.15a. Carbon blacks with a DBP value in the range of 0.9–1.2 mL/g are classified as medium-structure carbon, in which only parts of the particles are connected together to form branch-like structures, as shown in Fig. 7.15b. High-structure carbon blacks have the particles fused together to form grape-like structures, as shown in Fig. 7.15c. This structure has a larger specific surface area and tends to form a conductive network, which is very beneficial to improve the electrical conductivities of the composites[91,92].

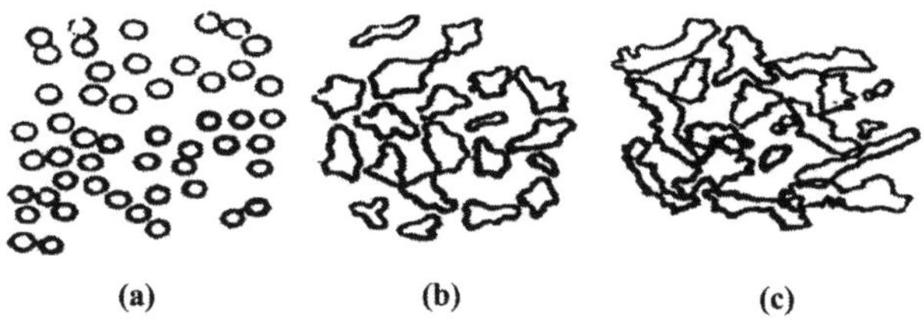

Figure 7.15 The schematic diagram of low- (a), medium- (b), and high-structure (c) carbon black.

Surface chemistry has great influence in the electrical properties and the interactions between carbon black particles and the matrix. Thermal treatments are also taken to improve the chemical activity of carbon black particles. The particles are firstly rinsed with acetone solution and after filtration they are calcined at 700°C for 1 h under a N_2 atmosphere. This process can not only remove the organic groups in the surface of the carbon blacks but also improve their chemical activities[93].

Dai et al. discussed the EMW absorbing characteristics of high-structure carbon black/cement composites and found that a small amount of high-structure carbon black can form a conductive network in the cement matrix and the percolation threshold zone contains only 0.36–1.34 vol% of carbon black. Both the electrical conductivity and dielectric loss tangents increase with increasing carbon contents in the matrix[90]. The composite exhibits high performance of microwave absorbing in the frequency range of

8–26.5 GHz. For a composite with a carbon black filling ratio of 1.34 vol%, the maximum reflectivity can reach –20 dB and the effective absorption bandwidth in which the reflectivity is superior to –10 dB is as wide as 11.6 GHz (from 14.9 to 26.5 GHz). Their results illustrate that since high-structure carbon black is rich in branch structures, it is convenient for the carbon particles to contact each other and form continuous chain-like conductive channels. Under an external electromagnetic field, the particles are apt to be polarized and present in the form of electronic polarization, ionic polarization, molecular polarization, and interface polarization[90]. Meanwhile, the incident microwave can generate an eddy current in the bulk composites due to the semiconductivity nature of carbon black and thus convert electromagnetic energy into heat energy.

The study of carbon fiber–reinforced cement composites presents similar electromagnetic absorbing performances. Studies of the microwave absorbing properties of carbon fiber/cement composites in the frequency range of 8–18 GHz conducted by Li et al. show that with a filling ratio of only 0.4 wt%, the composites can obtain a maximum reflectivity of –19 dB at 8.6 GHz. When the fiber content is over 0.6 wt%, the wave reflection at the surface of the composites prevails dominantly and thus the wave absorption properties decrease accordingly[94,95]. The improvement of electromagnetic absorbing properties of the carbon fiber/cement composites can be attributed, on the one hand, to the increased conductivity and, on the other hand, to the decreased porosity after the incorporation of carbon fibers[96].

The microwave absorbing properties of cement-based composites are not only dependent on the morphologies and filling fractions of the fillers but also dependent on the shapes of the composites due to the fact that various shapes and surface conditions reflect and absorb the incident wave differently. Laukaitis et al. investigated four types of carbon fiber–reinforced autoclaved aerated cement (AAC) slabs—slabs with large (80 mm height) pyramids, slabs with small (45 mm height) pyramids, uniform slabs, and slabs with small riffles on one surface—with the time domain method in the frequency range of 2–18 GHz. The results find that a uniform AAC slab without carbon fibers can have an absorbing performance of about –10 dB at 2–18 GHz, whereas a slab with small riffles on one side can improve the absorbing property to about –17 dB. The small riffles also induced "waving" of the microwave absorption versus frequency dependence[97]. The comparison of the AAC slabs with carbon fiber additions shows a similar phenomenon. The slabs having larger pyramids exhibit reflectivity ranging from –21 dB to –30 dB in the whole frequency range, whereas slabs with smaller pyramids show a reflectivity of –18 to –27 dB. These results are the effects of wave reflection from the riffles or pyramids and they correlate closely with the dimensions and shapes of the slabs.

Besides carbon black and carbon fiber, CNTs can also be effective fillers in cement matrices for microwave absorption. Due to their unique structure and dielectric properties, CNTs exhibit strong microwave absorption properties. In the tube configuration, CNTs have an electronic π-structure conjugate, which can lead to their unique transport behavior[98]. CNTs can also act as electric dipoles to resonate with the incident wave and thus produce a polarization current and attenuate the electromagnetic energy to heat or other kinds of energy. When the CNTs are mixed with cement paste, a considerable number of interfaces are formed between the CNTs and the cement matrix, thus causing dielectric polarization and relaxation arising from the interfacial polarization. Moreover, a hopping conduction can also occur between the CNTs if the content of the CNTs reaches a certain degree[99]. Wang et al. investigated the electromagnetic absorbing properties of MWCNT/cement composites in the frequency ranges of 2–8 GHz and 8–18 GHz and found that the content and dispersibility of the MWCNTs are the key factors that affect the microwave absorption properties of the composites[100]. With an addition of 0.6 wt% MWCNTs to the cement matrix, the composite attenuates the EMW remarkably in the 2–8 GHz range, as displayed in Fig. 7.16a. Three absorption peaks of –28 dB, –18 dB, and –16 dB can be observed at 2.9 GHz, 4.7 GHz, and 7.0 GHz, respectively. When the MWCNTs are increased to 9 wt%, the composite shows great absorption performances at 8–18 GHz, as shown in Fig. 7.16b. The reflectivity values in this frequency range are all superior to –8 dB and the effective absorption bandwidth below –10 dB reaches 7.1 GHz, which is much better than that of carbon fiber–reinforced cement-based composites[94].

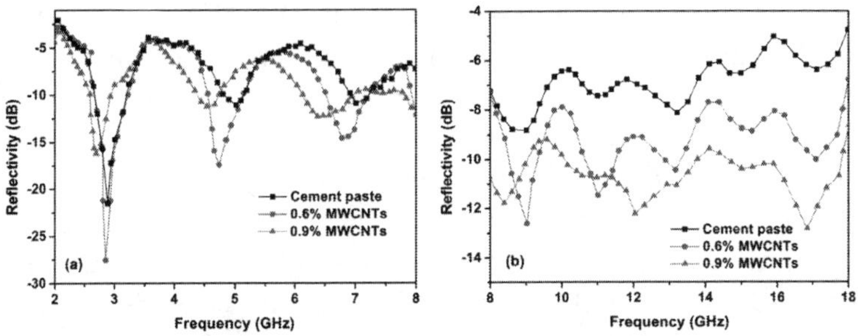

Figure 7.16 Influence of MWCNT content on the reflectivity of cement in the frequency range of 2–8 GHz (a) and 8–18 GHz (b).

7.3.2 Dielectric Loss Cement-Based Materials

Dielectric loss materials used in practical applications are usually metal oxides, such as ZnO, TiO_2, and MnO_2, which have appropriate electrical conductivity and can form dipoles in an electromagnetic field. ZnO and TiO_2 have been thoroughly studied as dielectric loss media in the past decades and their electromagnetic absorbing mechanisms have been discussed in detail. In the recent few years, MnO_2 has been investigated extensively as a kind of dielectric loss absorbent since its first report as an electromagnetic absorbing material in 2006[101-103]. Manganese oxide materials with various morphological and crystalline structures have been prepared with different physical and chemical methods, and their electromagnetic absorbing properties have been discussed.

Particulate titanium dioxides (TiO_2) are widely applied in cement-based materials because they can improve not only the microwave absorbing characteristics but also the hydration properties. TiO_2 can be used in Portland cement paste to accelerate its early-stage hydration and increase the hydration degree. Studies show that the size and dispersability of TiO_2 nanoparticles are critical factors influencing the hydration properties of cement paste. TiO_2-modified cements exhibit shorter final setting times and increased compressive strength[104–106]. An addition of 5 wt% nano-TiO_2 of 25 nm can accelerate the cement hydration for 2 h and increase the compressive strength by 20% due to its ability of pore refining and the hydration acceleration effects[106].

As to electromagnetic properties, Xiong et al. discussed the electromagnetic absorbing properties and mechanisms of TiO_2/cement composites in the frequency range of 12.4–18 GHz (Ku frequency band) with different kinds of TiO_2 fillings[107]. The findings show that nanosized TiO_2 addition can improve the microwave absorbing properties greatly due to the increased electrical conductivity after the incorporation of TiO_2 particles. Moreover, TiO_2 particles with nanosizes have more obvious effects than those with microsizes. For example, the TiO_2/cement composite with 5% 40 μm diameter TiO_2 addition gives a reflectivity of −6 dB in the frequency band tested. However, a composite with 5% 15 nm diameter rutile-type TiO_2 can provide a reflectivity of −10 to −16 dB. When substituting rutile TiO_2 with 40 nm diameter anatase-type TiO_2 with the same weight fraction, the reflectivity of the composite can gain a small improvement due to the higher magnetic loss of the anatase-type TiO_2[107]. Wang and Zhang's results also show that blending steel fibers with nano-TiO_2 can improve the electromagnetic absorbing properties greatly[108,109]. With 3 wt% TiO_2 and 3 wt% steel fiber in the cement paste, the composite can have a reflectivity over −9 dB in the whole frequency range of 8–18 GHz. And the effective absorption bandwidth in which the reflectivity is superior −10 dB reaches 7 dB[108].

Barium titanate ($BaTiO_3$) is a kind of dielectric material with high dielectric constants and is widely known as the basis for a large number of electrical and electronic applications due to its outstanding ferroelectric and piezoelectric properties. $BaTiO_3$ has long been used in cement to form piezoelectric ceramic–Portland cement composites to use for smart materials in the health monitoring and active vibration control of structures in civil engineering fields[110]. $BaTiO_3$ can also ameliorate the hydration properties of Portland cement and tune its dielectric loss and thus increase its electromagnetic properties[111]. A $BaTiO_3$/cement composite with 50 vol% $BaTiO_3$ shows low porosity with small pores, thus leading to increased mechanical strength. The dielectric loss tangents of the composite are observed to reduce with the increasing curing age from 0.598 to 0.373 at a curing age of 1 and 7 days, respectively[111]. The decreased dielectric loss tangent is beneficial to the electromagnetic impedance matching and thus can improve the microwave absorbing properties of the composites.

Since the first report as microwave absorbents in 2006, the microwave absorbing properties of MnO_2 materials have been investigated thoroughly in the matrices of epoxy, silicon dioxide, paraffin wax, etc. MnO_2 is also used as an auxiliary or second absorbent in composites to improve their electromagnetic characteristics. For example, nanosized MnO_2 particles have been used in carbon/epoxy composites for microwave absorption[112]. It is found that a 10 vol% addition of MnO_2 can improve the reflectivity of carbon black/epoxy composites greatly with carbon volume fractions of 10 vol% and 30 vol%. This result can be attributed to the fact that the additional introduction of MnO_2 improves the dispersion of carbon particles and the impedance matching at the surface of the composite. Moreover, the introduction of MnO_2 nanoparticles can increase the dielectric loss and polarization. The combination of the electric loss of carbon black and dielectric loss arissing from MnO_2 particles attenuates the incident EMW greatly[112,113]. With the further increase in MnO_2 contents, the electrical conductivity and the polarization of the composites decrease and thus lead to a reduction of the absorbing properties, as displayed in Fig. 7.17.

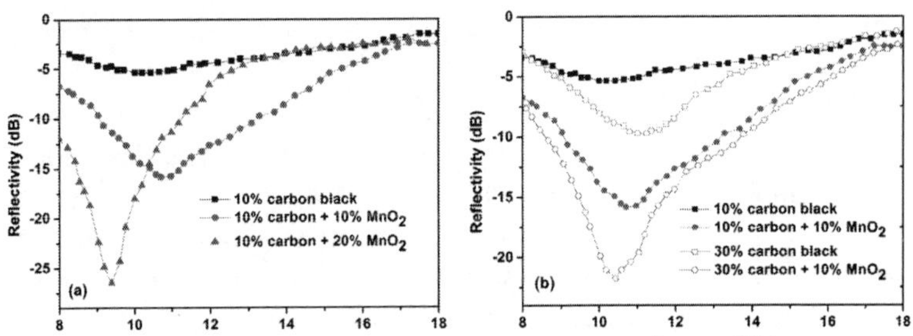

Figure 7.17 The effects of MnO_2 addition on the microwave absorbing properties of carbon black/epoxy composites with carbon black contents of 10 vol% (a) and 30 vol% (b).

MnO_2 has shown great application potential in carbon black/cement composites. Li et al. investigated the microwave absorbing properties of cement-based composites filled with carbon black and manganese dioxide in the frequency range of 2.6–18 GHz[114]. It is found that with a volume fraction of 4 vol%, carbon black can improve the reflectivity of the 20 mm thick cement plate from –5 dB to −10 dB. After 4 vol% MnO_2 was added, the reflectivity gets a further increase. The effective absorption bandwidth of the composite in which the reflectivity is superior to −10 dB reaches 12.4 GHz[114]. The improvement can be ascribed to the fact that the addition of MnO_2 ameliorates the dispersion of carbon black and improves the dielectric loss of the composite.

MnO_2 is also proven to be very effective to improve the microwave absorbing properties of expanded polystyrene (EPS)-filled cement matrix composites[115]. The electromagnetic characteristics of EPS/cement composites at 8–18 GHz found that after the addition of 75 g MnO_2 into 500 mL EPS/cement composites, the electromagnetic reflectivity can improve from −6 dB to about −9 dB. If the MnO_2 is coated on the surface of EPS spheres, the EPS/cement composites can give a reflectivity of −15 dB[115].

7.3.3 Magnetic Loss Cement-Based Materials

Most of the magnetic loss absorbents used in cement matrix composites are steel fibers, ferrites, and metal powders. For this kind of cement-based composites, Japanese researchers have carried out many studies and made great progress.

One Japanese building materials institute has successfully introduced ferrite tiles and steel mesh into an aramid fiber-reinforced concrete board or other decoration materials to fabricate a kind of microwave absorbing material. This kind of material can have a reflectivity over −14 dB in the frequency of 80–200 MHz and can be applied to a building as a microwave absorbing curtain wall[116,117].

Using cement and ferrite as the materials, Japan has fabricated a kind of curtain wall. It can absorb 90% of the incident waves and give an absorbing property of −20 to −30 dB in the frequency range of 100–200 MHz. It also has such advantages as being lightweight and shockproof and has been successfully applied in several buildings in Tokyo and Hiroshima[118].

Ferrite is a good kind of microwave absorbing component, especially in lower frequency bands, but it also has the disadvantages of a complicated fabrication procedure and high cost. In view of decreasing costs and simplification of the process, natural magnetic materials or wastes are often chosen as magnetic absorbents in cement composites[2]. Using magnetic sand as the magnetic component, a 30 mm thick sample can provide an absorption property of −8 dB in the frequency range of 75–100 GHz[119,120]. Natural magnetite is also proven to be an effective magnetic absorbent in cement-based composites. With a filling ratio of 15 and 20 wt%, a magnetite/cement composite with a thickness of 2.5 mm can give an absorbing peak of −15 dB at 9 GHz and −14 dB at 16 GHz, respectively[121]. The addition of magnetite increases the magnetic permeability and magnetic loss of the cement matrix and ameliorates the surface impedance matching of the incident wave and thus leads to the improvement of absorbing performances. Cement-based composites doped with metallurgical slag also illustrate good electromagnetic properties which can be considered for potential applications. A cement paste introduced with 5 wt% slag can give an average SE up to 39 dB in the X-band frequency range. For electromagnetic absorption, an absorption coefficient of 85% can be obtained after the incorporation of slag, which implies a reflectivity of about −16 dB[122].

Steel fiber is one of most commonly fiber fillers used in cement matrix composites. The diameter, length, aspect ratio, and environmental temperature and humidity are all key factors that influence the microwave absorbing performances of cement-based composites. Gao et al. investigated the influence of fiber length (14, 30, and 60 mm) and volume fraction (2, 5, and 7 vol%) on the electromagnetic absorbing properties in the frequency range of 2–18 GHz[123]. When the volume fraction is low, the reflectivity performances of the composites have not many differences with steel fibers of various lengths. When the volume fraction of the fiber is over 5 vol%, fibers with long lengths are easily to form conductive networks and thus cause high electrical conductivity, so reflection of the incident EMW prevails dominantly and leads to the reduction of absorption[123]. Curling the steel fiber to chiral helical structures can improve the microwave absorbing properties of cement composites in the higher frequency ranges. The results show that incorporating chiral steel fibers into Portland cement can provide an electromagnetic reflectivity of −6 dB for a 10 mm sample in the frequency range of 2–18 GHz. The reflectivity is still stable after hydrating for 370 days[124].

Fly ash, which emerged as a new kind of multifunctional material in the recent few years, is the by-product of coal combustion by electrical utilities. With the main component of oxides of silicon, aluminum, and iron, it is one of supplementary cementitious materials for the advantages of economy and durability[1]. The SiO_2 and Al_2O_3 are almost electromagnetically transparent due to their low electromagnetic parameters[125]. However, Fe_2O_3 and the residual carbon have great contributions to microwave absorption. The iron oxides are found to be coated on the surface of the spherical fly ash particles and make the fly ash have magnetic characteristics, which endows the fly ash with microwave attenuation properties. Due to its complicated dielectric loss and magnetic loss from carbon and the metal oxides, a 40 wt% high-iron fly ash–filled cement composite can provide a reflectivity value of −13 dB at 4.3 GHz and −12 dB at 7.3 GHz, as shown in Figs. 7.18a and 7.18b[126,127]. The microwave attenuation coefficients are also found to have a close relationship with the unburned carbon in fly ash. A 30 mm thick sample can show an attenuation of −8.5 dB at 1.2 GHz and −12 dB at 3.2 GHz with a carbon weight of 60% in fly ash[127].

Our previous study also shows that fly ash has the potential for microwave absorbing in cement-based composites; however, its pozzolanic activity and packing effect have a negative effect on the absorbing properties. Introducing EPS and MnO_2 can enhance the electromagnetic property to a certain degree[128]. The chemical compositions of fly ash are listed in Table 7.1, from which it can be seen that the main components of the fly ash are SiO_2 and Al_2O_3, contributing about 80% of the total weight. A high content of unburned carbon over 8 wt% can also be found, which will contribute to the electric and dielectric loss of the fly ash. The average content of 7.1 wt% Fe_2O_3 will lead to the magnetic loss of the product.

Table 7.1 Chemical compositions of fly ash in this work (wt%)

Component	SiO_2	Al_2O_3	Fe_2O_3	CaO	MgO	K_2O	Na_2O	SO_3	Ignition loss
Average content	50.6	27.1	7.1	2.8	1.2	1.3	0.5	0.3	8.2

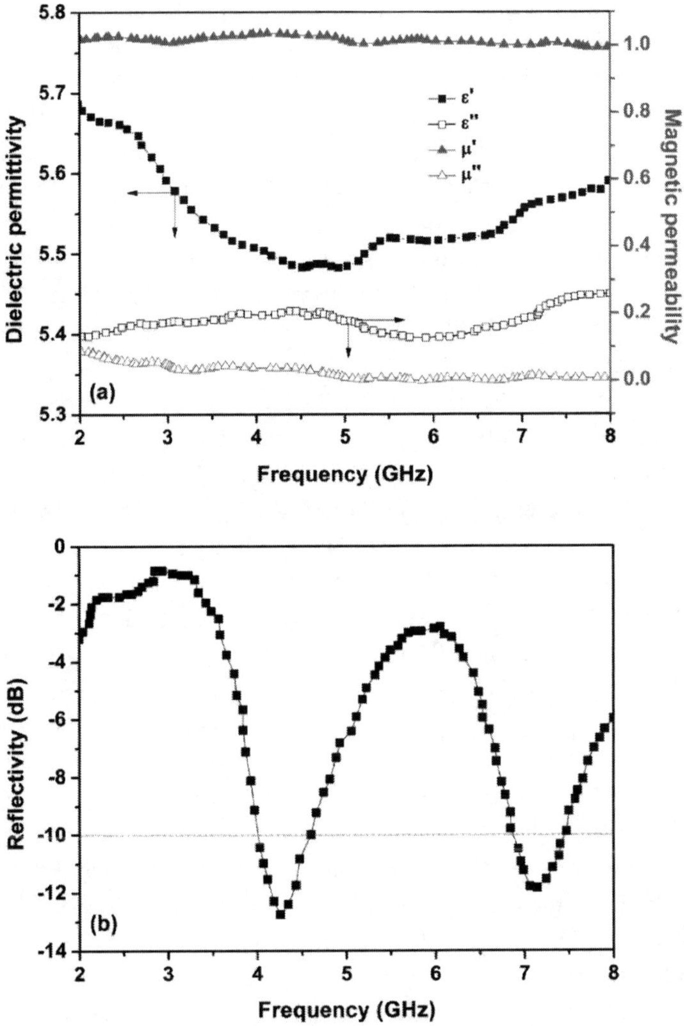

Figure 7.18 The electromagnetic parameters (a) and reflectivity (b) of fly ash/cement composites at 2–8 GHz.

The electromagnetic parameters in the frequency from 2 GHz to 18 GHz are shown in Fig. 7.19. It is clear that the ε' is almost a constant and ε'' has an obvious fluctuation with several peaks in the whole frequency range. The value of ε' is stable at about 3.2 and ε'' is in the range of 0.05–0.11. In contrast, the values of μ' and μ'' are much stable, in the range of 1.1–1.2 and 0.01–0.02, respectively. This can be concluded that fly ash mainly attenuates the incident EMW by both dielectric loss and magnetic loss. However, experimental results show that the introduction of 25 vol% fly ash into the cement matrix has a worse absorbing performance with only a reflectivity of about –6 dB in the whole frequency range. The introduction of fly ash decreases the numbers and sizes of the pores in the cement matrix and deteriorates the microwave absorbing effective paths. And also, due to the metal oxides and the unburned carbon in the fly ash, its introduction exacerbates the impedance mismatching between the samples and free space. All these factors lead to the weakening of the absorption properties of the cement-based composites[128].

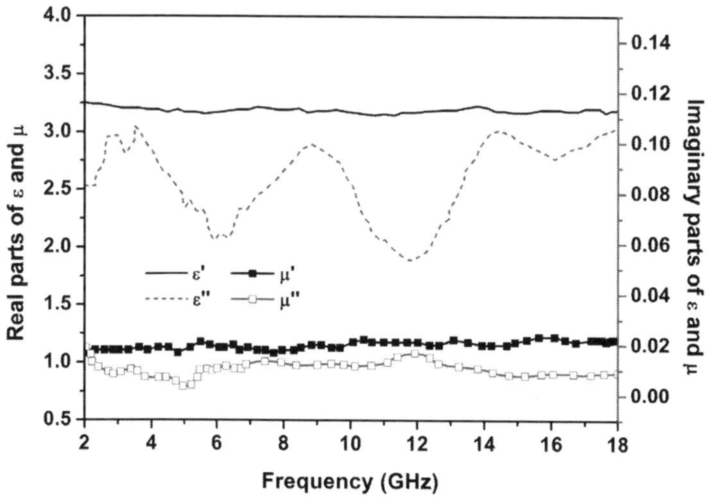

Figure 7.19 Electromagnetic complex permittivity and permeability of fly ash.

Nevertheless, microwave absorption is ameliorated by further incorporation of EPS and carbon or MnO_2 particulates. As illustrated in Figs. 7.20a and 7.20b, after 50 vol% EPS and 6 vol% carbon black or 6 vol% MnO_2 were employed in the composites, the microwave absorbing characteristics are improved greatly. The effective absorbing bandwidth with a reflectivity superior to –8 dB reaches 8 GHz and 12 GHz. This can be due to the improved impedance matching characteristics of the composites by introducing EPS[129].

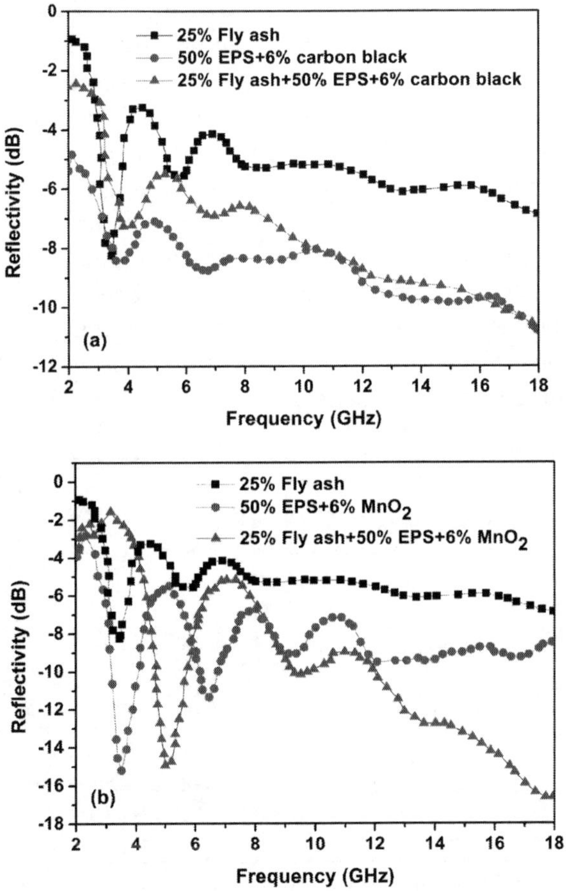

Figure 7.20 Electromagnetic reflectivity of fly ash/EPS/cement composites with carbon black (a) and MnO_2 (b) as the second absorbing component.

A composite with 6 vol% MnO_2 has a better absorbing performance than that with 6 vol% carbon black, as shown in Figs. 7.20a and 7.20b. This difference can be attributed to the high conductivity of carbon black. With 6 vol% carbon black filling into the cement composite, a continuous conductive network can be formed and the electrical conductivity turns out to be high enough to be reflective to the incident microwave wave, which results in the deterioration of absorbing properties. On the contrary, MnO_2 has much less conductivity and higher impedance than carbon black, so the filling of MnO_2 can tune the impedance matching very well and thus has a better absorbing performance than carbon black. From Fig. 7.20a it is also clear that the absorbing property of a fly ash/ EPS/cement/carbon black quaternary composite is even worse than that of a composite without fly ash. This result can also be attributed to the worsened impedance matching due to the conductivity of fly ash resulting from its carbon contents.

From Figs. 7.18 and 7.19 one can find that EPS and fly ash are not so conductive and both their dielectric permittivity and magnetic permeability are small. Especially, EPS is totally electromagnetic transparent. However, their introduction can improve the microwave absorbing properties of cement-based composites. According to electromagnetic absorption theory, two critical factors must be taken into consideration for the design of a microwave absorbing material. The first is the impedance matching. Only when the impedance matching is satisfied can the incident wave be transmitted into the interior of the absorbing material and then the absorption can be considered next. The second factor is that the incident wave must be attenuated or absorbed by the material through electric, dielectric, or magnetic loss[130].

EPS and fly ash, along with silica fume, are usually applied in cement-based composite materials as the second filler to improve the dispersion uniformity of the absorbent fillers and provide a transmission path for the incident wave. Results prove that their addition can improve the absorbing properties of the composites. Their influence on the electromagnetic absorption and absorbing mechanisms will be discussed in the next section.

7.3.4 Cement-Based Porous Composites

In the previous sections, we discussed the cement-based electromagnetic absorbing composites filled with conductive, dielectric or magnetic fillers. They all have a common point that the absorbing materials all attenuate the incident microwave by the absorbents in the cement composites, no matter whether it is by electric loss, dielectric loss, or magnetic loss. There is another type of filler which is almost electromagnetic transparent, but its addition can tune the incident impedance of the cement composite and make a homogeneous dispersion of the absorbing fillers and thus improve the electromagnetic absorbing performance of the material. EPS and silica fume are the typical ones of this kind and have been successfully employed in cement matrix composites for electric and electromagnetic applications. Especially EPS has a dielectric constant of $\varepsilon = 2 - j0$ with a dielectric loss tangent of 0.0001 and a magnetic permeability of $\mu = 1 - j0$, which means that it is transparent to EMWs. The cement matrix composites filled with these fillers are usually called porous materials and results show that these kinds of porous materials still demonstrate better microwave absorbing properties.

Cement is a kind of composite material constituted by several metal oxides and it has a certain electromagnetic absorbing property in the microwave frequency band. However, its microwave absorption mainly depends on the dielectric and magnetic loss of its metal oxides components, which leads to the result that its absorbing performance is usually too low to be applied directly. One common method is to increase its absorption by introducing conductive, dielectric, or magnetic fillers or loadings[2]. The introduction of these fillers can improve its absorbing property to some degree; however, the conductive or magnetic fillers all inevitably change the electromagnetic parameters (ε and μ) and impedance (η) of the composite. This leads to impedance mismatching between the composite and the free space and causes reflection at the surface of the composite material. So both the filling ratios and the absorption improvements are limited.

On the contrary, if an electromagnetic transparent filler, such as EPS, is introduced into the cement composite, though pure EPS particles have no direct contributions to microwave attenuation, the improved impedance matching can guide the incident wave into the interior of the composite and then be attenuated by the cement. Moreover, when coated by a layer of cement, the EPS particles also play reflection and scattering roles to the incident wave, as illustrated in Fig. 7.21. When the number of the EPS particles is large enough, the reflection and scattering may have great effects on the microwave attenuation. In addition, during the wave reflection and scattering in the interior of the composite, the cement components will also absorb and attenuate the waves. So the introduction of EPS has great potential in cement-based composites for electromagnetic absorbing application[129,131].

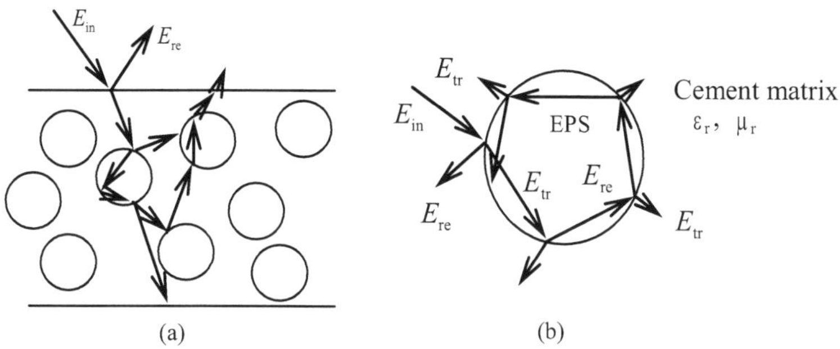

Figure 7.21 Sketch of the multireflection and scattering wave in the interior of the composite. (a) Wave transmission in the interior of the cement material plate and (b) illustration of the incident wave transmission in the interior of a single EPS particle.

7.3.4.1 The absorption mechanism analysis

To have better microwave absorption, the incident EMW must transmit into the interior of the microwave absorbing material and then be attenuated by the absorbing component of the material. Provided that there is an incident wave pinging on the surface of the material, no matter how complicated the absorbing process will be, there will only four interactions between the incident wave and the microwave absorbing material: incidence, reflection, transmission, and absorption. To improve the absorption, the reflection and transmission must be controlled and minimized. Suppose the interactions are illustrated as shown in Fig. 7.22, then the following equation must be satisfied:

$$W_{in} = W_{re} + W_{ab} + W_{tr} \quad (7.20)$$

where W_{in}, W_{re}, and W_{tr} are the power density of the incident wave, reflected wave, and transmitted wave, respectively, and W_{ab} is the power density absorbed by the microwave absorbing material.

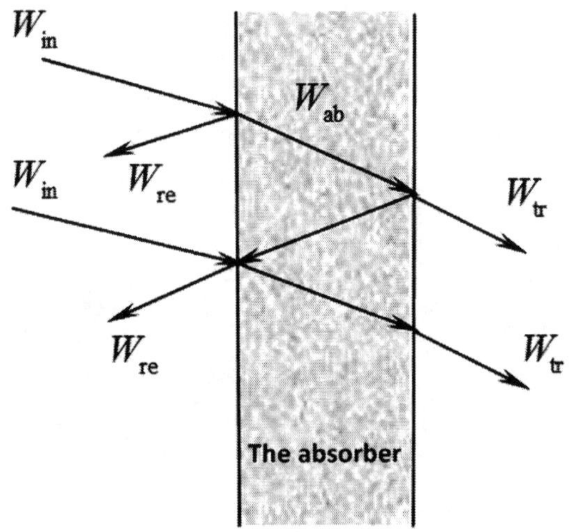

Figure 7.22 Sketch of power density distribution in a microwave absorber plate.

Suppose the microwave absorbing material is in a space with volume V and the surface enclosing the volume is S, then according to Maxwell's equation, one can get[132]

$$\nabla \times H = J + \frac{\partial D}{\partial t} \tag{7.21}$$

and a dot on each side of the equation with E gives

$$E \cdot \nabla \times H = E \cdot J + E \cdot \frac{\partial D}{\partial t} \tag{7.22}$$

We now make use of the vector identity,

$$\nabla \cdot (E \times H) = -E \cdot \nabla \times H + H \cdot \nabla \times E \tag{7.23}$$

Thus

$$H \cdot \nabla \times E - \nabla \cdot (E \times H) = J \cdot E + E \cdot \frac{\partial D}{\partial t} \tag{7.24}$$

According to Maxwell's equation,

$$\nabla \times E = -\frac{\partial B}{\partial t} \tag{7.25}$$

And therefore,

$$-\nabla \cdot (E \times H) = J \cdot E + E \cdot \frac{\partial D}{\partial t} + H \cdot \frac{\partial B}{\partial t} \tag{7.26}$$

However,

$$E \cdot \frac{\partial D}{\partial t} = E \cdot \frac{\varepsilon \cdot \partial E}{\partial t} = \frac{\partial}{\partial t}\left(\frac{\varepsilon E^2}{2}\right) \text{ and } H \cdot \frac{\partial B}{\partial t} = H \cdot \frac{\mu \cdot \partial H}{\partial t} = \frac{\partial}{\partial t}\left(\frac{\mu H^2}{2}\right) \quad (7.27)$$

And thus we get

$$-\nabla \cdot (E \times H) = J \cdot E + \frac{\partial}{\partial t}\left(\frac{\varepsilon E^2}{2} + \frac{\mu H^2}{2}\right) \quad (7.28)$$

After integrating throughout the volume, we get

$$-\int_{\text{vol}} \nabla \cdot (E \times H) dv = \int_{\text{vol}} J \cdot E\, dv + \frac{\partial}{\partial t}\int_{\text{vol}}\left(\frac{\varepsilon E^2}{2} + \frac{\mu H^2}{2}\right) dv \quad (7.29)$$

and applying the divergence theorem, we get the final expression as

$$-\oint_S (E \times H) \cdot dS = \int_{\text{vol}} (J \cdot E) dv + \frac{\partial}{\partial t}\int_{\text{vol}}\left(\frac{\varepsilon E^2}{2} + \frac{\mu H^2}{2}\right) dv \quad (7.30)$$

The first integral on the right is the total instantaneous ohmic power dissipated within the volume. The second term on the right is the total energy stored in the electric and magnetic fields, and the partial derivative with respect to time means the instantaneous power increase of the stored energy within this volume. The sum of the expressions on the right is the total power flowing into this volume. Thus the left integral over the closed surface S surrounding the volume V must be the total power flowing out of the volume[132].

The cross product $E \times H$ is known as the Poynting vector and is interpreted as an instantaneous power density in watts per square meter (W/m^2). Equation 7.30 is the so-called energy conservation equation in electromagnetic absorbing materials. Though Eq. 7.30 and Eq. 7.20 have totally different expressions, they express the same idea that the power density flowing into the material is equal to the sum of the power increment per unit time and the power attenuated by the material. Thus we can get the expression of Eq. 7.31:

$$W_{\text{in}} - (W_{\text{re}} + W_{\text{tr}}) = W_{\text{ab}} \quad (7.31)$$

After divided by W_{in} on both sides, we get

$$1 - (R + T) = A \quad (7.32)$$

where $R = W_{re}/W_{in}$, $T = W_{tr}/W_{in}$, and $A = W_{ab}/W_{in}$ are the reflection, transmission, and absorption coefficients of the incident, transmitted, and absorbed power, respectively.

From Eq. 7.32 one can see that the absorption coefficient A has a close relationship with both the reflection coefficient R and the transmission coefficient T, and only when $(R + T) \rightarrow 0$ can A get its maximum value. As to the EPS/cement composite, after the introduction of EPS into the cement matrix, T is increased and R is reduced, so there is the possibility of an increase in A.

7.3.4.2 Electromagnetic absorbing properties of EPS/cement composites

Figure 7.23a–d demonstrates the dependent electromagnetic absorption properties of cement pastes and EPS/cement composites with the EPS filling fraction of 40–70 vol% for a thickness of 10 mm, 20 mm, and 30 mm. The EPS/cement composites have shown improvement of electromagnetic absorbing properties, especially composites with 60 vol% EPS fillings. As illustrated in Fig. 7.23d, a 20 mm thick sample can have a reflectivity of –8 dB to –15 dB, and the effective absorption bandwidth reaches 6.2 GHz, which is much superior to that of the cement paste in Fig. 7.23a. However, when the EPS filling ratio increases to 70 vol%, the absorption decreases gradually, as shown in Fig. 7.23b. Thus it means that there is a limit value for the EPS volume fraction in cement-based composites.

EPS has a much small dielectric loss and no magnetic loss and it is almost electromagnetic transparent; however, its addition can improve the electromagnetic absorbing properties of the cement composite. This interesting phenomenon can be attributed to the energy conservation equation, as given in Eq. 7.30 or Eq. 7.32.

Cement is a kind of composite constituted of multiple phases and various metal oxides. After hydration, the various phases harden gradually and form a compact and dense mixture with adequate mechanical strength. There exists not much porosity in the interior after hydration and thus it lacks effective paths for EMW transmission. After the addition of EPS into the cement matrix, the transmission of the cement matrix is improved and the reflection is reduced greatly. When the sum of R and T reaches its minimum value, the maximum value of A will be obtained. For the same reason, silica fume is often chosen in carbon fiber, steel fiber, or CNT-filled cement matrix composites to improve the electromagnetic absorption of the composites. Silica fume, with a SiO_2 content over 80 wt%, has a much less electromagnetic loss and low electrical conductivity, but it can improve the incident impedance matching of the cement-based composites and ameliorate the dispersion of fibers or CNTs. Thus it can improve the electromagnetic shielding and absorbing properties of cement-based composites[133,134]. A surface treatment of silica fume with silane can improve the bond strength between the fibers and the cement matrix and the dispersion degree of conductive fillers; thereby it increases the SE and microwave absorbing properties of the composites[135,136].

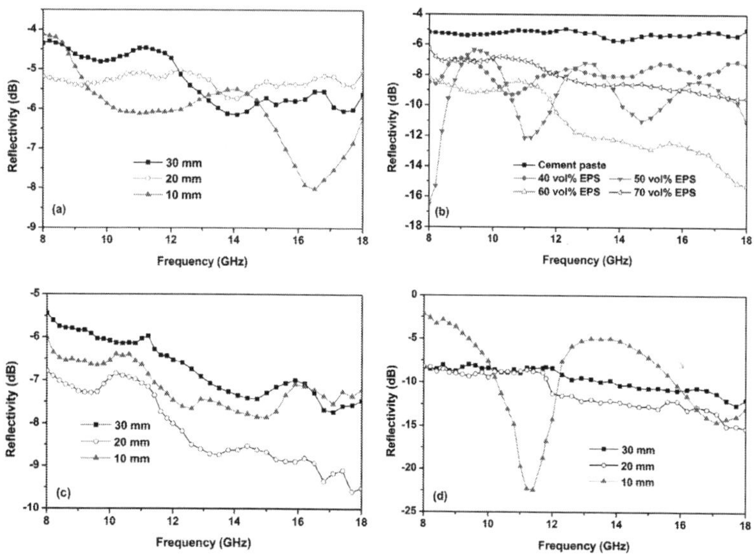

Figure 7.23 Electromagnetic reflectivity of cement paste (a), EPS/cement composites with a thickness of 20 mm and different EPS filling ratios (b), and different thicknesses of 10 mm, 20 mm, and 30 mm for 40 vol% (c) and 60 vol% EPS (d).

However, the volume fraction of EPS has its limit values. With the further addition of EPS, the transmission coefficient T will increase greatly and thus the value of A decreases, so its microwave absorbing properties decrease accordingly. As illustrated in Fig. 7.23b, when the EPS volume is increased to 70%, the absorption decreases greatly compared to that with 60%.

To improve the microwave absorption of EPS/cement composites, carbon black was tried by adding it to the composite. According to Eq. 7.32, after the incorporation of carbon black, there will be an increase in *R* and a decrease in *T*, so *A* is expected to increase with certain carbon black fillings. The EPS/cement composites with an EPS filling ratio of 60 vol% and thicknesses of 20 mm and 30 mm are chosen to investigate the influence of carbon black contents on the microwave absorbing performances of EPS/cement composites. As shown in Fig. 7.24, the same addition of carbon black has different effects on the microwave absorption of composites with different thicknesses.

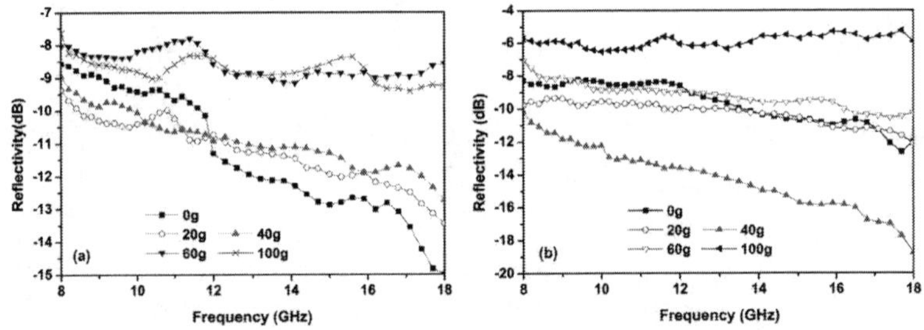

Figure 7.24 Electromagnetic reflectivity of EPS/carbon black/cement composites with 60 vol% EPS and different carbon black contents with a thickness of 20 mm (a) and 30 mm (b).

It is clear from Fig. 7.24 that when the thickness is 20 mm, an addition of 20 g and 40 g of carbon black can improve the reflectivity of the EPS/cement composite at lower frequencies, below 12 GHz. At higher frequencies, over 12 GHz, the reflectivity is still less than the sole EPS-filled composite. With the further addition to 60 g and 100 g, the reflectivity decreases to about −8 dB. When the thickness increases to 30 mm, the reflectivity variations with the carbon contents are different to those with 20 mm. As shown in Fig. 7.24b, when 20 g or 60 g of carbon black is added, the reflectivity of the EPS/cement composites has not much difference. The reflectivity in the whole frequency range of 8–18 GHz just varies from −8 to −11 dB. However, when 40 g of carbon black is added, the reflectivity curve is totally different. It has a reflectivity value of −10 dB at 8 GHz and increases gradually to −19 dB with a frequency of 18 GHz.

On the basis of Eq. 7.31 and Eq. 7.32, the microwave absorption of the EPS/cement composite is dependent on the combined effects of reflection and transmission. To have better absorption, the reflection and transmission must be confined. For composites with a thickness of 20 mm, the introduction of 20 or 40 g of carbon black improves the electrical conductivity of the composites and thus decreases the transmission T and increases the reflection R of the material. In addition, carbon black can also improve the electric and dielectric loss of the material. Thereby, the combination effects of these factors give a certain improvement of the reflectivity of the EPS/cement composite. With further filling of carbon black, the electrical conductivity is increased to such a degree that the reduction of T cannot offset the increase of R, and the reflection begins to dominate the interaction between the incident wave and the material at the surface of the EPS/cement composite. Thus it leads to the result that when carbon black is increased to 60 g and 100 g, the absorbing properties of the composite deteriorate greatly.

As to a composite with a thickness of 30 mm, it has a much longer transmission path than a 20 mm thick sample. When 40 g of carbon black is added, the attenuation of the incident wave overpasses the increased R caused by the rising conductivity and thus results in an improvement of reflectivity. When the contents of carbon black are increased to 60 g, the improved microwave attenuation by electric and dielectric loss comes from the carbon black, balancing its adverse effects cause by the increased R and decreased T, and thus the microwave absorption has a similar value as that of the EPS/cement composite without carbon black. With a further increase of carbon black contents, the composite turns much conductive and the microwave reflection R grows so high that the absorption decreases pronouncedly, as shown in Fig. 7.24.

7.3.4.3 Electromagnetic absorbing properties of double-layer cement composites

According to the energy conservation equation, a microwave absorbing material with high performance should first of all have good impedance matching with free space and thus confine the reflection of the incident wave. In other words, the absorber should provide effective transmission paths for the EMW. Secondly, it should have a high absorption capacity. However, for a single-layer absorber, these two demands are often contradictive and it is hard to be satisfied simultaneously. In this sense, double-layer electromagnetic absorbers can meet this demand. By a matching layer composed of low-absorbing media, the incident wave can penetrate into the absorber readily. Attached with an absorbing layer constituted of absorbing media, the incident wave can be attenuated effectively so as to improve the whole absorbing property of the absorber. The sketch of a two-layer absorber structure is shown in Fig. 7.25.

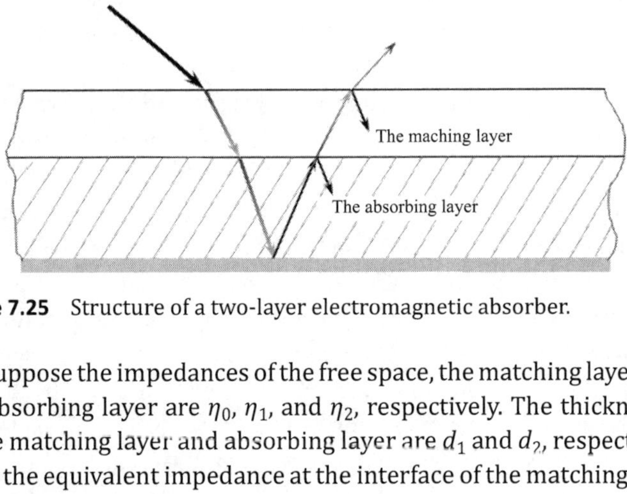

Figure 7.25 Structure of a two-layer electromagnetic absorber.

Suppose the impedances of the free space, the matching layer, and the absorbing layer are η_0, η_1, and η_2, respectively. The thicknesses of the matching layer and absorbing layer are d_1 and d_2, respectively. Then the equivalent impedance at the interface of the matching layer and the absorbing layer can be given by

$$Z_2 = \eta_2 \tanh(jk_2d_2) \tag{7.33}$$

Thus the input impedance of this bilayer absorber can be obtained:

$$Z_1 = \eta_1 \frac{Z_2 + \eta_1 \tanh(jk_1d_1)}{\eta_1 + Z_2 \tanh(jk_1d_1)} \tag{7.34}$$

From Eq. 7.33 and Eq. 7.34, one can get

$$Z_1 = \eta_1 \frac{\eta_2 \tanh(jk_2d_2) + \eta_1 \tanh(jk_1d_1)}{\eta_1 + \eta_2 \tanh(jk_2d_2) \tanh(jk_1d_1)} \tag{7.35}$$

In Eqs. 7.33–7.35, k_1 and k_2 imply the EMW numbers in the absorbing layer and the matching layer, respectively.

Thus one can see that as long as the electromagnetic parameters of the matching and absorbing layers are known, the reflectivity of the whole composite can be obtained.

With 60 vol% EPS filling cement as the matching layer and carbon black filling cement as the absorbing layer, two-layer electromagnetic absorbing composites are designed and their microwave absorption performances are characterized. Four kinds of samples are designed and the components of each sample are listed in Table 7.2.

Table 7.2 Design of two-layer microwave absorbing samples (vol%)

Sample	1	2	3	4
Matching layer	60%EPS	60%EPS	60%EPS	60%EPS
Absorbing layer	2.5%CB	6%CB	2.5%CB+60%EPS	6%CB+60% EPS

Figure 7.26 shows the electromagnetic reflectivity of the samples in the frequency range of 8–18 GHz. One can see from the curves that the reflectivity of samples with different absorbing layers has totally different absorbing performances[137]. When the absorbing layer contains only carbon black, samples 1 and 2 have reflectivity values of –9 dB to –13 dB at a frequency range less than 16 GHz. When EPS is introduced into the absorbing layer, samples 3 and 4 have improvements when the frequency is over 10 GHz. This result can be due to the improved transmission paths for the incident wave. As to sample 2, when the carbon content in the absorbing layer increase to 6 vol%, the electrical conductive network may be formed in the composite, which results in the fact that even though the incident wave can penetrate into the absorbing layer through the matching layer, the reflective phenomenon caused by the increased conductivity still prohibits the wave from being attenuated. However, when 60 vol% EPS is further introduced into the absorbing layer, the reflection in the absorbing layer is reduced greatly and the absorbing paths turn sufficient and thus results in a better absorbing performance.

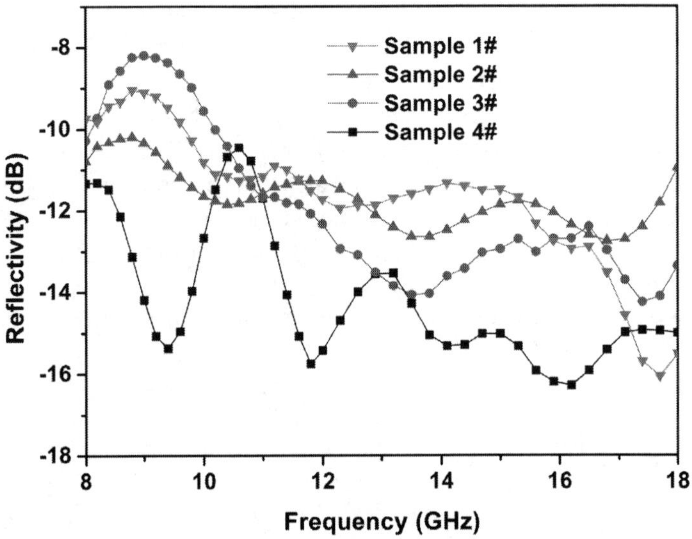

Figure 7.26 Electromagnetic reflectivity of two-layered microwave absorbers.

7.4 Summary

The development and application of cement matrices with EMI shielding and EMW absorbing characteristics are more and more urgently required to attenuate electromagnetic radiations in deteriorating electromagnetic environments. This emerging technology provides a new direction for the application of cement matrix materials. Cement paste filled with electric loss, dielectric loss, and magnetic loss media can provide EMI shielding and microwave absorbing performances through electric and ionic polarization, space charge polarization, interface and surface polarization, and the associated relaxation. These sorts of functional components improve the EMI shielding and microwave absorbing capacities; however, they also increase the reflection at the surface of the cement matrix composites at the same time. According to classical electromagnetic absorption theory, the absorption of the composite material can be ameliorated as long as the reflection and transmission are restricted. So, fillers with much less electromagnetic loss properties, such as EPS, silica fume, and fly ash, can be employed in cement matrix composites for microwave absorption due to their effective improvement in the transmission properties of the composites.

References

1. Cao JY, Chung DDL (2004). Use of fly ash as an admixture for electromagnetic interference shielding, *Cement Concrete Res*, **34**(10), 1889–1892.
2. Guan H, Liu S, Duan Y, Cheng J (2006). Cement based electromagnetic shielding and absorbing building materials, *Cement Concrete Compos*, **28**(5), 468–474.
3. Chung DDL (2000). Cement reinforced with short carbon fibers: a multifunctional material, *Compos Part B*, **313**(6–7), 511–526.
4. Akuthota B, Zoughi R, Kurtis KE (2004). Determination of dielectric property profile in cement-based materials using microwave reflection and transmission properties, in *MTC 2001: Instrumentation and Measurement Technology Conference*, Como, 327–332.
5. Liang JJ, Wang Y, Huang Y, Ma YF, Liu ZF, Cai JM, Zhang CD, Gao HJ, Chen YS (2009). Electromagnetic interference shielding of graphene/epoxy composites, *Carbon*, **47**(3), 922–925.
6. Deligeorgis G, Dragoman M, Neculoiu D, Dragoman D, Konstantinidis G, Cismaru A, Plana R (2009). Microwave propagation in graphene, *Appl Phys Lett*, **95**(7), 073107 (3 pp).
7. Allen MJ, Tung VC, Kaner RB (2010). Honeycomb carbon: a review of graphene, *Chem Rev*, **110**(1), 132–145.
8. Singh AP, Mishra M, Chandra A, Dhawan SK (2011). Graphene oxide/ferrofluid/cement composites for electromagnetic interference shielding application, *Nanotechnology*, **22**(46), 465701 (9 pp).
9. Yamane T, Numata S, Mizumoto T (2002). Development of wide-band ferrite fin electromagnetic wave absorber panel for building wall, *International Symposium on Electromagnetic Compatibility*, Minneapolis, **2**, 799–804.
10. Zhang X, Ding XZ, Ong CK, Tan BTG (1996). Dielectric and electrical properties of ordinary Portland cement and slag cement in the early hydration period, *J Mater Sci*, **31**(3), 1345–1352.
11. Kharkovsky SN, Akay MF, Hasar UC, Atis CD (2002). Measurement and monitoring of microwave reflection and transmission properties of cement-based specimens, *IEEE Trans. Instrum. Meas.*, **51**(6), 1210–1218.

12. Camp PR, Bilotta S (1989). Dielectric properties of Portland cement paste as a function of time since mixing, *J Appl Phys*, **66**(12), 6007 (7 pp).
13. Gorup K, Smit MK, Wittmann FH (1982). Microwave study of hydrating cement paste at early age, *Cement Concrete Res*, **12**(4), 447–454.
14. McCarter WJ, Garvin S (1989). Dependence of electrical impedance of cement-based materials on their moisture condition, *J Phys D*, **22**(11), 1773–1776.
15. Scuderi CA, Mason TO, Jennings HM (1991). Impedance spectra of hydrating cement pastes, *J Mater Sci*, **26**(2), 349–353.
16. Fu XL, Chung DDL (1995). Linear correlation of bond strength and contact electrical resistivity between steel rebar and concrete, *Cement Concrete Res*, **25**(7), 1397–1402.
17. Bogue RH (1947). *The Chemistry of Portland Cement*, Reinhold Publishing Corporation, New York.
18. Hansson ILH, Hansson CM (1983). Electrical resistivity measurements of Portland cement based materials, *Cement Concrete Res*, **13**(5), 675–683.
19. McCarter WJ, Curran PN (1984). The electrical response characteristics of setting cement paste, *Mag Concrete*, **36**(126), 42–49.
20. McCarter WJ, Afshar AB (1984). Some aspects of the electrical properties of cement paste, *J Mater Sci Lett*, **3**(10), 1083–1086.
21. McCarter WJ, Garvin S (1989). Admixtures in cement: a study of dosage rates on early hydration, *Mater Struct*, **22**(2), 112–120.
22. Abo-El-Enein SA, Ata AA, Hassanien A, Mikhai R Sh (1982). Kinetics and mechanism of slag cement hydration, *J Chem Technol Biotechnol*, **32**(7–12), 939–945.
23. Calleja J (1952). New techniques in the study of setting and hardening of hydraulic materials, *J Am Concrete Inst*, **48**(3), 525–536.
24. Calleja J (1953). Determination of setting and hardening time of high-alumina cements by electrical resistance techniques, *J Am Concrete Inst*, **50**(11), 249–256.
25. Taylor MA, Arulanandan K (1974). Relationships between electrical and physical properties of cement pastes, *Cement Concrete Res*, **4**(6), 881–897.
26. Mindess S, Young JF (1981). *Concrete*, Prentice-Hall Press.
27. Gu P, Xie P, Beaudoin JJ, Brousseau R (1992). A.C. impedance spectroscopy (I), a new equivalent circuit model for hydrated Portland cement paste, *Cement Concrete Res*, **22**(5), 833–840.

28. Fu X, Chung DDL (1998). Decrease of the bond strength between steel rebar and concrete with increasing curing age, *Cement Concrete Res*, **28**(2), 167–169.

29. Sanish KB, Neithalath N, Santhanam M (2013). Monitoring the evolution of material structure in cement pastes and concretes using electrical property measurements, *Constr Build Mater*, **49**, 288–2297.

30. McCarter WJ, Chrisp TM, Starrs G, Blewett J (2003). Characterization and monitoring of cement-based systems using intrinsic electrical property measurements, *Cement Concrete Res*, **33**(2), 197–206.

31. Zhang DW, Cao ZG, Fan LB, Liu SY, Liu WZ (2014). Evaluation of the influence of salt concentration on cement stabilized clay by electrical resistivity measurement method, *Eng Geol*, **170**, 80–88.

32. Bryson LS, Bathe A (2009). Determination of selected geotechnical properties of soil using electrical conductivity testing, *Geotech Test J*, **32**(3), 1–10.

33. Neithalath N, Persun J, Manchiryal RK (2010). Electrical conductivity based microstructure and strength prediction of plain and modified concretes, *Int J Adv Eng Sci Appl Math*, **2**(3), 83–94.

34. Chang FK (1999). *Structural Health Monitoring: A Summary Report*, paper presented in the Proceedings of the Second International Workshop on Structural Health Monitoring, Stanford, California.

35. Wen SH, Chung DDL (2003). Cement-based materials for stress sensing by dielectric measurement, *Cement Concrete Res*, **32**(9), 1429–1433.

36. Fu X, Lu W, Chung DDL (1998). Improving the strain-sensing ability of carbon fiber-reinforced cement by ozone treatment of the fibers, *Cement Concrete Res*, **28**(2), 183–187.

37. Yu X, Kwon E (2009). A carbon nanotube/cement composite with piezoresistive properties, *Smart Mater Struct*, **18**(5), 055010 (5 pp).

38. Han B, Yu X, Kwon E (2009). A self-sensing carbon nanotube/cement composite for traffic monitoring, *Nanotechnology*, **20**(55), 445501 (5 pp).

39. Konsta-Gdoutos MS, Aza CA (2014). Self sensing carbon nanotube (CNT) and nanofiber (CNF) cementitious composites for real time damage assessment in smart structures, *Cement Concrete Compos*, **53**, 162–169.

40. Le JL, Du HJ, Pang SD (2014). Use of 2D graphene nanoplatelets (GNP) in cement composites for structural health evaluation, *Compos Part B*, **67**, 555–563.

41. Kaynak A (1996). Electromagnetic shielding effectiveness of galvanostatically synthesized conducting films in the 300–2000 MHz frequency range, *Mater Res Bull*, **31**(7), 845–860.
42. Hayt WH, Buck JA (2001). *Engineering Electromagnetics,* 6th ed, McGraw-Hill Companies, Inc., New York.
43. Chung DDL (2000). Materials for electromagnetic interference shielding, *J Mater Eng Perform*, **9** (3), 350–354.
44. Chung DDL (2004). Review: electrical applications of carbon materials, *J Mater Sci*, **39**(8), 2645–2661.
45. Zornoza E, Catalá G, Jiménez F, Andión LG, Garcés P (2010). Electromagnetic interference shielding with Portland cement paste containing carbon materials and processed fly ash, *Mater Constr*, **60**(300), 21–32.
46. Bhattacharya S, Sachdev VK, Chatterjee R, Tandon RP (2008). Decisive properties of graphite-filled cement composites for device application, *Appl Phys A*, **92**(2), 417–420.
47. Callen BW (2005). High-performance composite powders for electronic devices, *Sulzer Tech Rev*, (1), 16–18.
48. Cao JY, Chung DDL (2003). Colloidal graphite as an admixture in cement and as a coating on cement for electromagnetic interference shielding, *Cement Concrete Res*, **33**(11), 1737–1740.
49. Chung DDL (2000). Flexible graphite for gasketing, adsorption, electromagnetic interference shielding, vibration damping, electrochemical applications and stress sensing, *J Mater Eng Perform*, **9**(2), 161–163.
50. Hong XH, Chung DDL (2015). Exfoliated graphite with relative dielectric constant reaching 360, obtained by exfoliation of acid-intercalated graphite flakes without subsequent removal of the residual acidity, *Carbon*, **91**, 1–10.
51. Chung DDL, Luo XC (1996). Electromagnetic interference shielding reaching 130 dB using flexible graphite, *Carbon*, **34**(10), 1293–1294.
52. Cao JY, Chung DDL (2003). Coke powder as an admixture in cement fro electromagnetic interference shielding, *Carbon*, **41**(12), 2427–2451.
53. Zhang XZ, Sun W (2012). Electromagnetic shielding and absorption properties of fiber reinforced cementitious composites, *J Wuhan Univ Technol: Mater Sci Ed*, **27**(1), 172–176.
54. Chung DDL (2001). Comparison of submicron-diameter carbon filaments and conventional carbon fibers as fillers in composite materials, *Carbon*, **39**(8), 1119–1125.

55. Fu XL, Chung DDL (1996). Submicron carbon filament cement-matrix composites for electromagnetic interference shielding, *Cement Concrete Res*, **26**(10), 1467–1472.

56. Fu XL, Chung DDL (1998). Submicron-diameter-carbon-filament cement-matrix composites, *Carbon*, **36**(4), 459–462.

57. Hudnut SW, Chung DDL (1995). Use of submicron diameter carbon filaments for reinforcement between continuous carbon fiber layers in a polymer-matrix composite, *Carbon*, **33**(11), 1627–1631.

58. Wu JH, Chung DDL (2003). Improving colloidal graphite for electromagnetic interference shielding using 0.1 μm diameter carbon filaments, *Carbon*, **41**(6), 1313–1315.

59. Baughman RH, Zakhidov AA, de Heer WA (2002). Carbon nanotubes–the route toward applications, *Science*, **297**(5582), 787–792.

60. Li N, Huang Y, Du F, He X, Lin X, Gao H, Ma Y, Li F, Chen Y, Elund PC (2006). Electromagnetic interference (EMI) shielding of single-walled carbon nanotube epoxy composites, *Nano Lett*, **6**(6), 1141–1145.

61. Singh AP, Gupta BK, Mishra M, Govind, Chandra A, Mathur RB, Dhawan SK (2013). Multiwalled carbon nanotube/cement composites with exceptional electromagnetic interference shielding properties, *Carbon*, **56**, 86–96.

62. Huang Y, Li N, Ma YF, Du F, Li FF, He XB, Lin X, Gao HJ, Chen YS (2007). The influence of single-walled carbon nanotube structure on the electromagnetic interference shielding efficiency of its epoxy composites, *Carbon*, **45**(8), 1614–1621.

63. Cao MS, Song WL, Hou ZL, Wen B, Yuan J (2010). The effects of temperature and frequency on the dielectric properties, electromagnetic interference shielding and microwave-absorption of short carbon fiber/silica composites, *Carbon*, **48**(3), 788–796.

64. Wen B, Cao MS, Hou ZL, Song WL, Zhang L, Lu MM, Jin HB, Fang XY, Wang WZ, Yuan J (2013). Temperature dependent microwave attenuation behavior for carbon nanotube/silica composites, *Carbon*, **65**, 124–139.

65. Wadhawan A, Garrett D, Perez M (2003). Nanoparticle-assisted microwave absorption by single-wall carbon nanotubes, *Appl Phys Lett*, **83**(13), 2683–2685.

66. Saravanan L, Liu JH, Miao HY, Wang LC (2013). Preparation and investigation of Co-dispersed MWCNT buckypaper for the microwave absorption, *Appl Mech Mater*, **479–480**, 20–24.

67. Singh BP, Saket DK, Singh AP, Pati S, Gupta TK, Singh VN, Dhakate SR, Dhawan SK, Kotnala RK, Mathur RB (2015). Microwave shielding

properties of Co/Ni attached to single walled carbon nanotubes, *J Mater Chem A*, **3**(25), 13203–13209.

68. Kim BJ, Bae KM, Lee YS, An KH, Park SJ (2014). EMI shielding behaviors of Ni-coated MWCNTs-filled epoxy matrix nanocomposites, *Surf Coat Technol*, **242**, 125–131.
69. Ji K, Zhao H, Zhang J, Chen J, Dai Z (2014). Fabrication and electromagnetic interference shielding performance of open-cell foam of a Cu-Ni alloy integrated with CNTs, *Appl Surf Sci*, **311**, 351–356.
70. Huang XH, Dai B, Ren Y, Xu J, Zhu P (2015). Preparation and study of electromagnetic interference shielding materials comprised of Ni-Co coated on web-like biocarbon nanofibers via electroless deposition, *J Nanomater*, **2015**, 320306 (7 pp).
71. Che RC, Peng LM, Duan XF, Chen Q, Liang XL (2004). Microwave absorption enhancement and complex permittivity and permeability of Fe encapsulated within carbon nanotubes, *Adv Mater*, **16**(5), 401–405.
72. Wang C, Han XJ, Xu P, Zhang XL, Du YC, Hu SR, Wang JY, Wang XH (2011). The electromagnetic property of chemically reduced graphene oxide and its application as microwave absorbing material, *Appl Phys Lett*, **98**(7), 072906 (3 pp).
73. Wen B, Cao MS, Lu MM, Cao WQ, Shi HL, Liu J, Wang XX, Jin HB, Fang XY, Wang WZ, Yuan J (2014). Reduced graphene oxides: light-weight and high-efficiency electromagnetic interference shielding at elevated temperatures, *Adv Mater*, **26**(21), 3484–3489.
74. Basavaraja C, Kin WJ, Kim YD, Huh DS (2011). Synthesis of polyaniline-gold/graphene oxide composite and microwave absorption characteristics of the composite films, *Mater Lett*, **65**(19–20), 3120–3123.
75. Singh K, Ohlan A, Pham VH, Balasubramaniyan R, Varshney S, Jang J, Hur SH, Choi WM, Kumar M, Dhawan SK, Kong BS, Chung JS (2013). Nanostructured graphene/Fe_3O_4 incorporated polyaniline as a high performance shield against electromagnetic pollution, *Nanoscale*, **5**(6), 2411–2420.
76. Yan DX, Ren PG, Pang H, Fu Q, Yang MB, Li ZM (2012). Efficient electromagnetic interference shielding of lightweight graphene/polystyrene composite, *J Mater Chem*, **22**(36), 18772–18774.
77. Chen ZP, Xu C, Ma CQ, Ren WC, Cheng HM (2013). Lightweight and flexible graphene foam composites for high-performance electromagnetic interference shielding, *Adv Mater*, **25**(99), 1296–1300.

78. Zhang HB, Yan Q, Zheng WG, He ZX, Yu ZZ (2011). Tough graphene-polymer microcellular foams for electromagnetic interference shielding, *ACS Appl Mater Interfaces*, **3**(3), 918–924.

79. Singh VK, Shukla A, Patra MK, Saihi L, Jani RK, Vadera SR, Kumar N (2012). Microwave absorption properties of a thermally reduced graphene oxide/nitrile butadiene rubber composite, *Carbon*, **50**(6), 2202–2208.

80. Chen J, Zhao D, Ge HY, Wang J (2015). Graphene oxide-deposited carbon fiber/cement composites for electromagnetic interference shielding application, *Constr Build Mater*, **84**, 66–72.

81. Mohammed A, Sanjayan JG, Duan WH, Nazari A (2015). Incorporating graphene oxide in cement composites: A study of transport properties, *Constr Build Mater*, **84**, 341–347.

82. El-Faramary NA, El-Hosiny FI (2000). Shielding of gamma radiation by hydrated Portland cement-lead pastes, *Radiat Meas*, **32**(2), 93–99.

83. El-Faramary NA, El-Hosiny FI (1999). The effect of autoclaving process and addition of silica fume on Portland cement in shielding gamma radiation, *Radiat Meas*, **29**(6), 619–623.

84. Shi Z, Chung DDL (1995). Concrete for magnetic shielding, *Cement Concrete Res*, **25**(5), 939–944.

85. Reineix A, Bounad A, Jecko B (1987). Electromagnetic pulse penetration into reinforced- concrete buildings, *IEEE Trans EMC*, **29**(1), 72–78.

86. Yi B, Chen B, Gao C (2002). Study on shielding effectiveness of reinforced-concrete in high power electromagnetic environment, *J Microw*, **18**(3), 52–58.

87. Antonini G, Orlandi A, D'elia S (2003). Shielding effects of reinforced concrete structures to electromagnetic fields to GSM and UMTS systems, *IEEE Trans Magn*, **39**(3), 1582–1585.

88. Liu SH, Dong XL, Liu JM, Duan YP (2013). *Electromagnetic Wave Shielding and Absorbing Materials*, Chemical Industry Press, Beijing.

89. Kao KC (2004). *Dielectric Phenomena in Solids–With Emphasis on Physical Concepts of Electronic Processes*, Elsevier Academic Press, California.

90. Dai YW, Sun MQ, Liu CG, Li ZQ (2010). Electromagnetic wave absorbing characteristics of carbon black cement-based composites, *Cement Concrete Compos*, **32**(7), 508–513.

91. Dishovsky N, Grigorova M (2000). On the correlation between electromagnetic waves absorption and electrical conductivity of carbon black filled polyethylenes, *Mater Res Bull*, **35**(3), 403–409.

92. Mather PJ, Thomas KM (1997). Carbon black/high density polyethylene conducting composite materials. *J Mater Sci*, **32**(2), 401–407.

93. Tzeng SS, Chang FY (2001). EMI shielding effectiveness of metal-coated carbon fiber-reinforced ABS composites, *Mater Sci Eng A*, **302**(2), 258–267.

94. Li KZ, Wang C, Li HJ, Li XT, Ouyang HB, Wei J (2008). Effect of chemical vapor deposition treatment of carbon fibers on the reflectivity of carbon fiber-reinforced cement-based composites, *Compos Sci Technol*, **68**(5), 1105–1114.

95. Li KZ, Wang C, Li HJ, Luo F, Hou D (2007). Reflectivity of carbon fiber-reinforced cement-based composites against electromagnetic waves, *Rare Metal Mater Eng*, **36**(10), 1702–1708.

96. Wang ZJ, Li KZ, Wang C (2014). Freezing-thawing effects on electromagnetic wave reflectivity of carbon fiber cement based composites, *Constr Build Mater*, **64**, 288–292.

97. Laukaitis A, Sinica M, Balevičius Levitas B (2008). Investigation of electromagnetic wave absorber based on carbon fiber reinforced aerated concrete using time-domain method, *Acta Phys Pol A*, **113**(3), 1047–1050.

98. Deng L, Han M (2007). Microwave absorbing performances of multiwalled carbon nanotube composites with negative permeability, *Appl Phys Lett*, **91**(2), 023119 (3 pp).

99. Chung DDL (2001). Electromagnetic interference shielding effectiveness of carbon materials, *Carbon*, **39**(2), 279–285.

100. Wang BM, Guo ZQ, Han Y, Zhang TT (2013). Electromagnetic wave absorbing properties of multi-walled carbon nanotube/cement composites, *Constr Build Mater*, **46**, 98–103.

101. Guan HT, Liu SH, Zhao YB, Duan YP (2006). Electromagnetic characteristics of nanometer manganese dioxide composite materials, *J Electron Mater*, **35**(5), 892–896.

102. Liu SH, Guan HT, Duan YP, Zhao YB (2006). Electromagnetic absorbing characteristics of manganese dioxide composites, *J Funct Mater*, **37**(2), 197–199.

103. Guan HT, Zhao YB, Liu SH, Lv SP (2006). Application of manganese dioxide to electromagnetic wave absorber: effective permittivity and absorbing property, *Eur Phys J–Appl Phys*, **36**(3), 235–239.

104. Lee BY, Kurtis KE, Jayapalan AR (2013). Effects of nano-TiO_2 on properties of cement-based materials, *Mag Concrete Res*, **65**(21), 1293–1302.

105. Ma BG, Li HN, Zhu YC, Han L, Li XG (2014). Influence of nano-SiO_2 and nano-TiO_2 on early hydration of calcium sulfoaluminate cement, *Key Eng Mater*, **599**, 39–45.

106. Zhang R, Cheng X, Hou PK, Ye ZM (2015). Influence of nano-TiO_2 on the properties of cement-based materials: hydration and drying shrinkage, *Constr Build Mater*, **81**, 35–41.

107. Xiong GX, Chen YR, Li JL, Ye YH, Zuo Y (2007). Discuss microwave absorption theory of the cement-based composite doping nano-TiO_2, *J Funct Mater*, **38**(5), 836–838.

108. Wang B, Zhang Y, Chen RJ (2014). Experimental and numerical simulation of absorbing property of cement-based material blending with nano-TiO_2 and steel fibers, *Appl Mech Mater*, **442**, 83–87.

109. Zhang Y, Wang XH, Luo JH (2014). Research on absorbing property of cement-based material blending with nano-TiO_2 and steel fibers, *Adv Mater Res*, **853**, 207–211.

110. Rianyoi R, Potong R, Yimnirun R, Guo RY, Bhalla AS, Chaipanichi A (2013). Electromechanical coupling coefficient of 1-3 connectivity barium titanate-Portland cement composites, *Integr Ferroelectr*, **148**(1), 138–144.

111. Rianyoi R, Potong R, Jaitanong N, Chaipanich A (2011). Influence of curing age on microstructure in barium titanate–Portland cement composites, *Key Eng Mater*, **484**, 222–225.

112. Duan YP, Zhang X, Ji ZJ, Liu SH (2008). Effect of scattering cross section of nanometer manganese dioxide on electromagnetic properties of carbon black, *Mater Sci Technol*, **16**(6), 741–744.

113. Zhang X, Liu SH, Duan YP, Guan HT, Wen B (2007). Electromagnetic characteristics of carbon black doped with nanometer manganese dioxide, *J Aeronaut Mater*, **27**(2), 58–61.

114. Li BY, Liu SH (2007). Research on absorbing property of cement-based material filled with composite absorbents, *J Funct Mater*, **38**(Suppl), 3075–3077.

115. Yang WW, Liu SH, Du JZ (2007). Research on the methods and properties of MnO_2 coating EPS absorbing spheres, *J Funct Mater*, **38**(Suppl), 3031–3034.

116. Kobayashi M, Kadashima Y, Nakagawa H (1998). Anti-radio wave transmission curtain wall used in buildings, *J Jpn Soc Compos Mater*, **24**(1), 32–34.

117. Kobayashi M, Kadashima Y, Nakagawa H (1998). Anti-radio wave transmission curtain wall used in tall buildings, *J Jpn Soc Compos Mater*, **24**(3), 110–113.

118. Chen B, Zhang D (2000). Research and application of new cement based composites, *New Build Mater*, **4**, 28–30.

119. Lou M, Kan T (2002). EM wave absorption construction materials containing magnetic media, *J Magn Mater Devices*, **33**(5), 13–15.

120. Li X, Kang Q, Zhou C (2004). Research on absorbing properties of magnetic concrete, *J Funct Mater*, **35**(Suppl), 854–856.

121. Guan BW, Ding DH, Xiong R, Yang T, Zhang JY (2015). Electromagnetic wave absorbing properties of cement-based composites doped with natural magnetite, *J Funct Mater*, **46**(2), 02019–02022.

122. Bantsis G, Mavridou S, Sikalidis C, Betsiou M, Oikonomou N, Yioultsis T (2012). Comparison of low shielding-absorbing cement paste building materials in X-band frequency range using a variety of wastes, *Ceram Int*, **38**(5), 3683–3692.

123. Gao XS, Han BG, Ou JP (2006). Absorption property experiment and stealthy effectiveness analysis of steel fiber cement based materials, *J Funct Mater*, **37**(10), 1683–1688.

124. Dai YS, Lu CH, Ni YR, Xu ZZ (2010). Microwave-absorbing properties of chiral steel fiber on cement-based composites, *J Funct Mater*, **41**(Suppl), 208–210.

125. Chen LY, Duan YP, Liu LD, Guo JB, Liu SH (2011). Influence of SiO_2 fillers on microwave absorption properties of carbonyl iron/carbon black double-layer coatings, *Mater Design*, **32**(2), 570–574.

126. Huang YB, Qian JS, Zhang JY (2009). Influence of high-iron fly ash on absorption properties of cement based materials, *J Funct Mater*, **40**(11), 1787–1790.

127. Liu HY, Tan HZ, Gao Q, Wang XB, Xu TM (2010). Microwave attenuation characteristics of unburned carbon in fly ash, *Fuel*, **89**(11), 3352–3357.

128. Li BY, Duan YP, Liu SH (2012). The electromagnetic characteristics of fly ash and absorbing properties of cement-based composites using fly ash as cement replacement, *Constr Build Mater*, **27**, 184–188.

129. Guan HT, Liu SH, Duan YP, Zhao YB (2007). Investigation of the electromagnetic characteristics of cement based composites filled with EPS, *Cement Concrete Compos*, **29**(1), 49–54.

130. Vinoy KJ, Jha RM (2011). *Radar Absorbing Materials: From Theory to Design and Characterization*, Kluwer Academic Publishers, Boston.

131. Du JZ, Liu SH, Guan HT (2006). Research on the absorbing characteristics of cement matrix composites filled with carbon black-coated expanded polystyrene beads, *Adv Cement Res*, **18**(4), 161–164.

132. Hayt WH, Buck JA (2002). *Engineering Electromagnetics*, 6th ed, McGraw-Hill Companies, Inc., New York.

133. Wen SH, Chung DDL (2001). Effect of admixtures on the dielectric constant of cement paste, *Cement Concrete Res*, **31**(4), 673–677.

134. Nam IW, Kim HK, Lee HK (2012). Influence of silica fume additions on electromagnetic interference shielding effectiveness of multi-walled carbon nanotube/cement composites, *Constr Build Mater*, **30**, 480–487.

135. Xu YS, Chuang DDL (2000). Improving silica fume cement by using silane, *Cement Concrete Res*, **30**(8), 1305–1311.

136. Cao JY, Chung DDL (2001). Improving the dispersion of steel fibers in cement mortar by the addition of silane, *Cement Concrete Res*, **31**(2), 309–311.

137. Zhang YF, Hao WJ, Li BY, Duan YP, Liu SH (2013). Double-layered cement composites with superior electromagnetic wave absorbing properties containing carbon black and expanded polystyrene, *Adv Mater Res*, **771–776**, 747–752.

Chapter 8

Structural Pyramid Materials

Application of pyramid absorbers in microwave anechoic chambers can be traced back to the 1960s[1]. Due to the good absorption performance at a high frequency and low price, polyurethane pyramid absorbing materials have been widely used.

The polyurethane foam matrix absorbing material has the following advantages: It does not require a specific-shape mold in production and can be cut to any shapes arbitrarily according to needs. The drawback is the dipping of absorbing agent needs manual or semimanual and semimechanical operations to complete. It is feasible to make a pyramid absorber with a low height, but if the pyramid absorber is high (such as 1.2–1.5 m), it becomes very difficult to apply: It is not easy to make the foam absorb enough absorbing agent solution and drying is more difficult and the cost is very high. Therefore, the production efficiency is low, the immersion amount of the absorbing agent cannot be controlled accurately, and the quality of the product is not stable[2]. With the deepening awareness of the harm caused by high-frequency electromagnetic waves, low-frequency electromagnetic waves are applied more widely. At the same time, the requirements of the pyramid's height is more and more high, and thus the drawback of a polyurethane foam matrix absorbing material is becoming more and more obvious.

A new type of absorbing material which is based on the flame-retardant plate of PVC and filled with ESP particles as the

Microwave Absorbing Materials
Yuping Duan and Hongtao Guan

ISBN 978-981-4745-10-9 (Hardcover), 978-981-4745-11-6 (eBook)
www.panstanford.com

carrier has been widely used. It can effectively solve the problems existing in the production of polyurethane foam materials. The electromagnetic properties of the absorbing material are better than those of polyurethane foam. This resonant absorber could be used in microwave chambers as a substitute for urethane pyramids, and the performance of the anechoic chamber will have certain improvement.

8.1 Design and Analysis of Pyramid Absorbers

8.1.1 The Pyramid's Height

To make sure that the electromagnetic waves could be reflected among pyramids adequately, the pyramids need to reach a certain height. This could make the reflection wave vectors counteract each other because the phase place is opposite, and so the reflection wave is reduced. The minimum height of pyramids should be designed for a wavelength of a minimum frequency. If the testing frequency band is 1–18 GHz, the pyramids' height should be above 300 mm at least.

8.1.2 Design of the Vertex Angle

The electromagnetic wave would be reflected or refracted on reaching the pyramids' surface, so the reflection needs to be analyzed first.

As is shown in Fig. 8.1, AB is the incident wave when BT and SC are normal. The pyramids' vertex angle is supposed as 2α, so the angle of incidence is $\theta = 90° - \alpha$.

In ΔBTQ, $\angle QBT = 90°$ and $\angle BQT = 2\alpha$, so $\angle BTQ = 90° - 2\alpha$.

In ΔBCT, $\angle TBC = \theta_1$; then $\angle BCT = 90° - \theta_2$

$$= 180° - (90° - 2\alpha) - \theta_1 = 3\alpha.$$

The second angle of incidence is gained as $\theta_2 = 90° - 3\alpha$; when the incidence reaches its *n*th time, $\theta_n = 90° - (2n-1)\alpha$.

When $\theta_n < 0$, the incidence of electromagnetic wave is toward the vertex of the pyramid. As the angle of incidence returning to the vertex is very small for the first time, the wave would reach the pyramid nearby when transmitted through the first pyramid. To make sure that the design is strict, it is assumed that the first reflection wave returns to the receiver, and the reflection times will be counted until the negative angle of incidence appears. The value of n is obtained from the formula $\theta_n = 90° - (2n-1)\alpha > 0^3$.

Given the values of each α, the reflection times n and each value of the incident angles θ_n can be obtained, as shown in Table 8.1.

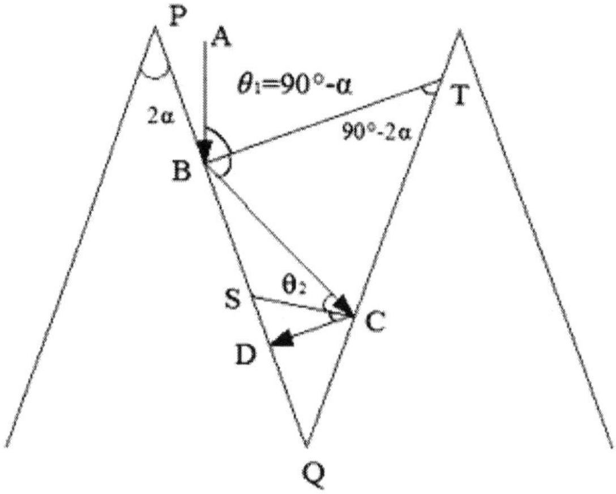

Figure 8.1 Sketch map of microwave reflecting.

When the electromagnetic wave meets the point E, the refraction wave EH transmits in the pyramid continuously, refracts at H, and the refraction wave HG transmits toward the base of the pyramids in the air, as shown in Fig. 8.2.

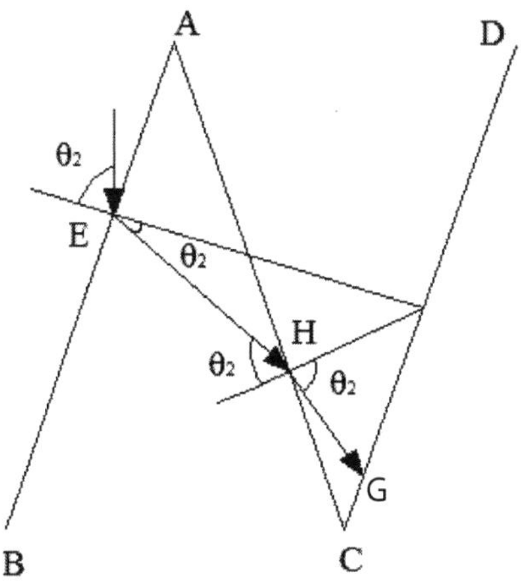

Figure 8.2 Sketch map of microwave refracting.

Table 8.1 Times of inflection and value of each angle of incidence when α is 10°–30°

α	30	29	28	26	24	22	20	18	17	16	15	14	13	12	11	10
θ_1	60	61	62	64	66	68	70	72	73	74	75	76	77	78	79	80
θ_2	0	3	6	12	18	24	30	36	39	42	45	48	51	54	57	60
θ_3	-60	-55	-50	-40	-30	-20	-10	0	5	10	15	20	25	30	35	40
θ_4					-78	-64	-50	-36	-29	-22	-15	-8	-1	6	13	20
θ_5							-90	-72	-63	-54	-45	-36	-27	-18	-9	0
θ_6										-86	-75	-64	-53	-42	-31	-20
θ_7													-79	-66	-53	-40
θ_8														-90	-75	-65
θ_9																-80

The electromagnetic wave would weaken when transmitted through the pyramids as it is absorbed, and the rest wave transmits toward the vertex of the pyramid after being reflected, as shown in Fig. 8.3.

In Fig. 8.3, BD⊥AB and HD⊥AH, so ∠BDH = 180° −($\theta_2 + \theta_3$) = 180° − 2α, and $\theta_3 = 2\alpha - \theta_2$. If $\theta_2 > 2\alpha$ then $\theta_3 < 0$, as shown in Fig. 8.4. So the demand $\theta_2 \le 2\alpha$ must be satisfied in the vertex angle's design, and the incident angle θ_1, which can satisfy $\theta_2 = 2\alpha$, is called the critical incident angle. Under this circumstance, the refraction law at the surface of the pyramid can be satisfied as follows[4]:

$$\frac{\sin\theta_1}{\sin\theta_2} = \sqrt{\mu_r \varepsilon_r} \tag{8.1}$$

In Eq. 8.1, $\theta_1 = \theta_n$ and $\theta_2 = 2\alpha$, and then

$$\frac{\sin\theta_n}{\sin 2\alpha} = \sqrt{\mu_r \varepsilon_r} = n \tag{8.2}$$

In Eq. 8.2, $2\alpha \le |\theta_n| \le 90°$. As the microwave incidence is toward the vertex of the pyramids, $\theta_n < 0$ is obtained. After substituting each θ_n value and its corresponding α into Eq. 8.2, the refractive index of each α can be obtained, as shown in Table 8.2.

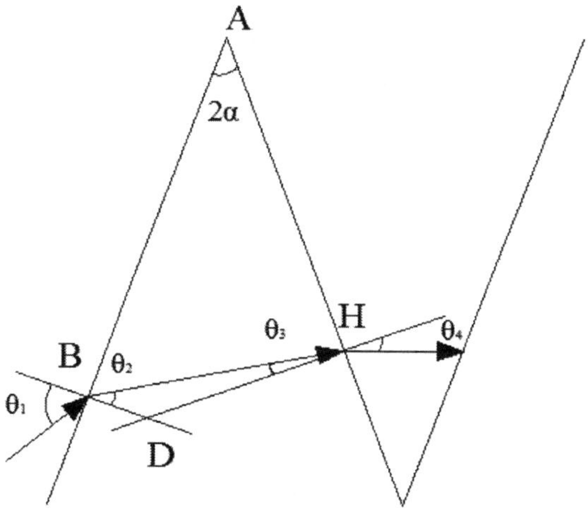

Figure 8.3 Sketch map of microwave incidence toward the vertex of the pyramids (1).

Figure 8.5 is gained from the data in Table 8.2. As shown in Fig. 8.5, series 1–6 corresponds to θ_1 – θ_6, and the six curves indicate the relation between α and n when n is from 4 to 9. The times of refractive index decrease as α increases. The value of the refractive index is limited as θ_n is restricted, and α and n do not correspond very well as n changes.

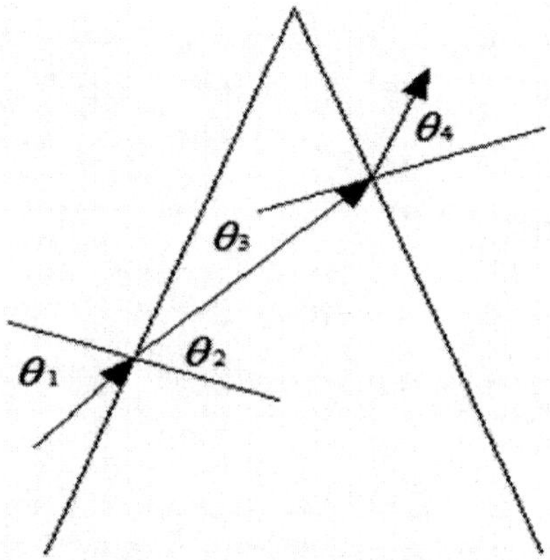

Figure 8.4 Sketch map of microwave incidence toward the vertex of the pyramids (2).

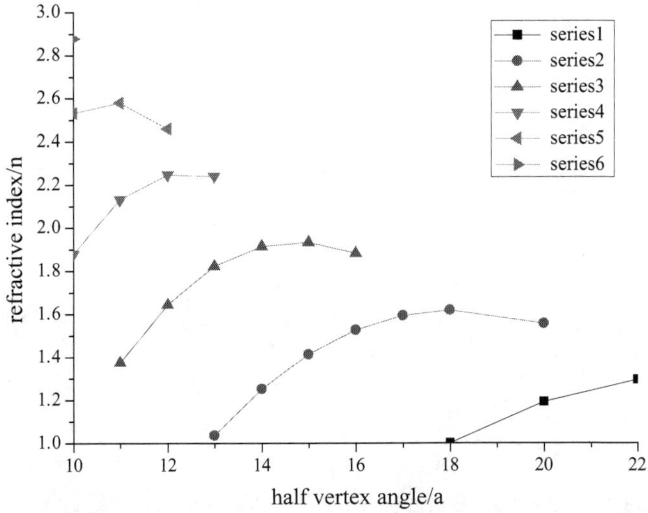

Figure 8.5 Sketch map of the relation between the refractive index and half vertex angle.

Table 8.2 Value of the refractive index n corresponding to the half vertex angle

α	22	20	18	17	16	15	14	13	12	11	10
n	1.294	1.192	1								
		1.556	1.618	1.593	1.527	1.414	1.252	1.036			
					1.882	1.932	1.914	1.822	1.645	1.375	
								2.239	2.246	2.132	1.879
									2.459	2.579	2.532
											2.879

From the result of Fig. 8.5, the trend of the curve is ascending on the whole and has a little drop after the peak value. This is determined by the property of the sine function. In $\frac{\sin\theta_n}{\sin 2\alpha} = \sqrt{\mu_r\varepsilon_r} = n$, the movement of the sine function is slow when θ_n is small, so sin θ_n is bigger than sin 2α. For this reason, the curve's trend is ascending, and it's opposite when θ_n is larger, so the peak value would appear in the curves.

When Fig. 8.5 is used, the relative dielectric constant and permeability should be tested first; then the refractive index is gotten from Eq. 8.2, and the corresponding angle is found from Fig. 8.5. This is the critical semiapex angle. The relation of the two parameters in the figure is complex, so the details are considered. For example, when the refractive index is 1.8, the corresponding semiapex angle is easily gotten. When n = 1.6, three values of α are obtained from the figure: one is 12 and others are near 18. These data on different curves indicate that θ_n is different. If it's permitted, α = 12 should be chosen as the larger of θ_n. Otherwise, α = 18 is suitable. As there are two values near 18, each could be chosen in principle; but the refractive index chosen cannot be larger than the material's real refractive index. And when the refractive index is chosen as 2.4, there are no suitable data from Fig. 8.5 because in Fig. 8.5 α is above 10°. And so the figure when α is less than 10° should be given by further work, if needed. When the vertex angle is not less than 20°, data must be chosen from Fig. 8.5 and the peak value of θ_7 could be used as it's the latest data to 2.4.

8.1.3 The Base's Height

If the base's height is too small, a part of the wave cannot be attenuated by the pyramid, the absorption property is weaken, and pyramid's barycenter is out of the base, which adds to the difficulty of fixing it to the wall; but an extremely large height has its disadvantages, too. It will add weight, so that the pyramid would lose its significance. The base's height is often determined as 1/3–1/4 of the whole height of the pyramid[5].

8.2 Resonant Absorber Based on Carbon-Coated EPS

8.2.1 Design of the Filling Method

A resonant absorbing pyramid uses expanded polystyrene (EPS) particles as the base material, which is coated with carbon instead of the traditional polyurethane mixed with carbon. The filling state of EPS particles decides the absorption property of the pyramid. There are three main types of distribution of EPS particles: random distribution, horizontal stratification, and vertical stratification distribution. They are used in different frequency bands to meet different matching conditions.

8.2.1.1 Horizontal stratification distribution

The relative surface area of EPS particles decreases with the increase in diameter, and the carbon content in the surface under the same process decreases accordingly. So from the top of the pyramid to the bottom, EPS particles should be arranged in descending order to meet the impedance matching condition, as shown in Fig. 8.6. If the EPS particles have the same size, different processes can change the thickness of the carbon layer in the surface, which can result in the change of carbon content. EPS particles should be arranged in increasing order from the top of the pyramid to the bottom to meet the impedance matching condition.

8.2.1.2 Cubic distribution

As shown in Fig. 8.7, the EPS particles with the largest diameter fill in the periphery, and then the medium-sized particles, while the center is filled with the smallest EPS particles. This method can ensure a smaller reflection of the incident wave in any direction, which is beneficial to improve the absorption property of the electromagnetic wave.

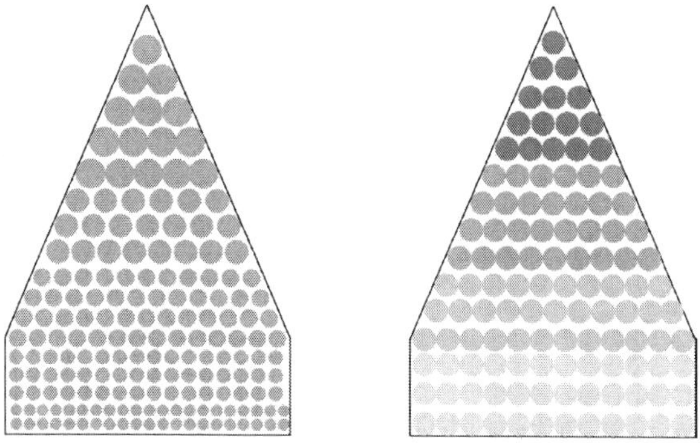

Figure 8.6 Horizontal stratification distribution of carbon. Pyramid filled with different size of EPS (left). Pyramid filled with different content of carbon (right).

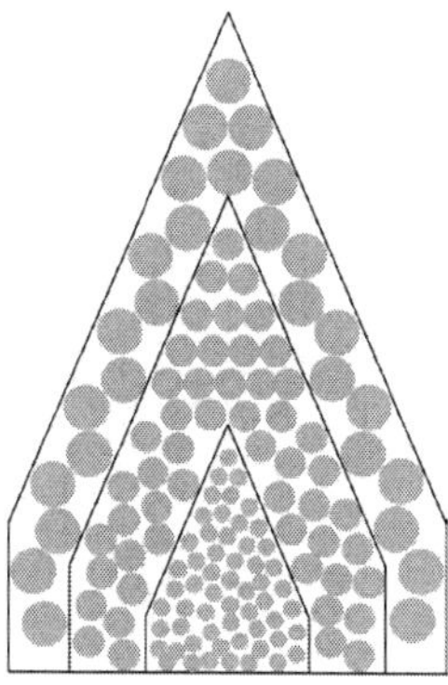

Figure 8.7 Cubic distribution of carbon.

8.2.2 Theoretical Analysis

8.2.2.1 Spherical resonant cavity model[6]

A high-frequency oscillating current is excitated and the electromagnetic wave is radiated when the microwave reaches the conductor's surface, which is reflection. Although the microwave couldn't transmit through the conductor, it could transmit through a certain thickness, which is called the skin depth. As EPS particles are spherical, they could be considered as the spherical resonant cavity model when their surface is coated with carbon. When the electromagnetic wave reaches the resonant sphere, most of it enters the spherical shell. Some of the microwave would reflect when reaching the other surface; the incidence and reflection rays could counteract when transmitting in the sphere. If the size of the resonant cavity is controlled appropriately, resonance would take place. Then the energy of the electromagnetic wave would be transformed to other energy in the process of resonance in the sphere[7].

From classical electromagnetic theory, the characteristic equation of the spherical resonant cavity is the Bessel equation[8]. To the transverse electric (TE) mode, it is

$$J_{n+\frac{1}{2}}(x)=0 \tag{8.3}$$

where x_{np} is used as the pth root of Eq. 8.3. To the TE$_{nmp}$ mode, the resonant frequency is

$$f_{nmp}=\frac{\omega_{nmp}}{2\pi}=\frac{k_{np}}{2\pi\sqrt{\mu\varepsilon}}=\frac{x_{np}}{2\pi a\sqrt{\mu\varepsilon}} \tag{8.4}$$

Its corresponding resonant wavelength is

$$\lambda_{nmp}=\frac{c}{f_{nmp}}=\frac{2\pi ac\sqrt{\mu\varepsilon}}{x_{np}}=\frac{2\pi a}{x_{np}} \tag{8.5}$$

where the suffixes n and m are the mode numbers of the electromagnetic wave along the X and Y directions.

In Eq. 8.5, a is the EPS particle radius and μ and ε are its magnetic permeability and dielectric permittivity, respectively. The values of x_{np} are listed in Table 8.3[9].

Table 8.3 Zero values (x_{np}) of $J_{n+\frac{1}{2}}(z)$

n/p	1	2	3	4
1	4.493	7.725	10.904	14.066
2	5.763	9.095	12.322	15.515
3	6.988	10.417	13.698	16.924
4	8.183	11.705	15.040	18.301
5	9.356	12.967	16.355	19.653

To Eq. 8.4, the value of x_{np} is fixed to a certain TE mode. The resonant frequency only relates to a, and its value reduces as a increases. So when EPS's radius increases, the corresponding resonant frequency is lower and its absorbing capacity is better at low frequency.

The electromagnetic wave, which is not absorbed in the resonant cavity, transmits into another resonant cavity, which is a transmission coupling. There is space among the spheres, so the reflection and incidence waves would interfere in the space because of the mismatch of two kinds of medium. If the space of different particles is at a quarter wavelength and both the amplitude and phases of the reflection and incidence waves are opposite, then the reflection and incidence waves would counteract and the electromagnetic wave would reach its maximal loss, which would increase material absorption. The absorber is composed of countless resonant cavities, so there would be too much loss which is contributed by the resonance of many resonant cavities and the counteracting of EPS particles when the electromagnetic wave transmits through the absorber. The result is that the material's absorbing capacity increases significantly.

8.2.2.2 Rectangular resonant cavity model

To meet the test's request, the base of the model is designed as 200 mm × 200 mm. When the EPS particle's filling height is l, the model could be supposed as a rectangular waveguide (length l). EPS particles coated with carbon are loss dielectric. The electromagnetic wave would be weakened by resonance when transmitted through the medium. From electromagnetic wave theory, this rectangular waveguide is considered as a rectangular medium resonant cavity with l as its length. Its resonant wavelength is

$$\lambda = \frac{2}{\sqrt{\left(\frac{m}{a}\right)^2 + \left(\frac{n}{b}\right)^2 + \left(\frac{p}{l}\right)^2}} \tag{8.6}$$

And the corresponding resonant frequency is

$$f = \frac{\sqrt{\left(\frac{m}{a}\right)^2 + \left(\frac{n}{b}\right)^2 + \left(\frac{p}{l}\right)^2}}{2\sqrt{\mu \times \varepsilon}} \tag{8.7}$$

In Eq. 8.7, ε is the effective dielectric permittivity of EPS which is coated with carbon, and its value can be obtained from the following formula:

$$\ln\varepsilon = V_1 \ln\varepsilon_1 + V_2 \ln \varepsilon_2 \tag{8.8}$$

where V_1 and V_2 refer to the volume content of carbon and EPS, respectively, and ε_1 and ε_2 are their dielectric constants. The mixed magnetic conductivity could be obtained similarly.

When the filling height l is 4 mm, $a = b > 1$ is concluded from the model's figure, and the transverse magnetic (TM) mode TM_{110} as the main mode of the resonant cavity is gained. When the volume ratio of carbon is 4%, the resonant frequency can be obtained as $f = 9.58$ GHz.

This frequency is the critical frequency of EPS with 4% carbon. Unless the frequency of the electromagnetic wave is above this value, resonance will not happen. The electromagnetic wave reflects in the resonant cavity when resonance happens and thus its loss increases. When the frequency is lower than the critical value, the loss of the electromagnetic wave comes from its multiple reflections in the resonant cavity. In this condition, the loss is relatively small.

8.2.2.3 Single sphere scattering and absorption analysis

The test sample is made up of numerous small spheres, so the wave scattering and absorption by a single EPS spherical particle are investigated.

The size of an EPS spherical particle is much smaller compared to the wavelength of the electromagnetic wave, so Rayleigh scattering theory can be used to describe the particle scattering of an electromagnetic wave. A spherical particle with radius a, permittivity ε_s, and permeability μ_s is located in the origin of the coordinate system, as illustrated in Fig. 8.8. A plane wave in the $\hat{z}$ direction is incident on the particle, $E = \hat{z} E_0 e^{ikx}$. Because the particle is very small, the scattered field can be seen as generated by the scattering of the source. The $\hat{z}$ direction of the electric field induces a dipole moment on the particle. As a result, the particle as a dipole antenna produces re-radiation. Its solution has the following forms:

$$E = \frac{-i\omega\mu Il e^{ikr}}{4\pi r}\left\{\hat{r}\left[\left(\frac{i}{kr}\right)^2 + \frac{i}{kr}\right]2\cos\theta + \hat{\theta}\left[\left(\frac{i}{kr}\right)^2 + \frac{i}{kr} + 1\right]\sin\theta\right\} \tag{8.9a}$$

$$H = \hat{\Phi}\frac{-ikIl e^{ikr}}{4\pi r}\left(\frac{i}{kr} + 1\right)\sin\theta \tag{8.9b}$$

The dipole moment Il is decided by E_0 and ε_s.

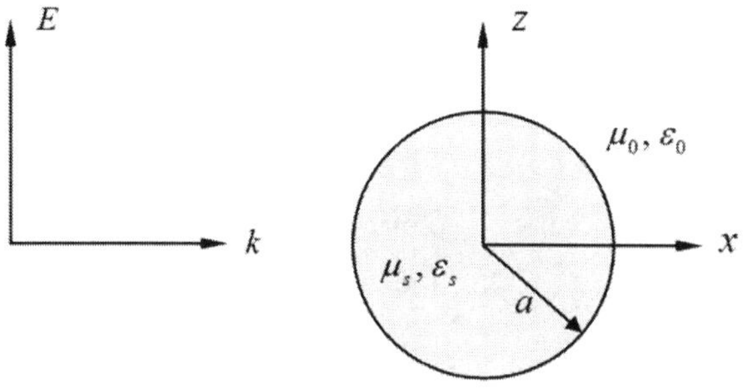

Figure 8.8 Rayleigh scattering of a sphere.

When $kr >> 1$, $E_\theta = -\left(\frac{\varepsilon_s - \varepsilon_0}{\varepsilon_s + 2\varepsilon_0}\right)k^2a^2E_0\frac{a}{r}e^{ikr}\sin\theta,\ H_\phi = \sqrt{\frac{\varepsilon}{\mu}}E_\theta$

The total scattered power of the sphere is

$$P_s = \frac{1}{2}\int_0^\pi r^2\sin\theta d\theta\int_0^{2\pi}d\phi E_\theta H_\phi^* = \frac{4\pi}{3}\sqrt{\frac{\varepsilon}{\mu}}\left(\frac{\varepsilon_s - \varepsilon_0}{\varepsilon_s + 2\varepsilon_0}k^2a^3E_0\right)^2$$

The scattering sectional area can be calculated as

$$\sum_s = \frac{P}{\frac{1}{2}\sqrt{\frac{\varepsilon}{\mu}}|E_0|^2} = \frac{8\pi}{3}\left(\frac{\varepsilon_s - \varepsilon_0}{\varepsilon_s + 2\varepsilon_0}\right)^2 k^4a^6 = \frac{128\pi^5}{3c^4}\left(\frac{\varepsilon_r^2 - \varepsilon_r}{\varepsilon_r + 2}\right)^2 f^4a^6 \tag{8.10}$$

As can be seen, the total scattered power of the sphere is proportional to the biquadratic of the wave number and the scattering of the high-frequency electromagnetic wave is much stronger than the low-frequency electromagnetic wave. The total scattered power of the sphere is also proportional to the sextic of the radius[10]. In the same way, the absorbing sectional area of the single sphere is as follows:

$$\sum_a = \frac{4\pi}{3} k a^3 \varepsilon_r'' \left(\frac{3}{\varepsilon_r + 2}\right)^2 = \frac{8\pi^2}{3c} a^3 f \sqrt{\varepsilon_r} \varepsilon_r'' \left(\frac{3}{\varepsilon_r + 2}\right)^2 \quad (8.11)$$

The formula is applied when a ≤ $0.05\lambda^{11}$, where $\varepsilon_r = 1 + \chi_e = \dfrac{3}{1-\dfrac{(\varepsilon_s-\varepsilon_0)x}{(\varepsilon_s+2\varepsilon_0)}} - 2$, x is carbon content and $\varepsilon_r'' = \dfrac{\varepsilon''}{\varepsilon_0} = \dfrac{\sigma}{\omega\varepsilon_0} = \dfrac{\sigma}{2\pi f \varepsilon_0}$.

$$\text{So} \sum_a = \frac{4\pi a^3 \sigma}{3c\varepsilon_0^2}\left[1-\frac{(\varepsilon_s-\varepsilon_0)x}{(\varepsilon_s+2\varepsilon_0)}\right]^2 \sqrt{\frac{3}{1-\dfrac{(\varepsilon_s-\varepsilon_0)x}{(\varepsilon_s+2\varepsilon_0)}}-2} \quad (8.12)$$

Conductivity can be obtained by the following formula:

$$\log\sigma = 39x - 41.81x^2 + 15.692x^3 - 12.8062 \quad (8.13)$$

The calculated scattering and absorbing cross sections of a single sphere are shown in Table 8.4 (radius of EPS is 1 mm, $\varepsilon_0 = 8.854 \times 10^{-12}$ F/m, f = 3 GHz).

The scattering and absorbing cross sections are enlarged with the increase in the carbon content. When the thickness of the carbon powder layers in the surface of the sphere is increased to exceeding the penetration depth of the electromagnetic wave in the carbon powder, the electromagnetic wave will not be able to pass through the sphere completely. At this time, the scattering and absorbing cross sections need to be revised.

The penetration depth should be calculated first. For a spherical shell composed of carbon powders, $\sigma = 3 \times 10^4$ S/m, $\varepsilon = 150\varepsilon_0$.

Due to $\sigma/\omega\varepsilon = \dfrac{3\times10^4}{2\pi\times3\times10^9\times150\times8.854\times10^{-12}} >> 1$, the penetration depth is

$$d_p = \sqrt{\frac{2}{\omega\mu\sigma}} = 53\mu m \quad (8.14)$$

When the thickness of the carbon powder layer exceeds the penetration depth, the effective cross section should be considered. So the effective interface coefficient is introduced:

$$k = \frac{\text{The effective volume of carbon}}{\text{The total volume of carbon}} = \frac{a^3 - (a - 2.65\times10^{-5})^3}{a^3 - 1} \quad (8.15)$$

Table 8.4 Scattering and absorbing cross sections of a single sphere (1)

Carbon content x	Permittivity ε_r	Radius α	log σ	Scattering cross section $\sum_s$	Absorbing cross section $\sum_a$	Effective cross section coefficient k
1	1.0303	1.0034	−12.42	1.0828×10^{-14}	6.018×10^{-19}	
2	1.0612	1.0068	−12.043	6.1038×10^{-14}	1.441×10^{-18}	
3	1.0928	1.0102	−11.6738	1.4883×10^{-13}	4.884×10^{-18}	
4	1.125	1.0137	−11.313	2.8615×10^{-13}	7.616×10^{-18}	
5	1.1579	1.0172	−10.96	4.897×10^{-13}	1.772×10^{-17}	
6	1.1915	1.0208	−10.6167	7.69×10^{-13}	3.913×10^{-17}	
7	1.2258	1.0245	−10.281	1.1234×10^{-12}	8.495×10^{-17}	
8	1.261	1.0282	-9.9538	1.5674×10^{-12}	1.807×10^{-16}	0.942
9	1.2967	1.032	-9.635	2.142×10^{-12}	3.789×10^{-16}	0.8342
10	1.333	1.0357	-9.308608	2.85×10^{-12}	8.19×10^{-16}	0.748

The values of k are shown in Table 8.4. To facilitate the expression, the logarithmic form of effective scattering and absorbing cross sections, $\log\sum_s^{/}$ and $\log\sum_a^{/}$, are shown in Table 8.5.

Table 8.5 Scattering and absorbing cross sections of a single sphere (2)

x	$\log\sum_s$	$\log\sum_a$	$\log k$	$\log\sum_s^{/}$	$\log\sum_a^{/}$
1	-13.965	-18.222		-13.965	-18.222
2	-13.214	-17.942		-13.214	-17.942
3	-12.827	-17.311		-12.827	-17.311
4	-12.543	-17.118		-12.543	-17.118
5	-12.31	-16.752		-12.31	-16.752
6	-12.114	-16.407		-12.114	-16.407
7	-11.949	-16.071		-11.949	-16.071
8	-11.802	-15.743	-0.026	-11.828	-15.769
9	-11.669	-15.421	-0.079	-11.748	-15.5
10	-11.545	-15.11	-0.126	-11.671	-15.236

Figure 8.9 is gained from the data in Table 8.5. As shown in Fig. 8.9, the effective scattering and absorbing cross sections are enlarging with the increase in carbon content, but the increasing rate is different. The increasing rate of the scattering cross section decreases with an increase in carbon content, but the absorbing cross section increases linearly.

For the absorbing material, the absorption capacity enhances with the increase in carbon content. But when the carbon content increases to a certain extent, the reflection will enhance and thus lead to a reduction in absorption. It also can be explained by the effective cross section. According to Eq. 8.14 and Table 8.5, when the carbon content is 7.5%, the thickness of the carbon powder layer is the penetration depth. When the carbon content is less than 7.5%, the electromagnetic wave can penetrate into the sphere. The absorbing cross section and the scattering cross section reflect the absorption capacity, so the absorption is gradually increasing. When the carbon content exceeds this value, absorption will still increase, but the scattered wave reflected back to the receiving horn antenna cannot be ignored with the increase in the depth of carbon powder layers. Reflection enhances with the increase of carbon content, so this time, the scattering cross section cannot completely reflect the scattering ability, and the electromagnetic wave scattering to space will be smaller than that in the figure. With the increase of carbon content, the deviation increases accordingly. So when the carbon content is more than 7.5%, the real scattering curve does not increase with the increase in carbon content but shows a downward trend. If both the absorbing and scattering cross sections are taken into consideration, 7.5% is the optimum carbon content where the maximum absorption peak appears.

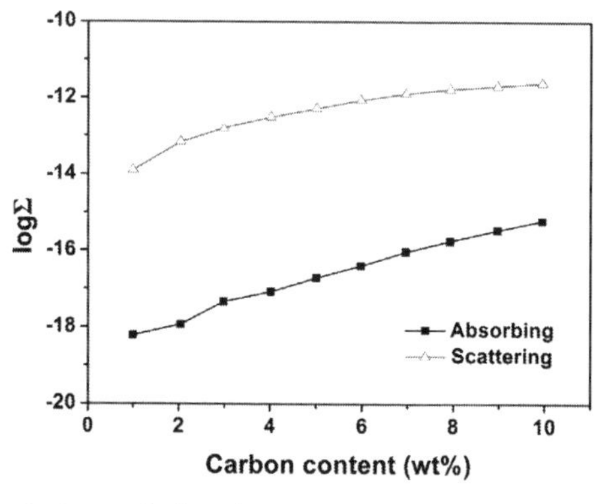

Figure 8.9 Variations of absorbing and scattering cross sections versus carbon contents.

8.2.2.4 Multiple scattering analyses

The effect of a single sphere on the electromagnetic wave is analyzed, but in the process of testing, the sample contains hundreds of thousands of small spheres. The influence on the electromagnetic wave is not simply the superposition of all the small spheres in the sample but also includes interaction between adjacent spheres. Under this circumstance, scattering theory will not be suitable. The test sample is considered as a whole, and the interior is a homogeneous mixture of small spheres and carbon powder. Multiple scattering and resonance of electromagnetic wave occur within the sample.

The wave attenuation of the pyramid absorbers consists of the scattering and absorption losses of the particles and the absorption loss in the matrix, which is written in the following form[12–14]:

$$I'_{\text{att}} = I'_{\text{sca}} + I'_{\text{abs}} + I'_{\text{mat}} \tag{8.16}$$

The wave attenuation of the particles includes the scattering loss I'_{sca} and the absorption loss I'_{abs} of the carbon-coated EPS particles, namely $I'_{\text{att}} = I'_{\text{sca}} + I'_{\text{abs}}$. The multiple scattering contains anisotropic scattering of the particles and multiple scattering among different particles. The anisotropic scattering of a particle is shown in Fig. 8.10, and it contributes to multiple scattering times among different particles.

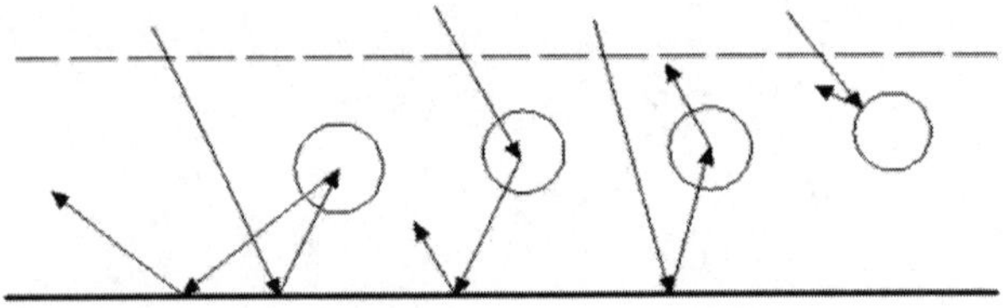

Figure 8.10 Physical phenomenon of anisotropic scattering by a particle.

Multiple scattering among different particles plays an important role in the wave loss of pyramid absorbers and obeys Twersky's theory[15]. Figure 8.11 represents a scattering field at r_a, and a scalar field at r_a is assumed as ψ^a. So ψ^a consists of the incident wave φ_i^a in the absence of any particles at r_a and the wave at r_a scattered from the other scatterers. This process is described in the following manner:

$$\psi^a = \phi_i^a + \sum_{s=1}^{N} u_s^a \phi_i^s + \sum_{s=1}^{N} \sum_{t=1,t\neq s}^{N} u_s^a u_t^s \phi_i^t + \sum_{s=1}^{N} \sum_{t=1,t\neq s}^{N} \sum_{m=1,m\neq t,m\neq s}^{N} u_s^a u_t^s u_m^t \phi_i^m + \sum_{s=1}^{N} \sum_{t=1,t\neq s}^{N} u_s^a u_t^s u_s^t \phi_i^s \ldots \tag{8.17}$$

Where ϕ_i^j is the scattering field by the scatterer at r_j and $u_s^a \phi_i^s$ is only a symbolic notation to indicate the field at r_a due to the scatterer at r_s when the wave ϕ^a is incident upon it. Other similar symbols represent the same meaning as mentioned above. The physical meaning of each term of Eq. 8.17 is as follows:

1. The first term is the incident wave ϕ_i^a at r_a.
2. The next term in this series, $\sum_{s=1}^{N} u_s^a \phi_i^s$, represents all the single scattering (see Fig. 8.11a).
3. The next summation, $\sum_{s=1}^{N} \sum_{t=1,t\neq s}^{N} u_s^a u_t^s \phi_i^t$, represents all the double scattering (as shown in Fig. 8.11b).
4. The third summation is triple, but the terms $t = s$ and $m = t$ are excluded, while the term $s = m$ is not excluded. Therefore, this summation can be written in two terms so as to include the terms containing the separate s, t, and m, plus the terms corresponding to $s = m$ (Fig. 8.11c,d):

$$\sum_{s=1}^{N} \sum_{t=1,t\neq s}^{N} \sum_{m=1,m\neq t,m\neq s}^{N} u_s^a u_t^s u_m^t \phi_i^m + \sum_{s=1}^{N} \sum_{t=1,t\neq s}^{N} u_s^a u_t^s u_s^t \phi_i^s \ldots \tag{8.18}$$

where the first triple summation is pictured in Fig. 8.11c. The second summation involves only two scatterers at r_s and r_t and is pictured in Fig. 8.11d.

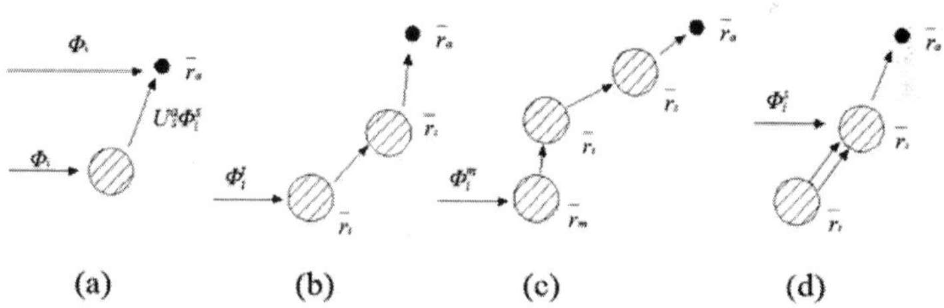

Figure 8.11 (a) Single scattering, (b) double scattering, (c) triple scattering through different particles, and (d) triple scattering when the propagation path goes through the same particle more than once.

In general, then, the complete field ψ^a at r_a, which is composed of the incident wave and all the multiple scattered waves, can be divided into two groups: One group, represented by the first summation in expression 4, consists of all the multiply scattered waves which involve chains of successive scattering going through different scatterers. This is illustrated in Fig. 8.12. Note that s is for all scatterers, and thus there are N terms for s; t should be for all the scatterers except s, and thus there are $N - 1$ terms for t. In a similar manner, there are $N - 2$ terms for m. The other group of terms represented by the second summation contains all the paths which go through the scatterer more than once, as shown in Fig. 8.12b.

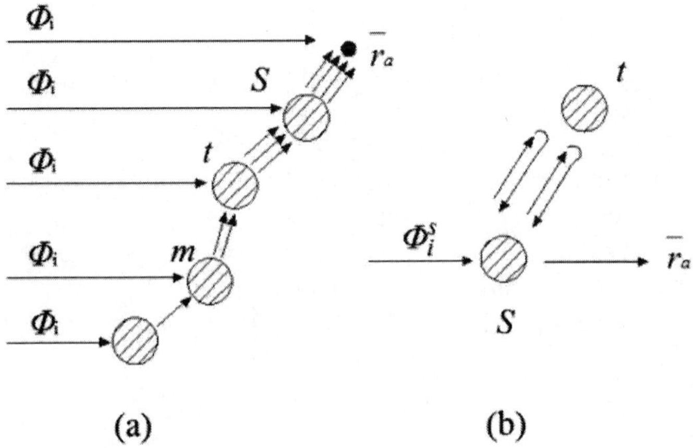

Figure 8.12 (a) Chains of successive scattering going through different scatterers and (b) chains of scattering paths which go through the same scatter more than once.

From the above analysis, it is obvious that the multiply scattering of the carbon-coated EPS particles in the pyramid absorbers is complicated. On the one hand, multiple scattering contributes to the incident wave attenuation; on the other hand, it increases absorption times of a single particle. With increasing carbon content of the carbon-coated EPS particles, the volume resistivity of the particles decreases, the absorption loss of a single particle increases, and the multiple scattering of the particles strengthens. So the wave attenuation of the pyramid absorbers improves greatly, which satisfies the experimental results.

8.3 Test for Microwave Absorbing Capacity

The physicochemical property of EPS is different from carbon, so the surface treatment is needed first. EPS is coated with carbon with special technology, and a fire retardant is used. After drying, it is filled for testing.

To examine the absorbing properties of the absorber, an orthogonal experimental design method is applied, and the detailed results are shown in Table 8.6. The filling height is fixed to five values: 2 cm, 4 cm, 6 cm, 8 cm, and 10 cm. The percentages of carbon are 1%, 2%, 3%, 4%, and 5%, and the diameters of EPS particles are 2 mm, 4 mm, and 6 mm. The test frequency is in the range of 8.2–12.4 GHz with an HP8720B scalar network analyzer using the bow test method[16]. The average absorption is noted as the test result in the frequency band.

It is concluded from the test results that the optimal percentage of carbon is 4%. When the percentage is 1%, its absorbing capacity is limited. This is due to the fact that the percentage of carbon is so low that a conductive network is not formed, and the loss is mainly because of carbon's damped vibration with the effect of the electromagnetic wave. This loss mode has little effect on the electromagnetic wave. As the percentage of carbon increases, a conductive network is formed gradually. When the percentage is 3%, the network has formed already and the loss mechanisms are as follows:

- The effect of dipoles. Considered as a dipole, the conductive powder's damped vibration is the main reason for the electromagnetic wave loss.
- Multiple reflection which weakens the electromagnetic wave.
- Leakage conductance effect[17] among the conductive powders.

The three mechanisms function together, so the electromagnetic wave is weakened obviously. When the percentage is 4%, each mechanism functions efficiently and the absorbing capacity reaches its maximum. When the percentage increases continuously to 5% or higher, there is excessive reflection and the absorption decreases.

The optimal value of the filling height is 4 cm as it relates to the percentage of carbon. When its height is below 4 cm, the absorbing capacity enhances with the increase in height. And when its height is above 4 cm, the carbon content is so high that the electromagnetic wave is reflected. The result is that the EPS particles filled at the base are useless as the electromagnetic wave could not reach there.

From Table 8.6, 2 mm is the optimal diameter. The reason is that when a 2-mm-diameter EPS is used, the EPS particle number is at its maximum and the resonant units are more than others in diameter. Although the frequency band is 8–12 GHz, which does not reach its resonant frequency from Eq. 8.4, the resonant model is also practicable. The electromagnetic wave is counteracted when it transmits in the resonant cavity, and its loss is obvious. As the particle number increases, the space in the resonant cavity increases, too. The loss of the electromagnetic wave increases when transmitting through the space. So the smaller the EPS diameter is, the better the absorbing capacity is.

Table 8.6 Experiment table designed by the orthogonal method

No.	Thickness *A*(cm)	Carbon *B* (vol%)	Particle size *C*(mm)	Average absorption(dB)
1	2	1	2	47.36
2	2	2	4	51.06
3	2	3	6	52.66
	2	4	–	
	2	5	–	
4	4	1	4	52.67
5	4	2	6	53.49
	4	3	–	
	4	4	–	
6	4	5	2	54.28
7	6	1	6	48.67
	6	2	–	
	6	3	–	
8	6	4	2	55.88
9	6	5	4	53.46
	8	1	–	
	8	2	–	
10	8	3	2	51.29
11	8	4	4	52.16
12	8	5	6	48.60
	10	1	–	
13	10	2	2	50.59

No.	Thickness A(cm)	Carbon B (vol%)	Particle size C(mm)	Average absorption(dB)
14	10	3	4	45.16
15	10	4	6	48.49
	10	5	-	
P1	151.08	148.70	259.40	
P2	160.44	155.14	254.51	
P3	158.01	149.11	251.91	
P4	152.05	156.53		
P5	144.24	156.34		
M1	50.36	49.57	51.88	
M2	53.48	51.71	50.90	
M3	52.67	49.70	50.38	
M4	50.68	52.18		
M5	48.08	52.11		
Ordering of influence	A,B,C			
Optimal combination	$A_2B_4C_1$			

Through analysis of the results of the orthogonal experiment, the optimal technical combination is as follows: The percentage of carbon is 4%, the diameter of EPS is 2 mm, and the filling height is 4 cm. Its absorbing capacity is shown in Fig. 8.13.

From Fig. 8.13, the following conclusions are made:

- The curve is relatively smooth when the frequency is low. The first absorption peak appears at 9.4 GHz, which is close to 9.58 GHz. This is the lowest resonant frequency, named critical frequency. The critical frequency decreases as the percentage of carbon increases from theory.
- The number of resonance peaks increases as the frequency increases when it is above 9.4 GHz, and the distance between two peaks becomes smaller, too. This is because the parameters (n, l) of the TE mode increase as the frequency increases. When the values of n and l are relatively large, the number of resonant peaks increases as Eq. 8.7. Resonance is

usually mixed with many modes, and it's difficult to distinguish them[18].

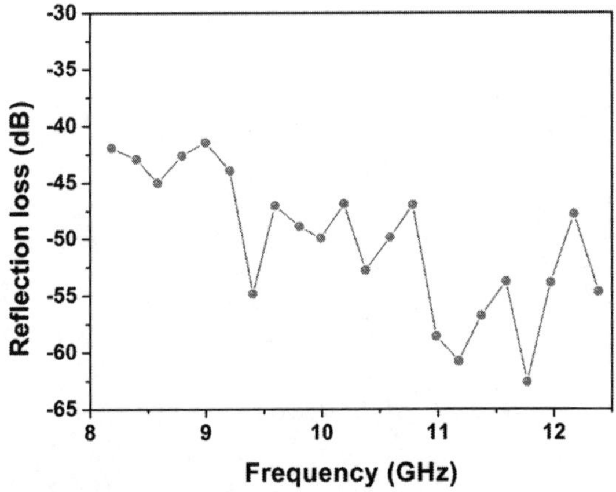

Figure 8.13 Absorbing capacity of the resonant cavity under the optimal technical combination.

References

1. Emerson WH (1973). Electromagnetic wave absorbers and anechoic chambers through the years, *IEEE Trans. Antennas Propag,* **21**(4), 484–490.
2. Wang, X, Zhu H (2001). Pyramidal absorbers made of hard foam polystyrene shell, *Chin J Radio Sci,* **16**(1), 41–44.
3. Wang, X, Zhu H (2000). Analysis and design for the absorbing materials used in microwave anechoic chamber, *J Microw,* **16**(4), 389–406.
4. Kondoh H (1986). An accurate FET modeling from measured S-parameters, *IEEE MTT-S Digest,* 380–387.
5. Lv S, Liu S (2007). Design and analyze of the figure of pyramid absorber used in microwave chambers, *Mater Sci Tech,* **4**, 33.
6. Stone AD (2000). Classical and wave chaos in asymmetric resonant cavities, *Physica A,* **288**, 1–4.
7. Liu S, Lv S, Zhao Y (2005). Study of the resonant absorber based of carbon coated EPS, *Mater Res Innov,* **9**(4), 114–116.
8. Zheng C, Li J (2004). Design and application of microwave absorbing materials, *Techno Space Navi Mater,* **5**, 1–5.

9. Fu J, Feng E (eds) (2001). *Advanced Electromagnetic Theory*, Xi'an Jiaotong University Press, China.
10. Jin AK (ed) (2003). *Electromagnetic Wave Theory*, Publishing House of Electronics Industry, Beijing, China.
11. Lshimaru A (ed) (1991). *Electromagnetic Wave Propagation, Radiation, and Scattering*, Prentice-Hall, New Jersey, USA.
12. Biwa S, Idekoba S, Ohno N (2002). Wave attenuation in particulate polymer composites: independent scattering/absorption analysis and comparison to measurements, *Mech Mater*, **34**(10), 671–682.
13. Biwa S, Watanabe Y, Ohno N (2003). Analysis of wave attenuation in unidirectional viscoelastic composites by a differential scheme, *Compos Sci Technol*, **63**(2), 237–247.
14. Biwa S (2001). Independent scattering and wave attenuation in viscoelastic composites, *Mech Mater*, **33**(11), 635–647.
15. Ishimaru A (ed) (1987). *Wave Propagation and Scattering in Random Media*, Academic Press, New York , USA.
16. Zhao J, Xue M (2001). Study of testing methods for absorption of RAM used in chamber with large incident angle, *J Microw*, **12**(4), 72–76.
17. Li L, Xu Z (2004). Physical mechanisms of the absorbing function in microwave-absorbing material and its composite design, *Adv Ceram*, **2**, 31–34.
18. Song Y, Xi C, Chen G (2002). Au-Au_2S nanoshell cavity resonance and parameters discussion, *Acta Optica Sinica*, **22**(11), 1392–1395.

Index